"Dr. Houston has done it again! He has written a[n entire] spectrum of causes, signs, evolution, and prevent[ion] for the most common reason to get sick and die, [the intan]gible quality of providing scientific rigor to the co[mplex] health and of bringing complex scientific concepts down to intuitive comprehension for the lay person. This is a wonderful book that should save you from days' worth of searching the web for solutions, explanations, and discussions about what causes heart attacks and what to do to stay clear of them. As always, he uses an eminently readable prose to accompany the reader from chapter to chapter on a journey of comprehensively updated and insightfully accurate understanding of coronary health and strategies for life prolongation. A recommended reading for those who aim to be enlightened to adopt preventive measures at home and empowered to discuss options at the doctor's office."

Sergio Fazio, MD, PhD
Adjunct Professor of Medicine
Stanford University Medical Center

"Dr. Houston once again provides incredible insight into novel treatment options for coronary heart disease. Eventually, conventional thinking will catch up to this cutting-edge knowledge."

Mimi Guarneri, MD, FACC
President Academy Integrative Health and Medicine (AIHM)
UCSD Associate Clinical Professor

"Dr. Mark Houston is a prolific author and has written many books on heart disease. This newest publication especially resonates with me as the doctor has taken coronary artery disease to a new level with his profound genetic analysis regarding cardiac pathology. Bringing specific genetic insights to the table in the analysis of early coronary artery disease and utilizing unique treatments that address our individual genetic profile will be the beginning of the 'NEW CARDIOLOGY' of the future. Dr. Houston is really years ahead of his time and provides some of the 'secret sauce' in the management of the world's number one killer, acute heart attack and coronary artery disease."

Stephen T. Sinatra, MD, FACC
Cardiologist and Author

"Heart disease is the number one killer of men and women, but Dr. Houston has a plan to end that. His new book provides the plan to prevent and reverse heart disease at its earliest stages. This book will be a must read for my patients."

Joel Kahn, MD, FACC
Clinical Professor of Cardiology
Kahn Center for Cardiac Longevity

"Dr. Houston continues to define the field of evidence-based Metabolic Cardiology and demonstrates a command of the physiologic complexity inherent in cardiovascular diseases. This book is a 'Must Have' for any practitioner managing patients with hypertension, hyperlipidemia, atherosclerosis, coronary heart disease, or heart failure. It is both a guide to understanding the primary causes of these conditions and offers concrete integrative, functional, and anti-aging solutions. His work has informed my own understanding of these topics and serves as an educational foundation for my graduate students and ongoing medical education events alike."

Andrew Heyman, MD, MHSA
Medical Director of Integrative Medicine
Department of Clinical Research and Leadership
School of Medicine and Health Sciences
George Washington University
Director of Academic Affairs
Metabolic Medical Institute and A4M

"I am incredibly pleased and honored to endorse Mark Houston's *The Truth About Heart Disease*. It has been my honor to participate in Dr Houston's Advanced Cardiology Course at A4M. This course has been instrumental in elevating the quality of medical care delivered by participants with its blend of traditional and integrative approaches to cardiac care. The level of wisdom and scientific backup in this course is unparalleled. With the update of *The Heart Book*, Dr. Houston has imparted much of the science and wisdom of his courses in a tome that has immense clinical utility. This textbook has become a classic, and the update only increases its clinical value. I heartily endorse not only this book but also Dr. Houston's many courses on Integrative Cardiology."

Douglas S. Harrington, M.D. FASPC, FACP, FASCP
Chairman
Predictive Health Diagnostics Company

The Truth About Heart Disease

You can prevent coronary heart disease yourself, but you need to have the knowledge of the risk factors, recognize the presenting symptoms, and take early action with aggressive and proper diagnostic testing. Start a prevention program for your heart health with *The Truth About Heart Disease*. In this book, Dr. Mark Houston provides you with scientific prevention and treatment programs to reduce your risk of coronary heart disease and myocardial infarction. These programs include optimal and proper nutrition, nutritional supplements, vitamins, antioxidants, anti-inflammatory agents, minerals, exercise, weight and body fat management, and other lifestyle changes. *The Truth About Heart Disease* will be of great value to all healthcare practitioners, cardiologists, and dietitians.

The Truth About Heart Disease

How to Prevent Coronary Heart Disease and Personalize Your Treatment with Nutrition, Nutritional Supplements, Exercise, and Lifestyle Tailored to Your Genetics

Mark C Houston, MD

CRC Press
Taylor & Francis Group
Boca Raton London New York

CRC Press is an imprint of the
Taylor & Francis Group, an **informa** business

First edition published 2023
by CRC Press
6000 Broken Sound Parkway NW, Suite 300, Boca Raton, FL 33487-2742

and by CRC Press
4 Park Square, Milton Park, Abingdon, Oxon, OX14 4RN

CRC Press is an imprint of Taylor & Francis Group, LLC

© 2023 Taylor & Francis Group, LLC

Reasonable efforts have been made to publish reliable data and information, but the author and publisher cannot assume responsibility for the validity of all materials or the consequences of their use. The authors and publishers have attempted to trace the copyright holders of all material reproduced in this publication and apologize to copyright holders if permission to publish in this form has not been obtained. If any copyright material has not been acknowledged please write and let us know so we may rectify in any future reprint.

Except as permitted under U.S. Copyright Law, no part of this book may be reprinted, reproduced, transmitted, or utilized in any form by any electronic, mechanical, or other means, now known or hereafter invented, including photocopying, microfilming, and recording, or in any information storage or retrieval system, without written permission from the publishers.

For permission to photocopy or use material electronically from this work, access www.copyright.com or contact the Copyright Clearance Center, Inc. (CCC), 222 Rosewood Drive, Danvers, MA 01923, 978-750-8400. For works that are not available on CCC please contact mpkbookspermissions@tandf.co.uk

Trademark notice: Product or corporate names may be trademarks or registered trademarks and are used only for identification and explanation without intent to infringe.

ISBN: 9781032230900 (hbk)
ISBN: 9781032230870 (pbk)
ISBN: 9781003275602 (ebk)

DOI: 10.1201/b22808

Typeset in Times
by Deanta Global Publishing Services, Chennai, India

Contents

About the Author xvii
Introduction...xix

Chapter 1   Overview of Coronary Heart Disease 1

    Some Informative Patient Cases with Coronary Heart Disease 1
    Patient Case 1 1
    Patient Case 2 1
    Patient Case 3 2
    Coronary Heart Disease in the United States 2
    Coronary Heart Disease Starts at an Early Age....................................3
    Five Effective Ways to Prevent Coronary Heart Disease in 80%
    of Patients 4
    A Brief Summary of the Coronary Heart Disease Prevention
    Program (CHDPP)...5
    Summary and Key Take Away Points..5
    References 6

Chapter 2   The Definitions and Real Reasons Behind CHD and MI 9

    Summary and Key Take Away Points..13
    References 13

Chapter 3   What Are the Symptoms and Signs of Coronary
    Heart Disease (CHD)?...15

    Summary and Key Take Away Points..16
    Bibliography 17

Chapter 4   Actual Patient Cases of CHD and MI from My Practice: Do Not
    Let This Happen to You 19

    Patient Case 1 19
    Comment 20
    Patient Case 2 20
    Comment 21
    Patient Case 3 21
    Comment 21
    Patient Case 4 21
    Comment 22
    Patient Case 5 22
    Comment 23
    Summary and Key Take Away Points..23

Chapter 5 The Arteries, the Endothelium, Endothelial Dysfunction, Glycocalyx, Glycocalyx Dysfunction, Nitric Oxide, and CHD 25

    Artery Structure and the Human Circulatory System 26
        The Lumen 26
        The Glycocalyx 30
        The Endothelium 33
        The Arterial Muscle (Media) 35
        The Artery Muscle or Media 35
        The Adventitia 35
    Summary and Key Take Away Points 35
    References 36

Chapter 6 What Do the Heart Arteries Do When They Are Damaged? The Infinite Insults and Finite Responses in CHD 37

    Introduction 37
    Inflammation 37
        What is Inflammation? 37
        How Do We Recognize and Measure Inflammation? 38
        How to Treat Vascular Inflammation 40
    Supplements 40
    Oxidation and Oxidative Stress 42
        How Is Oxidative Stress Measured in the Body? 43
        Antioxidants Fight Oxidative Stress 44
        How is Antioxidant Defense Measured? 45
        How to Treat Oxidative Stress 46
        Top Antioxidant Supplements 48
        Vascular Immune Dysfunction 49
    Final Conclusion 51
    Summary and Key Take Away Points 52
    Bibliography 53

Chapter 7 New Concepts in Coronary Heart Disease (CHD) 55

    Concept 1 55
    Concept 2 55
    Concept 3 56
    Concept 4 56
    Concept 5 56
    Concept 6 57
    Summary and Key Take Away Points 57
    Bibliography 57

Chapter 8 The Causes of Chest Pain: Is It a Heart Attack or Something Else? 59

    Review of CHD Symptoms 59

	Causes of Chest Pain in Addition to CHD or MI	59
	Summary and Key Take Away Points	60
	Bibliography	61
Chapter 9	CHD Genetics Part I: What Is in the Genes That Place You at Risk	63
	The Human Genome	63
	Summary and Key Take Away Points	65
	Bibliography	65
Chapter 9	CHD Genetics Part II: What Is in the Genes That Place You at Risk	67
	Nutrients	68
	Epigenetics	68
	Nutrition	69
	Mediterranean Diet	69
	Pritikin and DASH Diets	69
	Specific Nutrients	69
	Electrolytes	69
	Omega-3 Fatty Acids	70
	Monounsaturated Fats	70
	Genes Relevant to Cardiovascular Risk	70
	CHD and CHD Risk Factor Genes	70
	Genes Relevant to Hypertension	72
	Genes Related to Cholesterol and Diabetes Mellitus That Increase Risk for CHD and MI	73
	Summary and Key Take Away Points	74
	Bibliography	74
Chapter 10	What Are Your Genes Doing? Gene Expression in Coronary Heart Disease (CHD) and How to Treat	79
	Treatment of Your Genes and Gene Expression	79
	Summary and Key Take Away Points	81
	Bibliography	82
Chapter 11	The Gut and Heart Connection: Gut Microbiome	87
	Bad Food Choices Equal Bad Heart Arteries	89
	Problems with Medications That Reduce Stomach Acid and Cause Decreased Absorption of Vitamins and Minerals	91
	The Treatment for the Gut Microbiome	91
	Summary and Key Take Away Points	93
	Bibliography	94

Chapter 12 Nutrition Part I: The Importance of Nutrition in Coronary Heart Disease Prevention and Treatment 95

Introduction 95
Nutrition and CHD 95
    Mediterranean Diet 96
    Dietary Approaches to Stop Hypertension: DASH Diets 1 and 2 98
Dietary Fats 99
    Omega-3 Fatty Acids 99
    Monounsaturated Fats 100
    Saturated Fatty Acids 100
    Conclusions and Summary on SFAs 101
Saturated Fatty Acid Key Take Away Points and Conclusions 102
    Trans Fatty Acids 102
    Coconut Oil 103
    Milk, Milk Products, and Peptides 103
        Whey Protein 104
Refined Carbohydrates, Sugars, and Sugar Substitutes 105
Advanced Glycation End Products 105
Protein 106
    Vegetarian Diets and Plant-Based Nutrition 106
    Animal Protein Diets 106
    Soy Protein 107
    Fish 107
Dietary Acid Load and Protein 107
Specific Dietary and Nutritional Components and Caloric Restriction 108
    Caffeine 108
    Caloric Restriction 109
    Alcohol 109
    Gluten 110
    Nuts 110
    Dietary Sodium, Potassium, and Magnesium 111
Summary and Key Take Away Points 112
References 112

Chapter 13 Nutrition Part II: The Practice of Nutrition in Your Daily Life to Prevent and Treat CHD 125

Create Your Plate 126
Non-Starchy Vegetables 126
Clean-Sourced Proteins 126
Health-Promoting Fats 127
Starchy and Root Vegetables to Consider 128

Contents   xi

                Something Sweet 128
                Fruits 128
                Proteins and Foods to Say "Yes" To... 129
                    Pastured Poultry 129
                    Sweeteners 129
                    Flours... 129
                    Herbs.. 129
                    Starches 130
                    Fermented Foods 130
                    Nuts and Seeds 130
                Foods to Avoid or Eat in Moderation 130
                    Sweeteners to Avoid 131
                    Legumes to Avoid... 131
                    Oils to Avoid... 131
                Recipes 133
                Breakfast Ideas 142
                Summary, Conclusions, and Take-Away Points............................. 147
                Bibliography 147

Chapter 14 The Blood Vessel, Brain, and Immune System Connections........... 149

                The Autonomic Nervous System.. 149
                Treatment for an Overactive Sympathetic Nervous System............ 152
                Summary and Key Take Away Points.. 153
                Bibliography 153

Chapter 15 What Is Plugging Your Heart Arteries? Plaque Formation,
Types of Plaque, and Plaque Rupture... 155

                Why Do You Have a Myocardial Infarction?................................. 155
                Atherosclerosis 155
                Types of Plaques in the Coronary Arteries 159
                What Does a Coronary Artery Plaque Look Like? The
                Anatomy of a Plaque 160
                Types of Plaques: Hard and Soft Plaque 161
                What Makes a Plaque Prone to Rupture? These Characteristics
                are Listed below (Figures 15.8, 15.9A, and 15.9B)........................ 162
                Prevention and Treatment of Plaque 164
                Summary and Key Take Away Points.. 164
                Bibliography 164

Chapter 16 Nonobstructive Coronary Heart Disease and Coronary
Artery Vasospasm 167

                Introduction 167

	What Causes NO-CHD and CA-VS?	168
	Diagnosis of Coronary Spasm	168
	Treatment	169
	Summary and Key Take Away Points	169
	Bibliography	169
Chapter 17	Women and Coronary Heart Disease (CHD)	171
	Introduction	171
	Presenting Symptoms of CHD in Women	171
	CHD Risk Factors for CHD in Women	172
	Menopause and CHD in Women	172
	Summary and Key Take Away Points	173
	Bibliography	173
Chapter 18	Coronary Heart Disease Risk Factors: The Traditional Top Five Risk Factors and the Other 400	175
	Introduction	175
	Top Five CHD Risk Factors	175
	Hypertension	175
	Dyslipidemia	176
	Dysglycemia, Hyperglycemia, Insulin Resistance, Metabolic Syndrome, and Diabetes Mellitus	177
	Obesity and Body Fat	177
	Smoking	179
	Top 25 Modifiable CHD Risk Factors	179
	Hyperuricemia	180
	The Other 400 CHD Risk Factors	188
	Summary and Key Take Away Points	193
	Bibliography	193
Chapter 19	Testing: Labs, Noninvasive, and Invasive Testing	195
	Clinical Signs, Blood Tests, and Noninvasive Vascular Testing for Hypertension and Cardiovascular Disease	195
	The Arteries	195
	Endothelial Function and Dysfunction	196
	Carotid Artery Ultrasound	196
	The Eye and the Retina	198
	Coronary Artery Calcification	198
	Echocardiography (Heart Ultrasound)	199
	Aortic Ultrasound for Aneurysms and Kidney Size	200
	Central Blood Pressure and Aortic Stiffness (Augmentation Index)	201

Contents   xiii

                Electrocardiogram and Cardiopulmonary Exercise Testing 201
                Complete and Advanced Cardiovascular Laboratory Testing 201
                Summary and Key Take Away Points 201
                References 202

Chapter 20 How to Assess Your Risk Using CHD Scoring Tests 207

                Framingham Risk Score 207
                COSEHC Global Cardiovascular Risk Calculation 209
                Rasmussen Center CHD Scoring 210
                CHAN2T 3 CHD Scoring System 211
                PULS Cardiac Test (CHL) 211
                Gene Expression Testing 214
                    GES (Gene Expression Score) Corus CHD 214
                Measurements 214
                Summary and Key Take Away Points 216
                Bibliography 218

Chapter 21 The Integrative Coronary Heart Disease (CHD)
                Prevention Program 219

                Nutrition 219
                Nutritional Supplements 219
                    NEO 40 220
                    Arterosil 220
                    D Ribose Powder 221
                    Vitamin K2 MK 7 221
                    Aged Garlic: Kyolic Garlic Cardiovascular Formulation 224
                    Omega-3 Fatty Acids 225
                        The Cardiovascular Benefits of Omega-3 Fatty Acids 226
                        Clinical Research Studies 228
                        Safety of Omega-3 Fatty Acids 229
                    Dose, Ingredients, and Quality 229
                    Curcumin 229
                    Quercetin 231
                    Coenzyme Q10 231
                        Absorption, Pharmacokinetics, and Drug Depletions 231
                        Clinical Physiology and Functions 232
                        Cardiovascular Disease: CHF, CHD, Angina, and MI 232
                        Hypertension and Endothelial Function 236
                Summary 236
                    Mitoquinone: MITO Q 237
                    Multi-Nutrient Cardiovascular Protection Supplement 237
                        Dosage 237
                        Active Ingredients 237

Exercise, Cardiovascular Disease, and CHD 237
    The Aerobics, Build, Contour, and Tone Exercise Program 237
    The ABCs of Exercise with a Twist 239
    The Elements of ABCT 240
    The ABCT Elements in Detail 242
        Resistance Training 242
    What, How, and When to Lift 243
    Upping the Intensity with Supersets, Hybrids, and Rapid Sets 243
    ABCT Resistance Training Hints 243
        Aerobic Training in Intervals 244
    Maximum Heart Rate and Maximum Aerobic Capacity 244
    Always Combine Resistance and Aerobic Exercises 245
        Core Exercises 245
        Time-Intensive Exercise 245
        Nutrition, Water, and ABCT Energy Drink 246
        Exercise on an Empty Stomach 246
    Nutrition Before, During, and After the Exercise Session 247
    Getting Started with ABCT: Training Schedules and
    Descriptions of the Lifts 248
        The ABCT Training Schedules 248
        Chest Exercises 252
        Back Exercises 253
        Shoulder Exercises 254
        Arm Exercises 255
        Leg Exercises 255
ABCT Summary 257
ABCT Conclusion 258
Weight Management, Body Fat, and Visceral Fat 260
Summary of the CHD Nutrition Program 262
Tobacco Products, E-Cigarettes, and Vaping 263
Control all Risk Factors for Coronary Heart Disease 264
Summary and Key Take Away Points 265
Bibliography 265

Chapter 22 Medicines for CHD 277

    Blood-Thinning Medicines 277
    Statins 277
    Beta Blockers 278
    Nitrates 278
    Ranolazine 278

Chapter 23 Future Perspectives 281

    CHD Genetics and Informatics 281

Contents

        Stem Cells 281
        Nanotechnology 281
        Inflammation Treatments and Immunotherapy 281
        Vaccines 281
        Bibliography 282

Chapter 24 Grand Summary and Conclusions 283
        Eating a Healthy Heart Diet Is Easy, Tastes Good, and
        Can Prevent CHD 286

Sources 289

Index 293

About the Author

Mark C. Houston, MD, MS, MSc, FACP, FAHA, FASH, FACN, ABAARM, FAARM, DABC

Dr. Mark Houston graduated phi beta kappa and summa cum laude from Rhodes College, with a BA in chemistry and math. He graduated with the highest honors and the Alpha Omega Alpha honorary society distinction from Vanderbilt University Medical School. He completed his medical training at the University of California in San Francisco (UCSF) and then returned to serve as chief resident in medicine at Vanderbilt Medical Center, where he received the Hillman award for best teacher. Dr. Houston is the director of the Hypertension Institute and Vascular Biology, medical director of the Division of Human Nutrition and medical director of clinical research at the Hypertension Institute in Nashville, TN. He is on the faculty and director of the advanced cardiovascular medicine modules 16 A4M/MMI in the US and director of the cardiovascular medicine module 2 for A4M/MMI. He is a clinical instructor in the Department of Physical Therapy and Health Care Sciences at George Washington University (GWU) School of Medicine and Health Science. He served as an assistant professor of medicine and then as an associate professor of medicine at Vanderbilt University School of Medicine and was an associate clinical professor of medicine at Vanderbilt University School of Medicine (VUMS) from 1990 to 2012. He also served as an adjunct professor in metabolic medicine at the University of South Florida, Tampa (USF) Medical School (2014–2018).

He has four board certifications from the American Board of Internal Medicine (ABIM), the American Society of Hypertension (ASH) (FASH-Fellow), the American Board of Anti-Aging and Regenerative Medicine (ABAARM, FAARM), and the American Board of Cardiology (ABC) in hypertensive cardiovascular disease (DABC). He holds two master of science degrees in human nutrition from the University of Bridgeport, CT (MS), and another in metabolic and nutritional medicine (University of South Florida, School of Medicine-Tampa (MSc).

He was selected as one of the top physicians in the US in cardiovascular medicine in 2018 by the US Consumer Research Council. Dr. Houston was also named as one of the top physicians in hypertension in the US in 2008–2014 by the Consumer Research Council. He was honored by USA Today as one of the most influential doctors in the US in both hypertension and hyperlipidemia twice, in 2009 and 2010. He was selected for the Patient's Choice Award in 2010 and 2012 by Consumer Reports USA. He was selected as one of the top 100 physicians in the US by the American Health Council in 2017 and one of the top 50 functional and integrative medical doctors in the US in August 2017. He was also named one of America's best physicians in cardiology in 2018 by the National Consumer Advisory Board. In 2019, he was elected to the continental WHO'S WHO as a top doctor in the field of medicine as the medical director and founder of the Hypertension Institute.

Dr. Houston has presented over 10,000 lectures, nationally and internationally, and published over 250 medical articles and scientific abstracts in peer-reviewed medical journals, books, and book chapters. He is an author, teacher, clinician, and researcher.

He has written ten books that have been published:

Handbook of Antihypertensive Therapy
Vascular Biology for the Clinician
What Your Doctor May Not Tell You About Hypertension
Hypertension Handbook for Students and Clinicians
The Hypertension Handbook
What Your Doctor May Not Tell You About Heart Disease. Grand Central Publishing, 2012
Nutrition and Integrative Strategies in Cardiovascular Medicine. CRC Press, Sinatra and Houston, Editors, 2015
Vascular Biology and Cardiovascular Medicine for the Clinician. Mark Houston, Joe Lamb, and Anita Hays. Outskirts Press, 2019
Personalized and Precision Integrative Cardiovascular Medicine. Mark Houston, Mark, Editor and Contributor. Wolters Kluwer Publishers, Philadelphia and Chicago, 2020
Controlling High Blood Pressure Through Nutrition, Nutritional Supplements, Lifestyle, and Drugs. Houston, M and Bell, L. June 2021

Introduction

I have written this new book for personal as well as professional reasons. I have lost family members, friends, and patients to coronary heart disease (CHD) and heart attack or myocardial infarction (MI) over the past ten years. I do not want to lose anyone else. I do not want you to have CHD or MI or lose someone close to you. I want you to have at your fingertips all the information, facts, and education to help you and your physician prevent and treat CHD. I welcome you to read this book, which is the newest and most advanced scientific book on CHD and MI that has been written to date. A lot has happened in the past ten years since my last book on heart disease has been published. I know that you are bombarded with misinformation and confusing information about CHD. For example, which is the best nutrition program: paleo diet, keto diet, low fat, low carbohydrate, Mediterranean, DASH 2, or the newest fad diet? How much exercise and what type should you do? What nutritional supplements, antioxidants, and vitamins should you take, at what doses, and which brand is the best quality? What really causes CHD and MI, and how do you prevent them? This book on CHD is written for you, the nonmedical person to answer all of these questions and much more. I have explained CHD in terms that you will understand and that are easy and fun to read. Healthcare providers will also find this CHD book helpful. You will be able to apply the knowledge to reduce your risk of CHD and improve overall heart health and potentially save your life as well as those of your family and friends. Following are some of the features of this book:

1. All of the chapters are written in an easy-to-understand format, and all medical terms are defined for the reader.
2. Tables and figures are provided in each chapter to better illustrate what is described in the text and to provide concise, understandable, and valuable information.
3. All references are listed at the end of each chapter. If you wish to read more, the references will provide additional medical information.
4. A Summary and Key Take Away Points section is provided at the end of each chapter. This will give you a rapid and concise overview of the information in each respective chapter.
5. A Sources section is listed at the end of the book, which lists contact information for the best nutritional supplement companies, lab and testing companies, as well as other important sources mentioned in the text of the book. If you order from any of these companies please mention that you found their contact information in this book.
6. The names of specific nutritional supplements, if appropriate, are mentioned in the text should you wish to order them from the companies mentioned in the Sources section. Also, labs and testing companies are provided in the text and in the Sources section.

7. What this book will discuss. This book about the heart will discuss and review only CHD and MI. These will be defined in detail later in the book. Other diseases of the heart, such as heart failure, valvular heart disease, or abnormal rhythms of the heart, will not be discussed in this book. The reader is referred to other books on these topics.
8. The Prevention Program for CHD and MI. This book will discuss the role of nutrition, nutritional supplements, vitamins, antioxidants, minerals, exercise, weight and body fat control, smoking cessation, and other lifestyle changes to prevent CHD. There will be a very limited discussion on drug treatments (pharmacotherapies) in the prevention and treatment of CHD. Drug treatment is best discussed between you and your physician.

1 Overview of Coronary Heart Disease

SOME INFORMATIVE PATIENT CASES WITH CORONARY HEART DISEASE

There exists an enormous amount of misinformation on coronary heart disease (CHD) and myocardial infarction (MI) in medical books, medical publications, the newspapers, lay journals, magazines, and the news media. You are told to take a statin to reduce your cholesterol, and your heart is safe. You are told that if you get the top five treatable CHD risk factors to normal (high blood pressure, high cholesterol, diabetes mellitus, obesity, and tobacco cessation) you will not have an MI. These and other statements like this are very misleading and often not true. You must recognize the truth from the myths. Take your risk for CHD seriously and become educated and proactive patient with your heart health. Start now, as it is never too late. If you follow the recommendations in this book, you will dramatically reduce your risk for CHD and MI, and this may save your life. This is the best prevention and treatment program for CHD that is published to date. In this section, I will discuss some patient cases from my practice to give you a better understanding of the varied ways in which CHD can present. In Chapter 4, I present other patient cases with more details to help you understand CHD and MI.

PATIENT CASE 1

JR is a 52-year-old White male attorney with a wife and three children. He has had excellent health all of his life and regular examinations by his doctor. He has never had chest pain. His father died of an MI at age 55 years. JR was doing his daily run in the park on Sunday morning and collapsed. He was taken to the emergency room and pronounced dead on arrival. An autopsy showed he had sudden death from a massive MI.

PATIENT CASE 2

MK is a 48-year-old menopausal Black female school teacher, married with five children. She has high blood pressure and high cholesterol. She has complained of indigestion for two months and was treated with prescription medications for gastrointestinal reflux. She did not get any better. At 3 am on Saturday morning, she had severe substernal chest pain, shortness of breath, and sweating. She went to the hospital and was found to have an evolving MI. She survived.

PATIENT CASE 3

TM is a 62-year-old Hispanic male accountant with Type 2 diabetes for 30 years. He is overweight, has poor nutrition, eats fast food, and does not exercise. He is married with two boys. He started having chest pain and shortness of breath while climbing the steps in his house. He ignored his symptoms. His wife found him dead in bed on Monday morning. His autopsy showed a 100% blockage in the "widow-maker" artery of his heart, the left anterior descending artery (LAD). He had a fatal MI.

CORONARY HEART DISEASE IN THE UNITED STATES

All of these patients, just like millions of other Americans, have the same disease in common: CHD that resulted in an MI. In all of them, the CHD could have been prevented. You can prevent CHD in yourself, but you must know what the risk factors are, understand the symptoms that occur, get proper, early, aggressive diagnostic tests, and start a CHD prevention program. Taking action and responsibility for your heart health with all the information discussed in this heart book is key to prevention.

CHD, also referred to as coronary artery disease (CAD) is the number one cause of morbidity and mortality in the United States and is the most common form of heart disease (1, 2). In the United States, more than 2200 citizens die from CHD and MI *daily* (2–7). CHD accounts for approximately 610,000 deaths annually (estimated one in four deaths), is the leading cause of mortality in the United States, and is the third leading cause of mortality worldwide with 17.8 million deaths annually (7). Healthcare services for CHD are estimated to cost greater than 200 billion dollars annually in the United States. While CHD is a significant cause of death and disability, it is preventable (1–7). It is the result of plaque formation in the arteries supplying the heart. A plaque is a sticky substance, made up of fatty material, oxidized cholesterol and fats, inflammatory cells, white blood cells, immune cells, smooth muscle cells, and other substances, that builds up in the arteries and eventually explodes (called plaque rupture), causing an MI (Figure 1.1). CHD is used to describe a range of clinical disorders, such as asymptomatic atherosclerosis (cholesterol, white blood

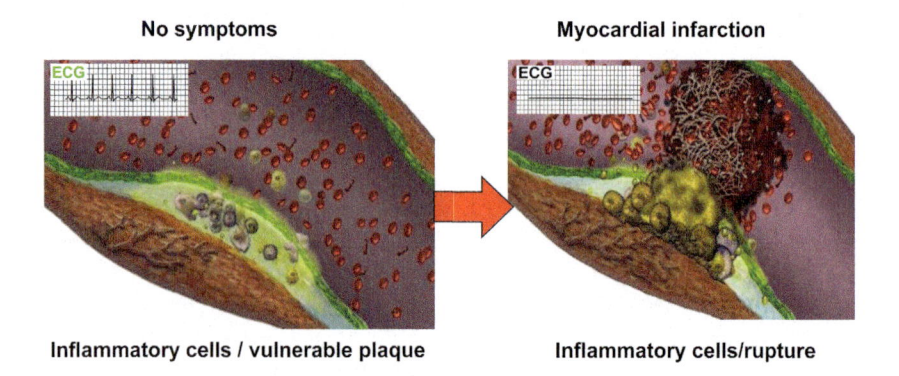

FIGURE 1.1 A coronary artery with a vulnerable plaque that ruptures causing MI.

cells, inflammation, and "glue-like material"), stable angina (pain in the chest), acute coronary syndrome, unstable angina, or MI as noted on an electrocardiogram (EKG) (6, 7). Approximately 80% of CHD can be prevented by optimal nutrition, exercise, weight and body fat control, stress reduction, and avoiding all tobacco products.

Clinical studies suggest that a limit has been reached in the ability to reduce the incidence of CHD using the traditional approaches (1, 8, 9). There are numerous external insults that damage the coronary arteries and the vascular system. The three finite responses of the coronary arteries and cardiovascular system to these infinite insults are *vascular inflammation*, *vascular oxidative stress*, and *vascular immune dysfunction*. These three finite coronary artery responses cause preclinical, functional, or structural damage to the heart and, eventually, clinical CHD. Laboratory measurement of the finite responses allows the physician to backtrack and determine the "why" (or the genesis) of CHD, remove the insult(s), and initiate optimal prevention and treatment methodologies to meet defined clinical noninvasive testing and laboratory goals.

Approximately 50% of patients continue to have CHD or MI, despite having presently defined "normal" levels of the top five CHD risk factors. The top five CHD treatable risk factors are hypertension, diabetes mellitus, dyslipidemia (abnormal cholesterol), obesity, and smoking (4, 6–9). The top five CHD risk factors, as well as many of the insult's top 25 CHD risk factors (there are actually about 400), must be redefined and treated early and aggressively. Simultaneously, the physician should measure and treat the three finite responses. Physicians should also embrace redefined CHD risk factors and CHD biomarkers by using new CHD risk-scoring systems, CHD and MI risk prediction tests, micronutrient testing, genetics, nutrigenomics, genetic expression testing, and noninvasive heart and vascular testing, which will allow cardiovascular medicine to become more personalized, precise, and integrated.

CORONARY HEART DISEASE STARTS AT AN EARLY AGE

Atherosclerosis and CHD begin in the fetus or at a very early age. Studies have shown the following (10–12):

1. **PDAY study:** patients at ages 15–19 at autopsy have atherosclerosis and CHD. About 60% have cholesterol deposits in the abdominal aorta (largest artery in the body and this is the abdomen part of the aorta). About 60% have fatty streaks in right coronary artery (heart artery).
2. **Korean War:** soldiers at autopsy had advanced CHD found at an average age of 22.
3. **Holman study:** patients were found to have fatty streaks in the aorta in the first decade of life. In addition, fatty streaks were found in the coronary arteries in the second decade of life, and fibrous plaques and CHD were noted in the second decade of life.

Cardiovascular disease starts very early in life (in utero) and is sub-clinical for 10 to 30 years or more prior to any CHD events. Endothelial dysfunction and glycocalyx

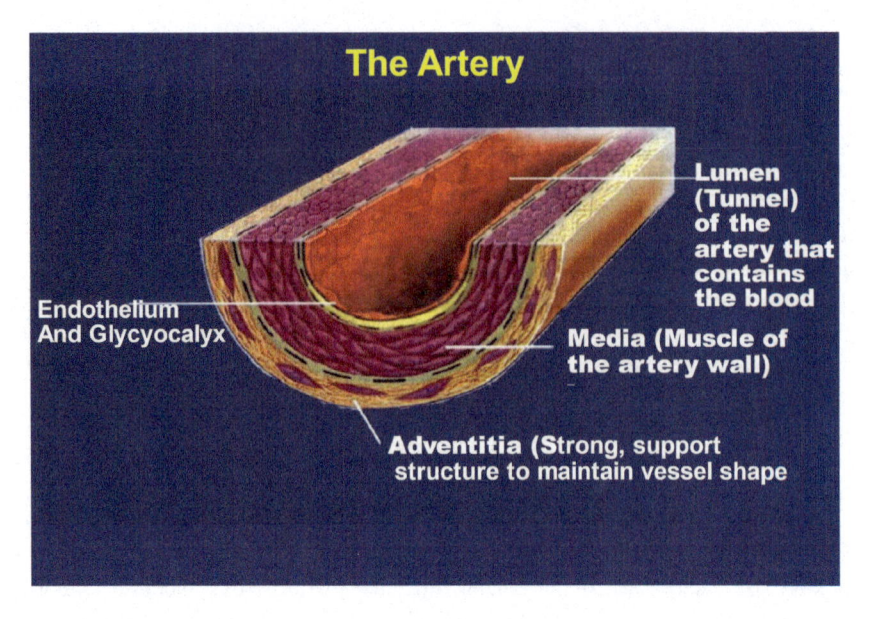

FIGURE 1.2 The coronary artery with the endothelium (the thin lining that separates the blood lumen from the arterial wall), the glycocalyx, the media (muscle) of the artery, and the adventitia.

dysfunction, which are abnormalities of the lining of the coronary arteries in the coronary arteries, are the earliest findings in CHD. This is followed by changes in arterial stiffness, loss of elasticity, cardiac and vascular muscle enlargement, and other changes in the coronary arteries and the cardiac muscle (Figure 1.2). Functional changes in the heart precede the structural changes or actual blockage with plaque in the arteries of the heart (8).

FIVE EFFECTIVE WAYS TO PREVENT CORONARY HEART DISEASE IN 80% OF PATIENTS

CHD can be prevented in about 80% of patients by implementing the following healthy lifestyle modifications listed below (Figure 1.3) (13, 14):

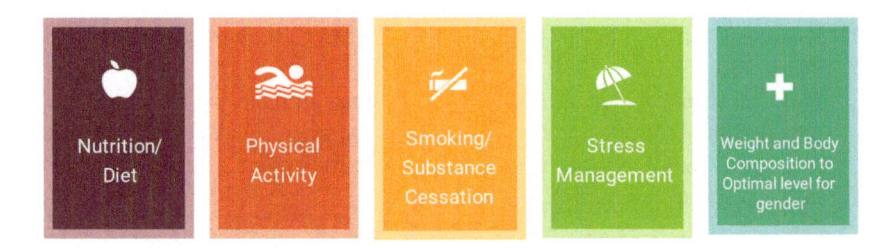

FIGURE 1.3 80% of Heart Disease is Preventable with 5 Healthy Lifestyle Modifications.

Overview of Coronary Heart Disease

- Optimal nutrition/diet with selective optimal nutritional supplements.
- Optimal exercise and physical activity (aerobic and resistance).
- Optimal weight, body, and visceral (belly) fat based on gender.
- Stress management.
- Smoking and tobacco (and smokeless tobacco) cessation.

All of these will be discussed in detail later in this book.

A BRIEF SUMMARY OF THE CORONARY HEART DISEASE PREVENTION PROGRAM (CHDPP)

1. Start an optimal nutrition program that is described in detail in this book (Chapters 12 and 13, related to calories, fats, carbohydrates (starches, refined carbohydrates, and complex carbohydrates), vegetable, fruit, and protein intake). Also consider caloric restriction and various types of fasting programs such as the Fasting Mimicking Diet (FMD) (Chapters 18 and 21).
2. Start an optimal and selective intake of high-quality nutritional supplements, vitamins, antioxidants, and minerals that are discussed in detail in this book (Chapters 18 and 21).
3. Start the recommended exercise program to include aerobics, resistance training, agility, and flexibility exercises (ABCT exercise program) (Chapters 18 and 21).
4. Maintain an ideal body weight, overall body fat, visceral (belly) fat, and body mass index (BMI) (Chapters 18 and 21).
5. Avoid all tobacco products and smokeless tobacco products (Chapters 18 and 21).
6. Reduce stress, anxiety, and depression (Chapters 18 and 21).

SUMMARY AND KEY TAKE AWAY POINTS

1. Coronary heart disease is the number one cause of death in the US and the third leading cause of death worldwide.
2. We have reached a limit in our ability to prevent and treat CHD with current concepts, prevention, and treatment strategies.
3. Coronary heart disease is preventable and treatable. It starts in utero or at a very early age.
4. Endothelial dysfunction and glycocalyx dysfunction are the earliest vascular findings in CHD and precede the clinical symptoms and events such as MI by decades.
5. Coronary heart disease is the result of atheromatous changes and plaque formation in the coronary arteries.
6. Approximately 80% of the cases of coronary heart disease can be prevented by achieving optimal nutrition, exercise, weight and body fat, reducing stress and anxiety, and avoiding all tobacco products and smokeless tobacco.

7. The three finite responses of the coronary arteries and cardiovascular system to the infinite insults are *vascular inflammation, vascular oxidative stress,* and *vascular immune dysfunction.*
8. The top five CHD treatable risk factors are hypertension, diabetes mellitus, dyslipidemia (abnormal cholesterol), obesity, and smoking. These must be redefined, measured accurately, and treated early and aggressively.
9. It is also important to measure and treat the top 25 CHD risk factors.
10. However, there are over 400 known risk factors.
11. Genetics and family history are very important and will be discussed in detail in Chapters 9 and 10.
12. Start the Coronary Heart Disease Prevention Program (CHDPP).

REFERENCES

1. Yusuf S, Hawken S, Ounpuu S, et al. Effect of potentially modifiable risk factors associated with myocardial infarction in 52 countries (the INTERHEART study): Case-control study. INTERHEART Study Investigators. *Lancet* 2004;364:937–952.
2. Houston MC. *What Your Doctor May Not Tell You About Heart Disease. The Revolutionary Book that Reveals the Truth Behind Coronary Illnesses and How You Can Fight Them.* New York: Grand Central Life and Style, Hachette Book Group; 2012.
3. O'Donnell CJ, Nabel EG. Genomics of cardiovascular disease. *New England Journal of Medicine* 2011;365:2098–2109.
4. Houston MC. Nutrition and nutraceutical supplements in the treatment of hypertension. *Expert Review of Cardiovascular Therapy* 2010;8:821–833.
5. ACCORD Study Group, Gerstein HC, Miller ME, et al. Long-term effects of intensive glucose lowering on cardiovascular outcomes. *New England Journal of Medicine.* 2011;364:818–828.
6. Regmi M, Siccardi MA. Coronary artery disease prevention. 2020 Aug 10. In: *StatPearls* [Internet]. Treasure Island: Stat Pearls Publishing; 2021 Jan. PMID: 31613540.
7. Brown JC, Gerhardt TE, Kwon E. 2021. Risk factors for coronary artery disease. Jun 5. In: *Stat Pearls* [Internet]. Treasure Island: Stat Pearls Publishing; 2021 Jan. PMID: 32119297.
8. *Personalized and Precision Integrative Cardiovascular Medicine.* Houston, M. Editor and Contributor. Philadelphia and Chicago: Wolters Kluwer Publishers; 2020.
9. *Controlling High Blood Pressure Through Nutrition, Nutritional Supplements, Lifestyle and Drugs.* Houston, M and Bell, L. Boca Raton: CRC Press; June 2021.
10. Strong JP, Malcom GT, McMahan CA, Tracy RE, Newman 3rd WP, Herderick EE, Cornhill JF. Prevalence and extent of atherosclerosis in adolescents and young adults: implications for prevention from the pathobiological determinants of atherosclerosis in youth study. *JAMA.* 1999;281:727–35.
11. Enos WF, Holmes RH, Beyer J. Coronary disease among United States soldiers killed in action in Korea: Preliminary report. *JAMA.* 1986;256:2859–62.
12. Holman, JR, Strong, JC. The natural history of atherosclerosis: the early aortic lesions as seen in New Orleans in the middle of the of the 20th century. *American Journal of Pathology.* 1958;34:209.

13. American Heart Association Nutrition Committee. Diet and lifestyle recommendations revision 2006: a scientific statement from the American Heart Association. *Circulation.* 2006 Jul 4;114(1):82–96 and Circulation. 2009 Mar 3;119(8):1161-75; Implementing American Heart Association pediatric and adult nutrition guidelines: a scientific statement from the American Heart Association Nutrition Committee of the council on nutrition, physical activity and metabolism, council on cardiovascular disease in the young, council on arteriosclerosis, thrombosis and vascular biology, council on cardiovascular nursing, council on epidemiology and prevention, and council for high blood pressure research.
14. Arnett DK, Blumenthal RS, Albert MA, Buroker AB, Goldberger ZD, Hahn EJ, Himmelfarb CD, Khera A, Lloyd-Jones D, McEvoy JW, Michos ED, Miedema MD, Muñoz D, Smith SC Jr, Virani SS, Williams KA Sr, Yeboah J, Ziaeian B. ACC/AHA guideline on the primary prevention of cardiovascular disease: A report of the american college of cardiology/American heart association task force on clinical practice guidelines. *Circulation.* 2019 Sep 10;140(11):e596–e646.

2 The Definitions and Real Reasons Behind CHD and MI

Coronary heart disease (CHD) is due to a reduction in the blood supply due to obstruction by a plaque in one or more of the coronary arteries to the heart muscle (myocardium) that results in a decreased delivery of fresh blood with oxygen and nutrients (1, 2) (Figure 2.1). The medical term for this is obstructive CHD. Depending on the severity of the blockage you can have angina, unstable angina, acute coronary syndrome (pending MI), MI, fatal arrhythmias, such as ventricular tachycardia, ventricular fibrillation, or sudden death. The main coronary arteries are the left main (LM), left anterior descending (LAD), left circumflex (LCX) and right coronary arteries (RCA) (Figure 2.1). These are relatively small arteries, and over time, they become obstructed with plaque, may constrict due to vasospasm, become diseased, and blood flow is decreased to the heart. Most cases of CHD are due to an obstruction of plaque in the artery. The plaque is made up of fatty material, oxidized cholesterol and fats, inflammatory cells, white blood cells, immune cells, smooth muscle cells, and other substances (1, 2) (Figures 2.2 and 2.3). However, in some cases, the arteries narrow or constrict; this is called "coronary artery spasm". In both of these circumstances, you will experience several symptoms such as chest pain (angina), shortness of breath, fatigue, palpitations sweating, nausea, dizziness, lightheadedness or syncope (passing out). If the plaque ruptures and forms a blood clot inside the lumen that blocks 100% of the artery, the result is a myocardial infarction (Figure 2.4) (1–3). The vast majority of patients do not have any symptoms prior to their first MI (Figure 2.5).

What actually causes coronary heart disease? The various pathways to CHD are many and complicated. However, most medical and lay publications will make you believe that there are only five CHD risk factors (hypertension, diabetes mellitus, dyslipidemia (abnormal cholesterol), obesity, and smoking). This is very misleading and is not accurate. These top five must be redefined based on new information and the top 25 CHD risk factors need to be reassessed as well. There are actually over 400 risk factors for CHD (4–7). All of these CHD risk factors will be discussed in detail later in this book.

What you need to know now is that the end results of all these CHD risk factors are only three finite responses (2) (Figure 2.6) in the arteries:

1. Inflammation.
2. Oxidative stress.
3. Vascular immune dysfunction.

DOI: 10.1201/b22808-2

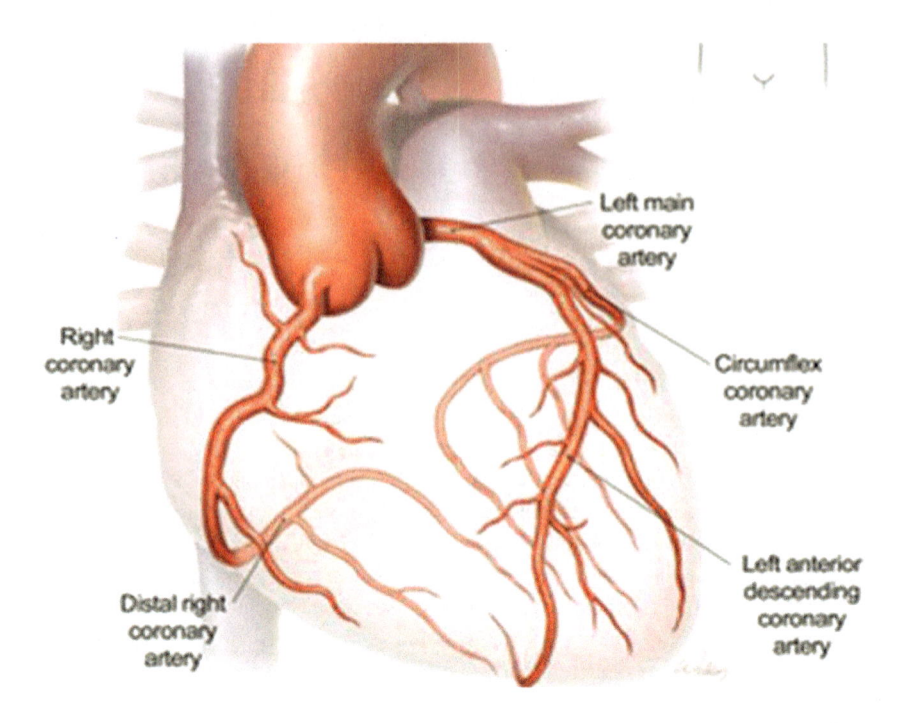

FIGURE 2.1 The main coronary arteries: left main (LM), left anterior descending (LAD), left circumflex (LCX) and right coronary arteries (RCA), and all the smaller arterial branches.

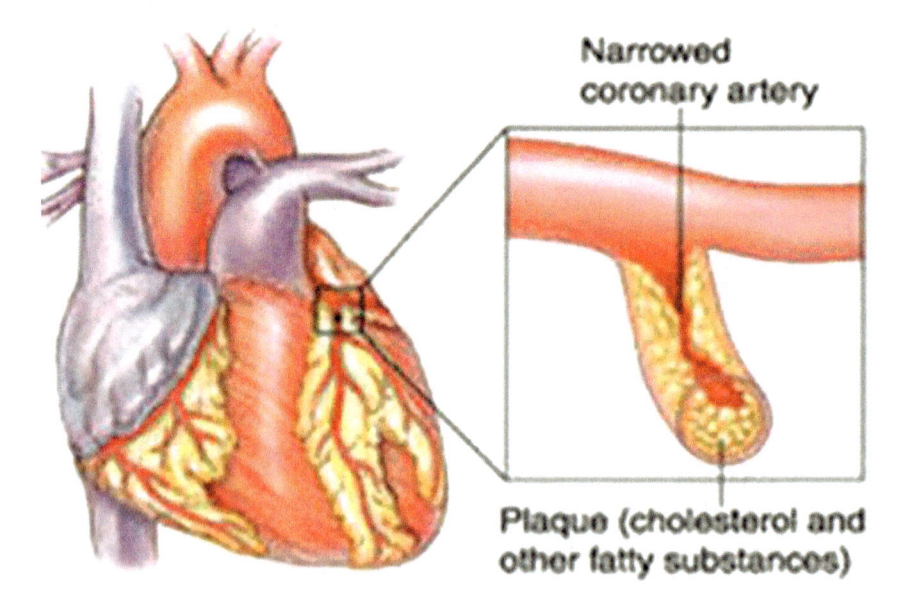

FIGURE 2.2 Coronary artery plaque (yellow) showing a severe narrowing in the LAD and reduction in blood flow.

Definitions and Reasons Behind CHD and MI   11

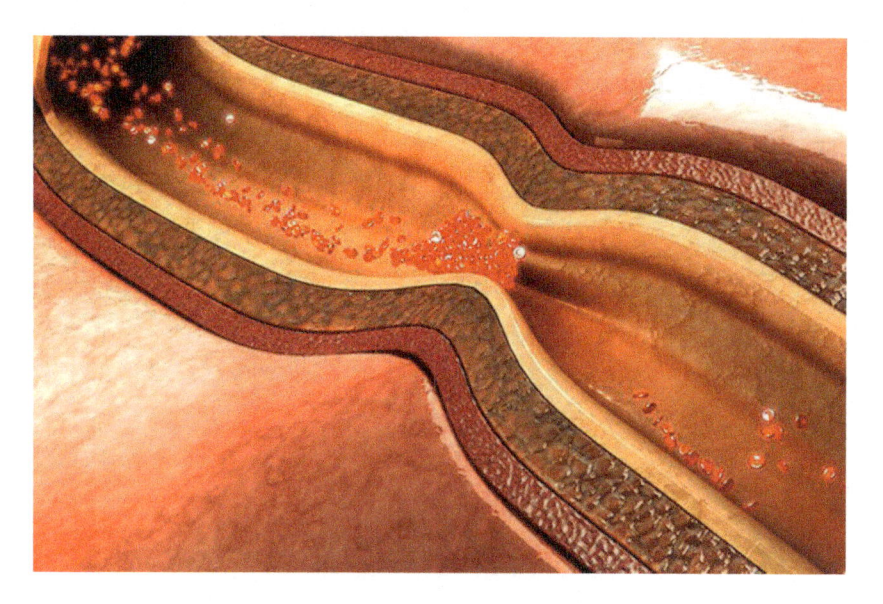

FIGURE 2.3 Coronary artery spasm showing the narrowing of the artery as the muscle wall constricts into the lumen of the artery, reducing blood flow (shown in red).

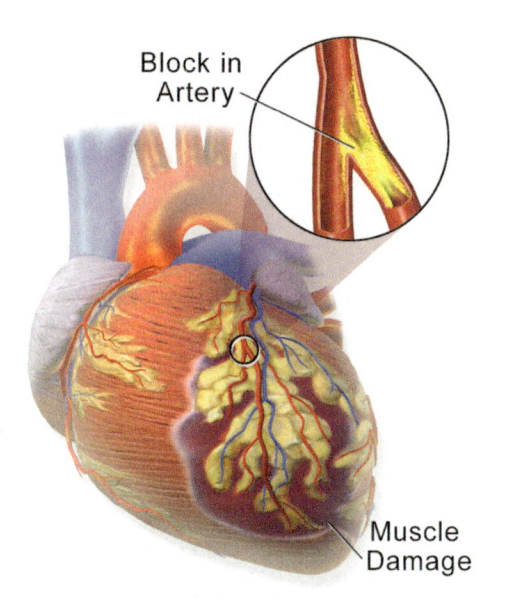

Heart Attack

FIGURE 2.4 Acute myocardial infarction with 100% blockage of the LAD from a ruptured plaque (yellow in image detail at the top) and a clot in the artery. The damage to the myocardium is shown by the whitish-yellow color in the lower part of the image.

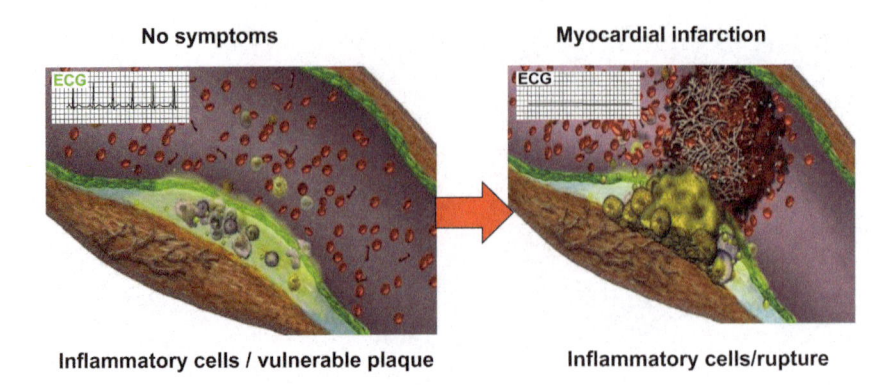

FIGURE 2.5 Seventy-five percent of myocardial infarctions are caused by unstable plaque rupture (yellow) and a blood clot (thrombus, red) in a coronary artery with no previous angina symptoms.

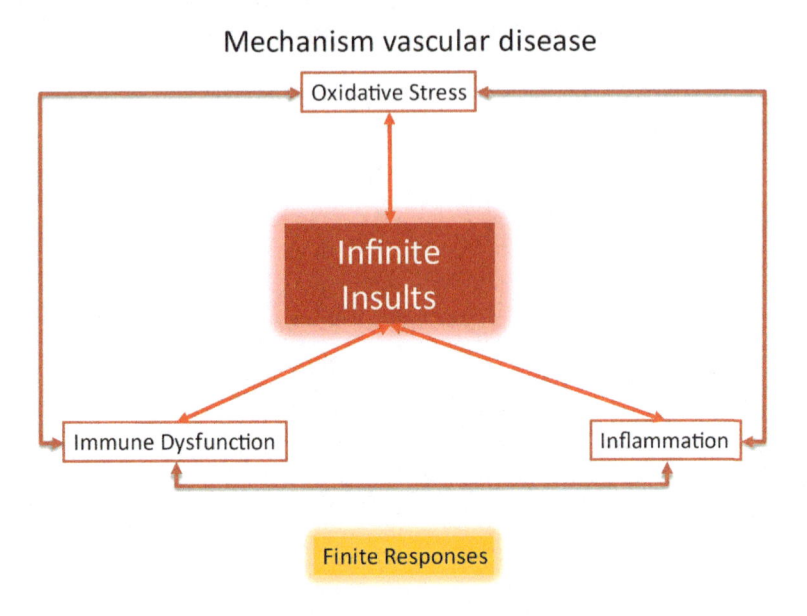

FIGURE 2.6 Infinite insults result in the three finite responses that cause CHD: vascular inflammation, vascular oxidative stress, and vascular immune dysfunction.

Once you identify one or more of these three finite vascular responses, it is important to track down the real causes or the genesis of the responses by going "backward on the path" to the origin of these finite responses. Once identified, then these infinite insults must be removed and treated. I will teach you all about this new concept and how to find and follow these pathways to prevent you from developing CHD or an MI.

SUMMARY AND KEY TAKE AWAY POINTS

1. Coronary heart disease is due to a reduction in the blood supply from obstruction with a plaque in one or more of the coronary arteries to the heart muscle (myocardium) that results in a decreased delivery of fresh blood with oxygen and nutrients. Less commonly, vasospasm may cause CHD.
2. The plaque is made up of fatty material, oxidized cholesterol and fats, inflammatory cells, white blood cells, immune cells, smooth muscle cells, and other substances.
3. The top five CHD risk factors (hypertension, diabetes mellitus, dyslipidemia (abnormal cholesterol), obesity, and smoking) must be redefined based on new information, and the top 25 CHD risk factors need to be assessed as well. There are actually over 400 risk factors for CHD.
4. There are an infinite number of insults to the coronary artery to cause CHD but only three finite responses: inflammation, oxidative stress, and vascular immune dysfunction.
5. Seventy-five percent of myocardial infarctions are caused by unstable plaque rupture and a blood clot (thrombus) in a coronary artery without previous angina symptoms.

REFERENCES

1. Arnett DK, Blumenthal RS, Albert MA, Buroker AB, Goldberger ZD, Hahn EJ, Himmelfarb CD, Khera A, Lloyd-Jones D, McEvoy JW, Michos ED, Miedema MD, Muñoz D, Smith SC Jr, Virani SS, Williams KA Sr, Yeboah J, Ziaeian B. ACC/AHA guideline on the primary prevention of cardiovascular disease: A report of the American College of Cardiology/American Heart Association Task Force on clinical practice guidelines. *Circulation*. 2019 Sep 10;140(11):e596–e646.
2. Houston MC. *What Your Doctor May Not Tell You About Heart Disease. The Revolutionary Book that Reveals the Truth Behind Coronary Illnesses and How You Can Fight Them.* New York: Grand Central Life and Style, Hachette Book Group; 2012.
3. Matta A, Bouisset F, Lhermusier T, Campelo-Parada F, Elbaz M, Carrié D, Roncalli JJ Coronary artery spasm: New insights. *Interventional Cardiology*. 2020 May 14;2020: 45–63.
4. Houston MC. Nutrition and nutraceutical supplements in the treatment of hypertension. *Expert Review of Cardiovascular Therapy* 2010;8:821–83.
5. Brown JC, Gerhardt TE, Kwon E. Risk factors: For coronary artery disease. Jun 5. In: *Stat Pearls* [Internet]. Treasure Island: Stat Pearls Publishing; 2021 Jan. PMID: 32119297.
6. *Personalized and Precision Integrative Cardiovascular Medicine.* Houston, Mark. Editor and Contributor. Philadelphia and Chicago: Wolters Kluwer Publishers; 2020.
7. *Controlling High Blood Pressure Through Nutrition, Nutritional Supplements, Lifestyle and Drugs.* Houston M. and Bell, L. Editors. Boca Raton: CRC Press; June 2021.

3 What Are the Symptoms and Signs of Coronary Heart Disease (CHD)?

Early recognition of both typical and atypical symptoms and signs of coronary heart disease (CHD) can be lifesaving. Getting to the emergency room, hospital, or your doctor as soon as symptoms start will allow for rapid and appropriate examination, testing, and treatment. Let us review these symptoms and signs in detail.

The symptoms and signs of CHD and MI are very similar. The medical terms used are angina, unstable angina, and acute coronary syndrome of MI. The most common are listed below (Figures 3.1 and 3.2).

1. Chest pain (angina). This is usually a heaviness, pressure, fullness, tightness, discomfort, stabbing, or squeezing in the middle of the chest (sternum) or in the left chest area. It is often worse with exertion. The pain may be atypical, however, and mimic heartburn. Patients may describe actual burning or feel a cramp, ache, or even sharper pain. The chest pain may radiate to the neck, jaw, arm, shoulder, or back (Figure 3.1).
2. Shortness of breath (dyspnea). Shortness of breath can range from very mild to severe, depending on how much the heart is damaged from reduced blood and oxygen supply. It is often worse with exertion, especially if going upstairs or walking on an incline. If you have heart failure, your breathing can be worse when lying down.
3. Sweating. Often there can be widespread cold sweating, even in the absence of a hot environment or exertion.
4. Nausea or vomiting. This can also be mild to severe, depending on the degree of heart damage due to reduced blood and oxygen supply to the heart muscle.
5. Abdominal pain (epigastric), especially in the upper abdomen between the lower ribs.
6. Fatigue. A sudden or gradual change in your energy level, with more lethargy than expected without a good cause, is frequent.
7. Dizziness, lightheadedness, fainting, or passing out (syncope).
8. Palpitations with skipping of heartbeats and fast or slow heart rate.
9. Numbness in the arms or hands.

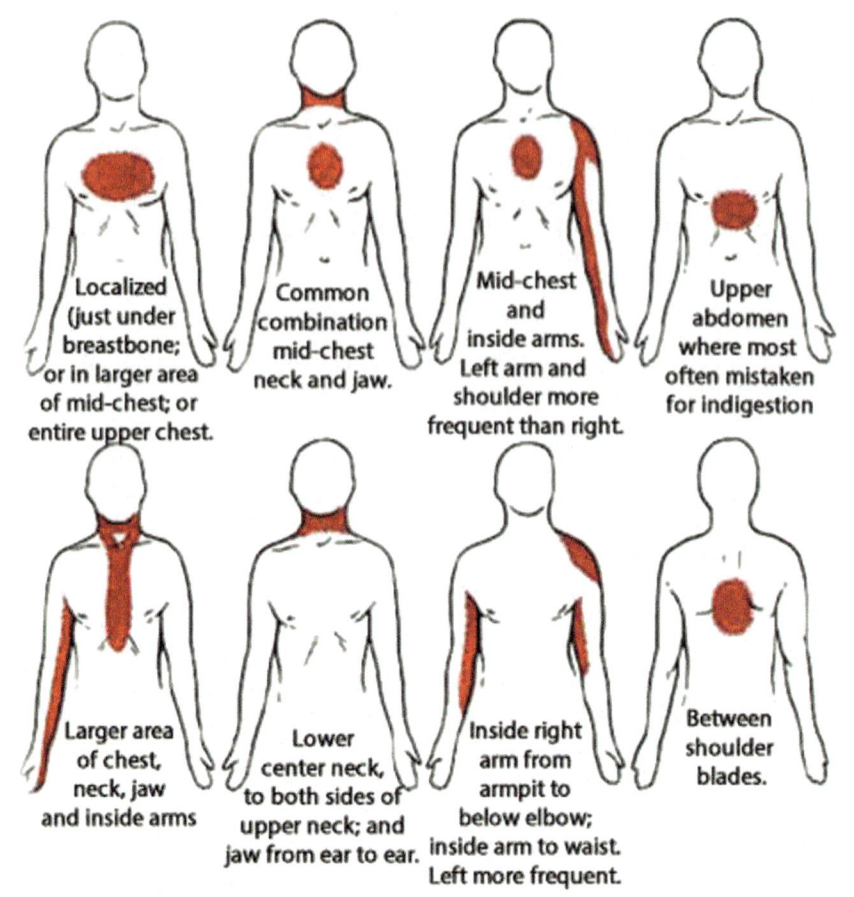

Heart Attack Pain Diagram

FIGURE 3.1 Various combinations of chest pain (angina and MI) and areas of radiation to other areas of the body.

About 21% of women and 10% of men had no symptoms at all with an MI or severe CHD. Men and women may respond differently to the symptoms of an MI, particularly if the symptoms are vague.

SUMMARY AND KEY TAKE AWAY POINTS

The symptoms and signs of CHD and MI can be mild or severe, typical or atypical, and will vary depending on your gender. If you develop any of these then go to the nearest emergency room or hospital. Time is critical in dealing with an MI and may truly save your heart and your life!

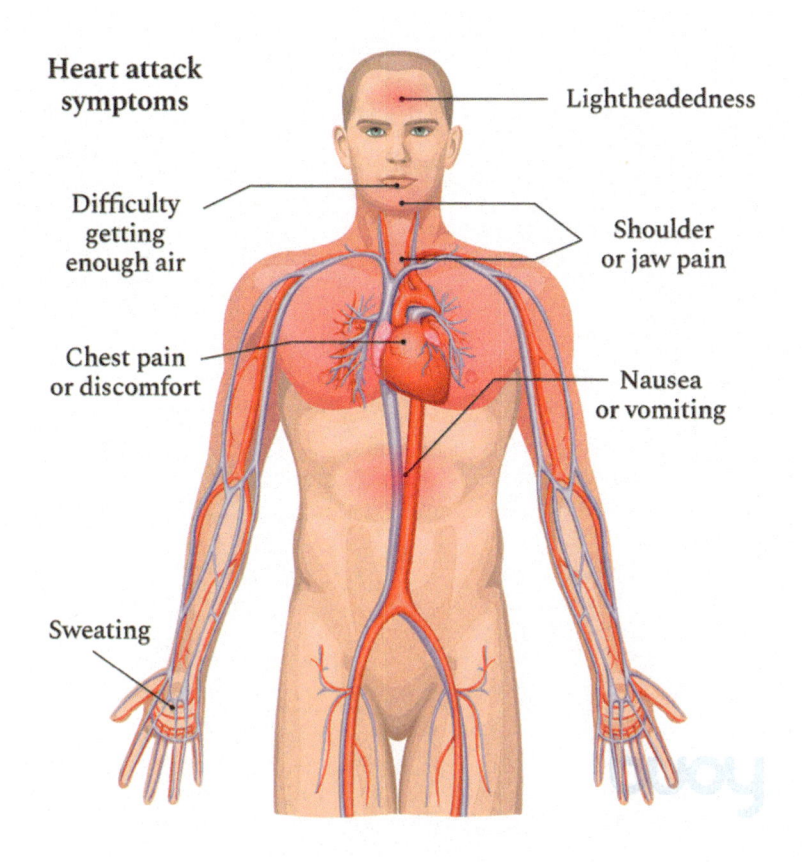

FIGURE 3.2 Some of the symptoms and signs of angina and MI.

BIBLIOGRAPHY

Khamis RY, Ammari T, Mikhail GW. Gender differences in coronary heart disease. *Heart*. 2016 July 15;102(14):1142–9.

4 Actual Patient Cases of CHD and MI from My Practice
Do Not Let This Happen to You

The following patient cases are from my cardiovascular practice in Nashville, TN, over the past several years. Many of these stories you may recognize in yourself, a family member, or a friend. I hope these real patient cases will help you to understand CHD and MI better and provide you with information that will arm you with practical actions and concepts that could save your life.

PATIENT CASE 1

JW is a 49-year-old Asian female who visited our office complaining of left chest pain that radiated to her left arm. She had these symptoms for over six months, and they were getting worse. The chest pain happened both at rest and with mild exertion, such as walking or going up steps or an incline. In addition, she had some shortness of breath with exercise. She had a family history of CHD in her father at age 72. She did not have other risk factors for CHD, such as high blood pressure, high cholesterol, diabetes mellitus, and she was not overweight and did not smoke.

She had seen four other physicians and was told that she was anxious and stressed, and she was given medication to reduce her stress that did not improve her symptoms.

One week before her visit to my office, she had severe chest pain and went to the emergency department at a local hospital. She was given an electrocardiogram, a chest X-ray, and blood tests that were all normal. She continued to have pain and was then given a special heart test called a nuclear medicine scan. This test involves an injection of a special material into the veins that will then image the heart to see if there is any decrease in blood supply that would indicate a blockage to the heart arteries; it was normal.

JW was told her heart was normal and she did not have CHD or an MI, and was discharged. She was not given any treatment and was told to continue her normal activities and not to worry.

She consulted with me later that week in my office. She had a normal physical examination. I did numerous tests that showed she had stiffness in the coronary arteries. She had a normal cardiac stress test (treadmill test). She was then placed on aspirin 81 mg per day, some nitrates, and a natural compound of beets that increases nitric oxide to improve the dilation of the arteries.

I suspected that she had "coronary artery spasm" based on her history and the tests. This is a relatively common cause of chest pain in women due to constriction and spasm of the coronary arteries, but there is no plaque or blockage in the arteries.

I scheduled JW for a coronary arteriogram, where a special dye is injected into the arteries of the heart from an artery in the leg or the wrist. The coronary arteries did not show any plaque or blockage. However, she was administered a drug during the arteriogram that proved she had coronary artery spasm when the artery constricted, which produced chest pain.

She was treated with another drug called amlodipine, which dilates the coronary arteries. She remained on aspirin, nitrates, and the natural beet compound to dilate the arteries.

JW has not had any more chest pain or shortness of breath for the past six months and feels well.

COMMENT

This patient case clearly demonstrates how women often have their symptoms of heart disease either ignored or incorrectly diagnosed. A normal electrocardiogram does not tell you that you do not have CHD. The normal nuclear medicine scan cannot completely eliminate the possibility of obstructive CHD, and cannot properly diagnose coronary artery spasm. In fact, nuclear medicine scans can miss about 15% of serious and significant blockages in the coronary arteries. She could have had an MI in the future if her diagnosis had not been made and the correct treatment started.

PATIENT CASE 2

MS is a 41-year-old Black male with chronic high blood pressure and high cholesterol for 10 years who has been treated with medications to lower the blood pressure and cholesterol. For the past three months, he has had progressive loss of energy and fatigue without explanation. He is an attorney, married with three children, and under a lot of stress. He says that he has no chest pain or shortness of breath and, other than the fatigue, has no other complaints.

His physical exam was normal, except for a mild increase in his blood pressure of 140/86 mm Hg (normal is 120/80 mm Hg) and an elevated resting heart rate of 88 beats per minute. His electrocardiogram was normal. He had a cardiac stress test that was abnormal, suggesting that he may have CHD with blockage in some of the coronary arteries.

MS underwent a coronary arteriogram and was found to have a blockage of 95% in his left anterior descending artery (LAD). A stent was placed in the artery; he was given proper medication and has done well for the past year with no more fatigue.

COMMENT

This patient presented with an "atypical" symptom of fatigue that was caused by CHD. The severe blockage in his left anterior descending artery, which supplies about two-thirds of the heart muscle, did not allow adequate delivery of oxygen, blood, and nutrients to the heart muscle. This resulted in a decreased delivery of oxygen, blood, and nutrients to all the arteries in the body, as the heart function was poor. The result of all of this was fatigue. In addition, the fast heart rate and increased blood pressure could be related to the blockage, as they returned to normal after the stent.

PATIENT CASE 3

HW is a 60-year-old White male who has smoked one pack of cigarettes per day for the past 30 years. Over the past one year, he has had progressive severe sternal mid-chest pressure and heaviness while exercising that radiates to his left neck and left arm. If he stops to rest for a few minutes, the chest pain goes away. He also gets very short of breath with exercise and often has to stop and rest for several minutes. He has been taking some medications for indigestion, as he thought this would help. However, there was no improvement in any of his symptoms.

His family history showed that both his father and mother had MIs before the age of 55 years and died.

HW's physical exam showed some wheezing in his lungs from smoking. His blood tests were normal. His cardiac stress test was abnormal, and he had chest pain and shortness of breath after walking on the treadmill for only two minutes. The test indicated that he had severe CHD in multiple areas of the heart.

He underwent a coronary arteriogram. Three arteries were involved: the left anterior descending was 85% blocked; the right coronary artery was 90% blocked; and the left circumflex artery was 92% blocked.

He underwent surgery with a coronary artery bypass graft (CABG) to all three arteries. This was successful; he went home on the proper medications and has had no chest pain or shortness of breath for two years.

COMMENT

This patient had the classic or common symptoms of CHD. His risk factors were his family history and 30 years of smoking. His genetic testing showed that he had four genes for CHD and MI that he inherited from his parents. Because he had three vessels that were severely blocked, he needed a CABG. If he had not had this CHD diagnosed, he would have had a major MI within one year and would likely have died.

PATIENT CASE 4

AC is a 42-year-old Hispanic female who came to see me with shortness of breath. She could walk on level ground for about 50 feet before she got short of breath.

However, walking up one flight of steps gave her severe shortness of breath. There was no chest pain. She had mild fatigue. She had no history of lung problems or asthma and does not smoke. She was overweight and has had type 2 diabetes mellitus for 25 years that was poorly controlled. She did not exercise, often did not take her diabetes mellitus medications, and did not eat healthy foods. She is a single mother with three boys and works two jobs.

Her physical exam was normal except for a mild increase in her blood pressure of 128/90 mm Hg (normal is 120/80 mmHg).

Her lab tests showed high blood sugar and elevated hemoglobin A1C (which measures average blood sugar levels).

AC's electrocardiogram at rest was normal. Her treadmill test was normal, and she has normal lung function. She had a nuclear medicine heart scan that showed decreased oxygen and blood supply to the bottom of her heart. A coronary arteriogram was done. It showed "microvascular angina", a disease of the small arteries in the heart, due to her diabetes mellitus. The large arteries of the heart did not have any blockages. She was started on proper medications, aspirin, weight management, better nutrition, and control of her blood sugar. Over the next six months she improved with the resolution of her shortness of breath.

COMMENT

This patient was found to have "microvascular angina", which means that the very small arteries of the heart are diseased and do not dilate well with exercise. As a result, the heart muscle does not get enough blood, nutrients, or oxygen. This is what caused the shortness of breath. This type of angina can only be treated with medication, lifestyle changes, and control of the CHD risk factors. There is no surgery for this disease.

PATIENT CASE 5

LH is a 57-year-old White male with high blood pressure for 15 years, high cholesterol for 20 years, and type 2 diabetes for 11 years, and he had been a chronic smoker since the age of 18. He took many medications for each of his medical diseases, but he missed a lot of them. He was overweight, ate fast and processed food, did not exercise, and got only five hours of sleep each night. He worked as a stockbroker for a large company, was under a lot of stress, and was anxious. He was single and had little time to socialize with friends.

He complained of sternal "pressure-like" chest pain with minimal exertion, such as walking across the room. The pain radiated to his back, left arm, and neck. He became very short of breath, started to sweat, and got dizzy and light-headed. He said that he has had extreme fatigue for over a year. Sometimes while going up steps, he became nauseated.

His physical exam showed a weight of 250 pounds, and he was only 5 feet 10 inches tall. His blood pressure was elevated at 146/95 mm Hg. His (bad) LDL cholesterol was high at 160 mg/dL, and his fasting blood sugar was 156 mg/dL, with an increased hemoglobin A1C of 7.5 units.

His electrocardiogram was abnormal. Other tests indicated very stiff coronary arteries, and he had a large amount of calcium in one artery of his heart, the left main (LM). His calcium score was 980 in that artery; a normal calcium score is zero.

His exercise stress test was very abnormal, suggesting severe CHD.

He underwent a coronary arteriogram and was found to have a 95% blockage in the left main artery of his heart. This artery supplies 100% of the blood supply to the heart.

LH had a CABG to the left main artery and did well with the resolution of all his symptoms. He was started on medications for his heart, blood pressure, diabetes mellitus, and high cholesterol. He saw a nutritionist for weight loss. He stopped smoking. He was sent to a counselor for stress reduction, sleep management, and lifestyle changes. He started cardiac rehabilitation and a supervised exercise program.

COMMENT

This patient had all the classic symptoms and signs of severe CHD and angina. He had the top five CHD risk factors. The obstruction of 95% of his left main coronary artery would have resulted in a massive MI and death within a few months if it had not been diagnosed and treated.

SUMMARY AND KEY TAKE AWAY POINTS

I have discussed several patients from my cardiovascular practice who demonstrate the common, typical, or classic signs and symptoms of CHD as well as the less common or atypical signs and symptoms. These presentations are variable with chest pain, shortness of breath, fatigue, dizziness, light-headedness, sweating, nausea, and more. You have seen examples of coronary artery spasm, microvascular angina, severe obstructive CHD and their clinical presentations, diagnostic testing, and surgical or medical treatments. I hope that these examples will increase your awareness of CHD and angina so you can recognize the symptoms early and seek medical evaluation by your physician.

5 The Arteries, the Endothelium, Endothelial Dysfunction, Glycocalyx, Glycocalyx Dysfunction, Nitric Oxide, and CHD

The arteries are the vessels or conduits (like pipes) that carry blood, oxygen, and nutrients from the heart to all the cells, tissues, and organs throughout the body. All of the arteries in the body have basically the same structure and function. The structure of an artery consists of five parts. All the arteries in the body have the same structure but vary in size. The coronary arteries are relatively small and are subject to more stress due not only to their size but also due to their location over the heart myocardium and the constant stress of heart contractions (Figure 5.1) (1–13). The five parts of all arteries are listed in order from the inside of the artery (lumen) where all the blood elements are found (red blood cells, white blood cells, platelets, proteins, albumin, globulins, clotting factors, nutrients) to the outer supporting structure of the artery called the "adventitia" (Figures 5.1–5.10).

1. Lumen (Figure 5.4).
2. Glycocalyx (Figures 5.5 and 5.5).
3. Endothelium (Figures 5.9 and 5.10).
4. Artery muscle (Media) (Figure 5.1).
5. Adventitia (Figure 5.1).

Note: In Figure 5.1, the glycocalyx and the endothelium are very thin, adjacent to one another, and closely connected and almost appear as one structure. All arteries have the same basic structure and function but the various parts, as well as the entire structure, can vary in size and thickness. The larger arteries have more artery muscle relative to the endothelium and glycocalyx, but due to their larger size they actually have thicker endothelial cells and glycocalyx. The smaller arteries have relatively more endothelial lining and glycocalyx but the thickness is less than that in the larger arteries. Arteries are the largest as they exit the heart (the aorta) and become

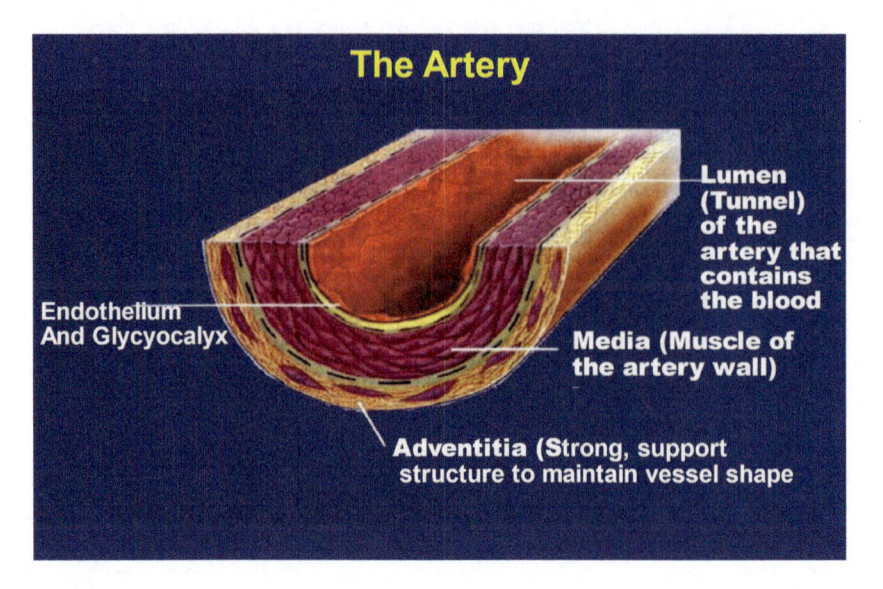

FIGURE 5.1 The artery structure and the human circulatory system: the lumen, glycocalyx, endothelium, muscle wall (media), and adventitia.

smaller as they go to the distant parts of the body, to the tissues and cells. Terms used for the variable sizes of the arteries include macrovasculature (large arteries), conduit arteries (medium-sized arteries), and microvasculature (small or resistance arteries) (Figure 5.2).

ARTERY STRUCTURE AND THE HUMAN CIRCULATORY SYSTEM

Oxygen-poor blood returns from the body from the veins (blue) and enters the lungs, and then oxygen-rich blood is pumped out of the heart into the aorta (red) to the rest of the body in the arteries (red).

THE LUMEN

The lumen is the inside of the artery (Figure 5.4), which would be similar to the inside of a water hose or a pipe. Inside the lumen is the blood which contains red blood cells, white blood cells, platelets, plasma with various types of proteins (albumin, globulins, clotting factors, nutrients and other blood components). The red blood cells carry oxygen to the cells, tissues, and organs throughout the body. Low red blood cell (RBC) count (anemia) can result in poor oxygen delivery to all of the vital organs such as the heart, brain, and kidneys. In unusual cases, the RBC is elevated and can cause "slugging" in the arteries because the blood is too thick. This can lead to MI or a stroke. White blood cells (WBCs) are involved in protecting us from infections, help to repair injuries, and provide an immune response to

Human circulatory system

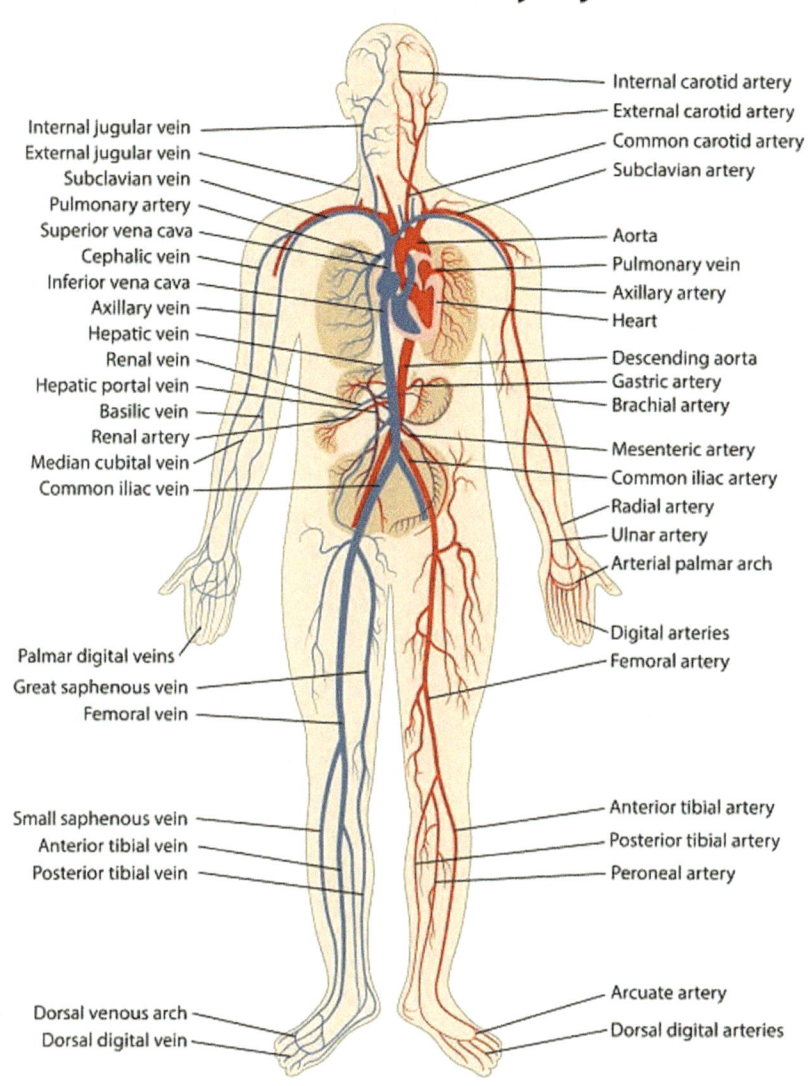

FIGURE 5.2 The human circulatory system showing the heart, lung, veins, and arteries.

outside invaders. The WBCs are involved in inflammation as part of this immune response and protect us against infections and even cancer. The WBCs include neutrophils, monocytes, lymphocytes, eosinophils, and basophils (Box 5.1). The platelets and the coagulation factors help to form clots (thrombus) in the artery lumen when we bleed. Proteins such as globulins are involved in immune function

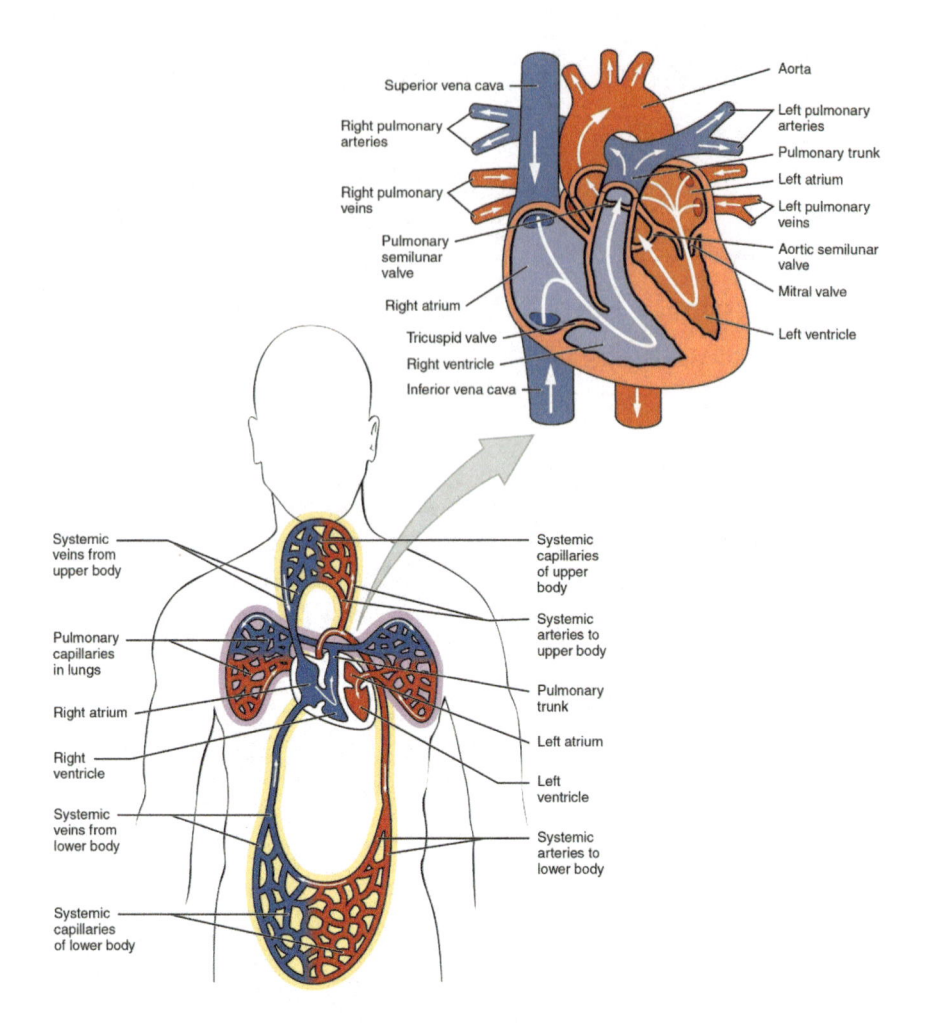

FIGURE 5.3 Oxygen poor blood returns from the body from the veins (blue) and enters the lungs, then oxygen rich blood is pumped out of the heart into the aorta (red) to the rest of the body in the arteries (red).

and protect against infections. The plasma albumin helps to maintain the blood volume and flow. Lastly, all of the nutrients that are absorbed from our diet, following digestion in the gastrointestinal tract, are delivered into the blood for delivery to all of the cells, tissues, and organs. Many of the immune WBCs are involved in CHD as they may cause damage to the endothelium, the glycocalyx, and the artery muscle. Remember: inflammation is one of the three finite responses of the artery that induces CHD and MI (Chapters 2, 5, and 6) (1–13). Vascular inflammation, oxidative stress, and immune dysfunction are closely related to the various WBC populations and what they do.

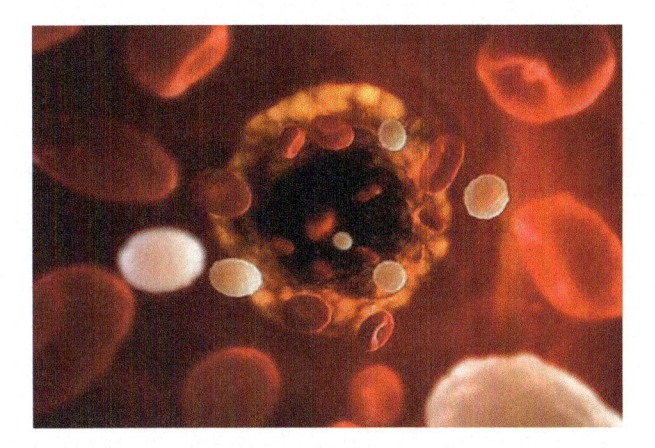

FIGURE 5.4 The lumen contains all the blood elements, such as red blood cells, white blood cells, platelets, proteins, albumin, globulins, clotting factors, and nutrients.

BOX 5.1   Types of White Blood Cells and Their Function

Monocytes comprise about 5–7% of WBCs and are essential in the immune system. They come in two forms: B cells and T cells. Unlike other WBC that provide nonspecific immunity, B and T cells have specific purposes. B lymphocytes (B cells) are responsible for humoral immunity, which is the immune response that involves antibodies. B cells produce the antibodies that "remember" an infection. They stand ready in case your body is exposed to that pathogen again. T cells recognize specific foreign invaders and are responsible for directly killing them. "Memory" T cells also remember an invader after an infection and respond quickly if it is seen again. B lymphocytes play a key role in the effectiveness of many current vaccines. In some cases, such as tuberculosis and pertussis vaccines, T lymphocytes are the main players. They have a longer life span than many WBCs and help to break down/kill bacteria. Both B and T cells are involved in vascular immune function, inflammation, oxidative stress, atherosclerosis, and CHD. For example, a modified form of the bad type of cholesterol (LDL) cholesterol is oxidized cholesterol (ox LDL). It is considered a foreign invader and is attacked by the macrophages and various types of WBC causing inflammation, oxidative stress, immune dysfunction, atherosclerosis, plaque, and CHD (more on this later in Chapters 6 and 15).

Lymphocytes are the garbage trucks of the immune system. Around 20–40% of WBCs in your bloodstream are lymphocytes. Their most important function is to clean up dead cells in the body. They create antibodies to fight against bacteria, viruses, and other potentially harmful invaders.

Neutrophils make up roughly half of the WBC population. They are usually the first cells of the immune system to respond to invaders, such as bacteria or viruses. As first responders, they also send out signals alerting other cells in the immune system to come to the scene. Neutrophils are the main cells found in pus. Once released from the bone marrow, these cells live for about eight hours. Your body produces roughly 100 billion of these cells every day. They kill and digest bacteria and fungi and other harmful organisms.

Basophils account for about 1% of white blood cells. These cells are perhaps best known for their role in asthma. However, they are important in mounting a nonspecific immune response to pathogens (organisms that can cause disease). When stimulated, these cells release histamine, among other chemicals. This can result in inflammation and narrowing of the airways in the lung.

Eosinophils also play a role in fighting off bacteria. They are very important in responding to parasitic infections (such as worms) as well. They are perhaps best known for their role in triggering allergy symptoms. Eosinophils can go overboard in mounting an immune response against something harmless. For example, eosinophils mistake pollen for a foreign invader. Eosinophils account for no more than 5% of the WBCs in your bloodstream. However, there are high concentrations of eosinophils in the digestive tract.

THE GLYCOCALYX

The endothelial glycocalyx (EGC) ("sugar coating") is a microscopically thin gel-like layer that coats the entire luminal side of the vascular endothelium and provides a nonadherent shield, like a protective coat (Figure 5.5). It works in direct communication with the endothelium, like an "air traffic control tower". The EGC is a negatively charged sulfate and carbohydrate-rich mesh of membrane containing soluble molecules including glycoproteins, proteoglycans, and glycosaminoglycans (a group of sugars bound to proteins), and creating a slippery gel-like layer in a dynamic equilibrium with blood, which continuously affects composition and thickness. EGC can be damaged due to high blood pressure, various toxic compounds, and enzymes. EGC thickness increases with the vascular diameter in arteries ranging from 2 to 3 μm (micrometer) in small arteries to 4.5 μm in carotid arteries. It may be thicker than the endothelium (Figure 5.6) (14, 15).

The EGC serves as a selective barrier to prevent various substances from moving from the blood into the endothelium and the subendothelial layer that can create damage to the artery and cause CHD. The components and functions of the EGC are listed below (also see Figures 5.6 and 5.7):

- Proteoglycans (seven types) with long-chain glycosaminoglycans (GAG) –side chains (five types) (heparan sulfate (50–90%), chondroitin sulfate, dermatan sulfate, keratan sulfate, and hyaluronic acid), and glycoproteins that bind to the endothelial cell as backbone molecules.

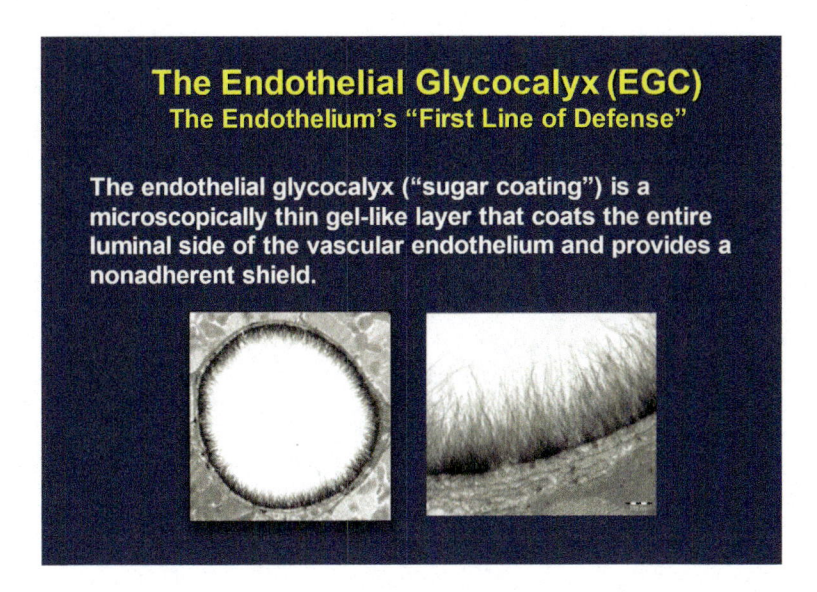

FIGURE 5.5 The endothelial glycocalyx (EGC), showing the thin hairlike projections into the lumen from the endothelial lining.

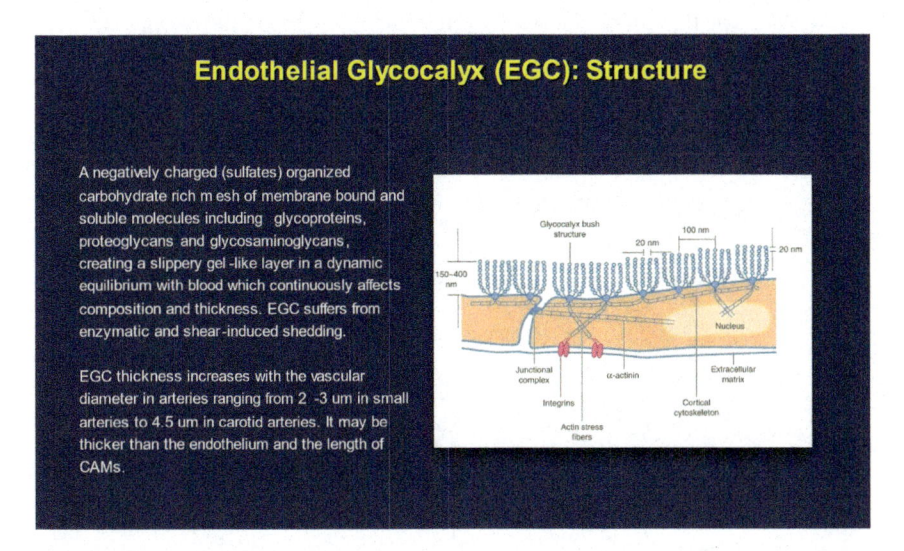

FIGURE 5.6 Endothelial glycocalyx structure.

- On top are soluble proteoglycans (SP), thrombomodulin (TM), extracellular superoxide dismutase (ec-SOD), and anti-thrombin III (AT-III).
- EGC limits access to endothelial cell membranes of circulating plasma components, such as lipoproteins (LDL, Lipoprotein(a), etc.), activated platelets, and sticky leukocytes (Figure 5.8).

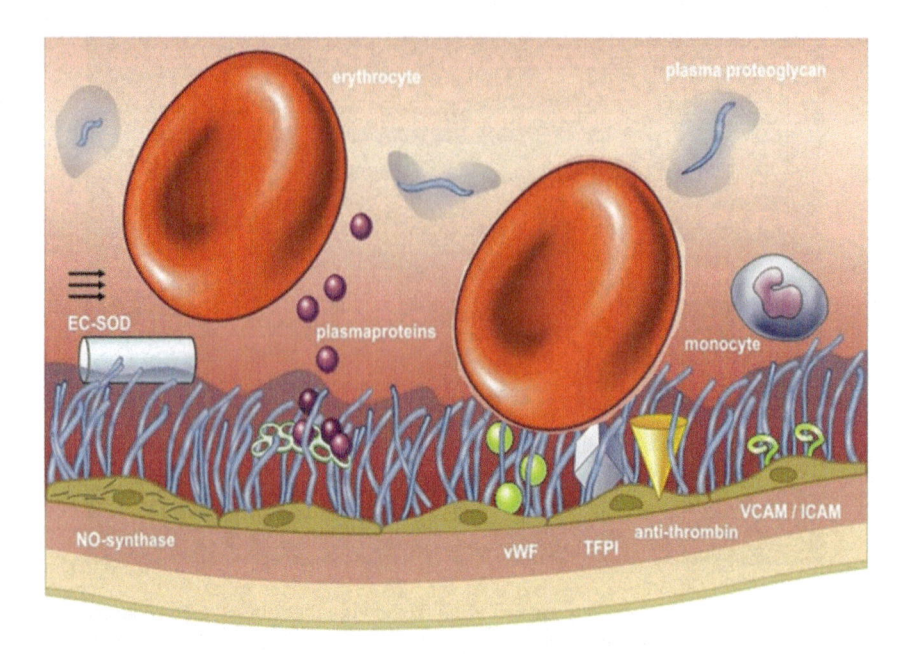

FIGURE 5.7 The EGC, with blue hairlike projections into the arterial lumen, RBC, WBC and proteins, nitric oxide (NO), clotting factors, and antioxidant enzymes.

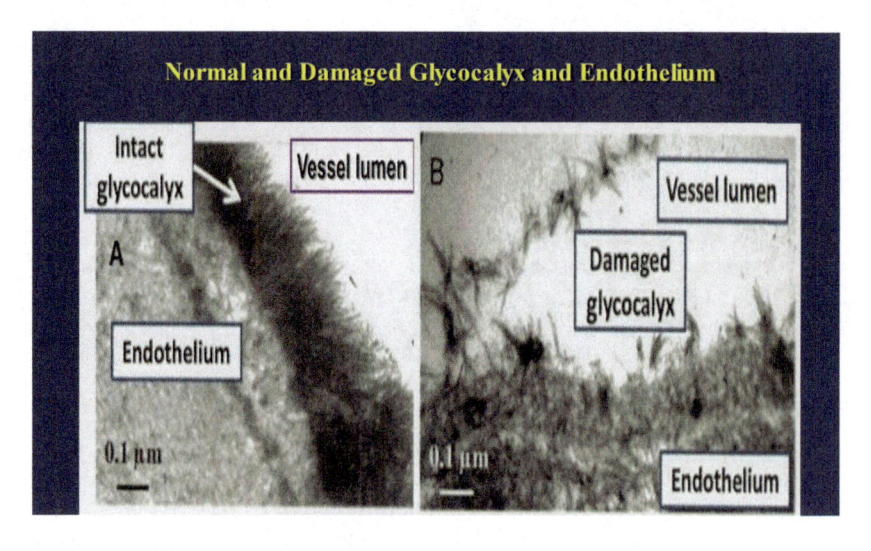

FIGURE 5.8 The normal and intact EGC on the left and the damaged EGC on the right. Note the disruption of the structure of the thin hairlike projections.

Arteries, Endothelium, Glycocalyx, NO, CHD

- EGC provides anti-inflammatory and antioxidant effects and thrombo-resistance (reduces clotting), decreases leakage from the blood into the artery, and slows the process of atherosclerosis. It responds to stress by triggering the synthesis and release of nitric oxide from endothelial cells to maintain normal vascular tone and prevent damage to the EGC (glycocalyx dysfunction (GD)) (Figure 5.8).

THE ENDOTHELIUM

The endothelium is one of the most important parts of the artery, and its role in protecting us from CHD is crucial. It is like the air traffic control tower that communicates to all the substances and the blood cells in the lumen as well as to the arterial muscle (1–13). It works in direct communication with the glycocalyx. The endothelium is a thin layer of cells (Figure 5.9) that separates the blood from the muscle of the artery (the media) (Figure 5.1). The left-side panel in Figure 5.9 is a normal endothelium, and the right-side panel presents a damaged endothelium. There are many things that can damage the endothelium, such as hypertension, high cholesterol, high blood sugar, diabetes mellitus, smoking, obesity, high homocysteine, and more. Both the structure and function of the endothelium are changed in the right panel of Figure 5.9. These changes are abnormal and cause the endothelium to not function in its usual fashion. The medical term for this is endothelial dysfunction (ED). The endothelium serves as a barrier (like a solid plank fence) to prevent things in the blood from crossing over into the subendothelial layer and the arterial muscle (1–13). However, just as a damaged fence allows undesired passage from one side to the other, when the endothelium is damaged, some of the endothelial cells become separated and the membrane will start to leak substances from the blood (see the right-side panel in Figure 5.9).

The endothelium lies in a very key area and determines what happens in the blood, the subendothelial layer, and the artery muscle (Figure 5.10) (1–13). The alterations in the blood will make the platelets stick together and form blood clots, make the

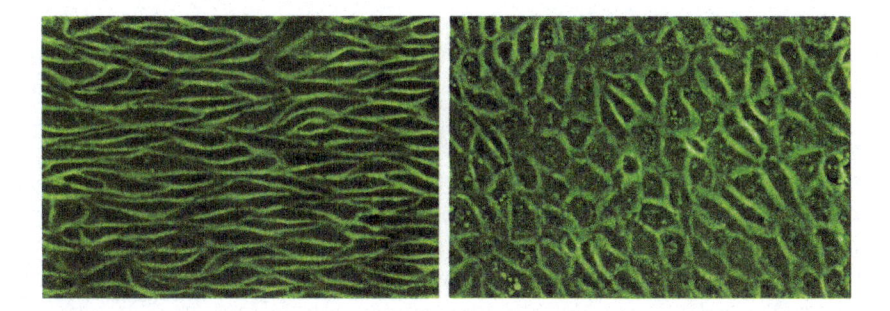

FIGURE 5.9 Normal and damaged endothelium showing normal endothelial cells on the left, which have normal structure and function. The right panel shows abnormal endothelial cells with disrupted structure and function (ED).

The Endothelium Maintains Vascular Health

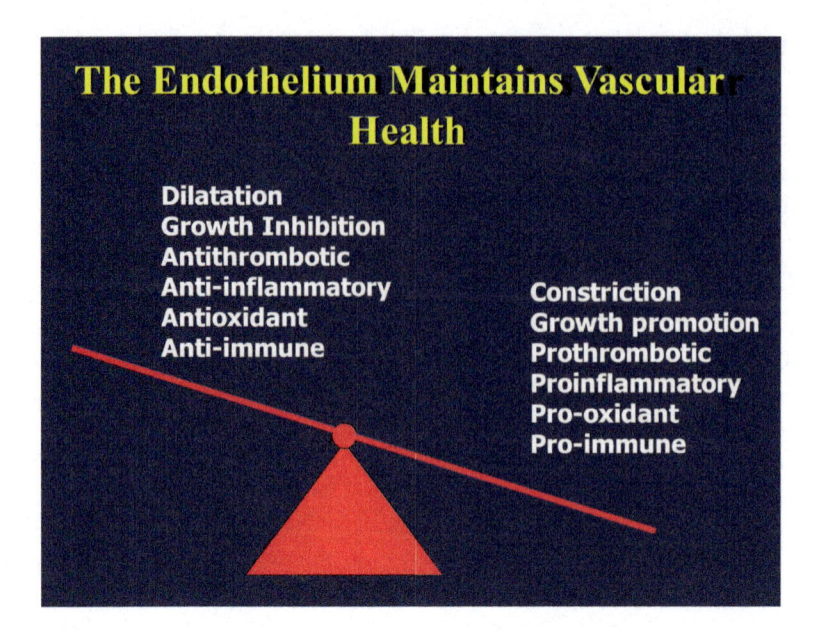

Dilatation
Growth Inhibition
Antithrombotic
Anti-inflammatory
Antioxidant
Anti-immune

Constriction
Growth promotion
Prothrombotic
Proinflammatory
Pro-oxidant
Pro-immune

FIGURE 5.10 The vascular endothelium maintains the balance of vascular health by making numerous compounds that regulate blood pressure and vasodilation, vascular inflammation, oxidative stress, immune function, risk of clotting thrombosis), and growth in the heart muscle and arteries.

white blood cells attach to the endothelial lining, and cause inflammation, oxidative stress, and immune dysfunction of the artery. This starts the process of total arterial damage with the accumulation of cholesterol, fats, and other substances, leading to fatty deposits, fatty streaks, complex fat, and inflammatory cell accumulation—and then the formation of a plaque with atherosclerosis, CHD, and MI.

The alterations to the arterial muscle include leaking of the blood vessel with the loss of proteins into the cells, tissues, and organs. If this happens in the kidney, then the proteins spill into the urine (proteinuria), which is very abnormal and predicts future kidney disease. Another alteration is the constriction or narrowing of the artery with reduced blood flow and hypertension. Finally, there may be abnormal growth, thickening, and stiffness of the artery which leads to arteriosclerosis and subsequent CHD and MI.

The endothelium acts like an endocrine organ which makes numerous compounds that regulate blood pressure, vascular inflammation, oxidative stress, immune function, and risk of clotting and growth in the heart muscle and arteries. The functions determine future CHD, MI, and overall arterial health. One of these compounds is *nitric oxide*, which helps to dilate the artery and lower blood pressure. Nitric oxide is a short-lived gas that also provides numerous other health benefits to the artery, reducing blood pressure and the risk of CHD. The endothelium must function normally to prevent CHD (1–13). The endothelium is the largest organ in the body (2–5, 11–13). It is the size of six and a half tennis courts or about 14,000 square feet (2–5, 11–13).

If the endothelium becomes damaged and does not make as much nitric oxide but rather produces other compounds that damage the arteries, then there is vasoconstriction, increase in blood pressure, increase in clotting thrombosis), abnormal growth of the heart and arterial muscle, and increase in vascular inflammation, oxidative stress, and immune dysfunction. Thus, damage to the endothelium increases the risk for CHD, MI, and generalized atherosclerosis, with plaque buildup as well as hardening of the arteries (stiffness or arteriosclerosis) (2–5, 11–13). This endothelial dysfunction may precede the development of CHD by decades (2–5, 11–13).

As the artery becomes stiff and loses its elastic stretch, the blood pressure goes higher and the artery can leak, burst, or form a clot. This results in an MI or stroke (1–13).

THE ARTERIAL MUSCLE (MEDIA)

The Artery Muscle or Media

The media is the muscle of the artery (Figure 5.1). This would be similar to the outside of a water hose. The muscle of the artery can dilate or constrict, which will change the flow in the artery and the blood pressure. Depending on the size of the artery, there may be more or less muscle present. For example, the aorta is the largest artery in the body and has a significant amount of muscle. However, the coronary smaller arteries and the brain have less muscle and more endothelium. If the blood pressure is elevated, it will cause all of the arteries to become thicker and stiffer which may seriously reduce the blood flow in the smaller arteries to the heart and cause anginal chest pain and CHD. Also, the arteries could primarily become stiff and thick due to other reasons, such as high blood sugar, diabetes mellitus, high cholesterol, smoking, obesity, and genetics.

THE ADVENTITIA

The adventitia (Figure 5.1) is the outermost layer of the artery that surrounds the muscle, providing support to the artery and blood supply to the artery through small arteries called the "vasa vasorum".

SUMMARY AND KEY TAKE AWAY POINTS

1. Arteries have five parts: lumen, glycocalyx, endothelium, muscle (media), and adventitia.
2. The endothelium and the glycocalyx are like air traffic control systems of the coronary arteries (as well as other arteries) and determine what happens in the blood. The subendothelial layer and the artery muscle control clotting, leakage, contraction of the muscle, hypertension, vasodilation, growth, inflammation, oxidative stress, immune dysfunction of the artery, atherosclerosis, arteriosclerosis, and CHD.
3. The endothelium makes nitric oxide and many other substances that help to reduce blood pressure, atherosclerosis, arteriosclerosis, CHD, and MI.

4. Glycocalyx dysfunction (GD) and endothelial dysfunction (ED) of the coronary arteries may precede the development of CHD by decades.
5. There are many treatments for endothelial dysfunction and glycocalyx dysfunction that will be discussed in subsequent chapters of this book (Chapters 13 and 21).

REFERENCES

1. Whelton PK et al. ACC/AHA/AAPA/ABC/ACPM/AGS/APhA/ASH/ASPC/NMA/PCNA guideline for the prevention, detection, evaluation, and management of high blood pressure in adults: A report of the American College of Cardiology/American Heart Association Task Force on clinical practice guidelines. *Hypertension.* 2018 Jun;71(6):e13–e115.
2. Houston M. The role of nutrition and nutraceutical supplements in the treatment of hypertension. *World Journal of Cardiology* 2014;6(2):38–66.
3. Houston M. Nutrition and nutraceutical supplements for the treatment of hypertension: Part 1. *Journal of Clinical Hypertension* 2013;15:752–7.
4. Houston M. Nutrition and nutraceutical supplements for the treatment of hypertension: Part II. *Journal of Clinical Hypertension* 2013;15:845–51.
5. Houston M. Nutrition and nutraceutical supplements for the treatment of hypertension: Part III. *Journal of Clinical Hypertension* 2013;15:931–7.
6. Borghi C, Cicero AF Nutraceuticals with a clinically detectable blood pressure-lowering effect: A review of available randomized clinical trials and their meta-analyses. *British Journal of Clinical Pharmacology*, 2017;83(1):163–71.
7. Sirtori CR, Arnoldi A, Cicero AF. Nutraceuticals for blood pressure control. Review. *Annals of Medicine* 2015;47(6):447–56.
8. Cicero AF, Colletti A. Nutraceuticals and blood pressure control: Results from clinical trials and meta-analyses. *High Blood Pressure & Cardiovascular Prevention.* 2015;22(3):203–13.
9. Turner JM, Spatz ES. Nutritional supplements for the treatment of hypertension: A practical guide for clinicians. *Current Cardiology Reports.* 2016;18(12):126. Review.
10. Caligiuri SP, Pierce GN. A review of the relative efficacy of dietary, nutritional supplements, lifestyle and drug therapies in the management of hypertension. *Critical Reviews in Food Science and Nutrition* 2016 Aug 5:0. [Epub ahead of print].
11. Houston MC, Fox B, Taylor N. *What Your Doctor May Not Tell You About Hypertension. The Revolutionary Nutrition and Lifestyle Program to Help Fight High Blood Pressure.* New York: AOL Time Warner, Warner Books; September 2003.
12. Houston M. Treatment of hypertension with nutrition and nutraceutical supplement: Part 1. *Alternative and Complimentary Medicine.* 2019;24:260–75.
13. Houston M. Treatment of hypertension with nutrition and nutraceutical supplement: Part 2. *Alternative and Complimentary Medicine.* 2019;25:23–36.
14. Van den Berg, Vink & Spaan. The endothelial glycocalyx protects against myocardial edema *Circulation Research* (2003), 92: 592–4.
15. Fu BM et al. Systems. *Biology and Medicine* (2013);5:381–90.

6 What Do the Heart Arteries Do When They Are Damaged? The Infinite Insults and Finite Responses in CHD

INTRODUCTION

In response to a large number of insults due to our genetics, environment, and nutrition, the human body reacts in three specific ways: vascular inflammation, vascular oxidative stress, and vascular immune dysfunction. I will define and explain all of these in this chapter (Figure 6.1). Apart from targeting the abnormalities in cholesterol, blood pressure, blood glucose, obesity and tobacco cessation, and controlling all the other CHD risk factors, treatments that can improve and regulate chronic vascular inflammation, vascular oxidative stress, and vascular immune response may prove to be very promising strategies in the management of CHD.

INFLAMMATION

WHAT IS INFLAMMATION?

At one time, it was thought that CHD and MI were caused only by excess cholesterol, fats, and other types of cells that blocked the coronary arteries. This blockage inside the artery, called a "plaque", eventually got large enough to stop the blood flow and cause an MI. However, it is now known that vascular inflammation plays a very important and primary role in causing CHD and MI from the beginning. Inflammation is the body's natural response to prevent infection and repair damage and heal the injured tissue. The classic signs of inflammation are redness, pain, heat, and swelling. For example, if you cut your hand, bacteria enter the body and are recognized as abnormal foreign invaders to which the body produces a defense by sending WBCs, immune cells, and red cells to the damaged area. Various chemicals are produced by these cells that kill the bacteria, the inflammation resolves, and the cut hand will heal. This is normal, short-lived, helpful, and required to resolve this acute problem and prevent long-term inflammation.

However, if the inflammation and immune cells do not win and kill the bacteria, the result is chronic inflammation, and these immune system cells and chemicals now

Mechanism vascular disease

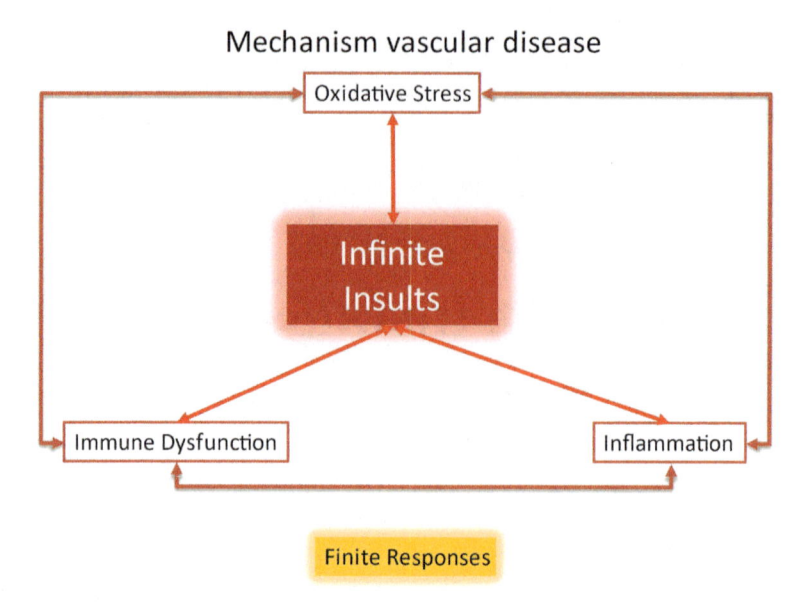

[Diagram: "Infinite Insults" at center, connected bidirectionally to three "Finite Responses": Oxidative Stress, Immune Dysfunction, and Inflammation.]

FIGURE 6.1 The three finite vascular responses.

attack the body's normal cells and damage them. In other words, the normal process becomes abnormal and the body's cells become damaged as "innocent bystanders". Chronic inflammation is one of the most important causes of CHD and MI as well as other diseases such as arthritis, diabetes mellitus, hypertension, and cancer.

In the blood vessel, the initial "cut" or damage occurs at the lining of the artery from the lumen called the glycocalyx and endothelium (Chapter 5). This is a single layer of cells that protects the artery and makes many chemicals and substances to fight CHD, MI, and chronic inflammation. If this layer starts to "leak" then immune cells (such as T cells and macrophages), inflammation chemicals, bad cholesterol particles (LDL-P) and oxidized LDL (oxLDL), as well as other substances, go into the subendothelial layer and the artery wall, start to grow, and expand, resulting in plaque formation, CHD, and MI (Figure 6.2).

How Do We Recognize and Measure Inflammation?

Inflammation in the arteries can be easily measured with several blood tests. One of the most common and best validated in scientific studies is the high-sensitivity C reactive protein (HS-CRP). Inflammatory chemicals from all over the body, such as adipose tissue macrophages (a type of white blood cell), and also from the heart and arteries, travel to the liver where HS-CRP is made and released into the blood stream (Figure 6.3). The precursors for HS-CRP are called "proinflammatory cytokines" and include interleukin I B (IL-1B), interleukin 6 (IL-6), and tumor necrosis factor alpha (TNF alpha). The HS-CRP can be elevated in CHD due to poor nutrition, too many refined carbohydrates and starches, high trans fat and some saturated fat intake, hypertension, high cholesterol and other blood lipids, diabetes and high blood

What Do Arteries Do When They Are Damaged?

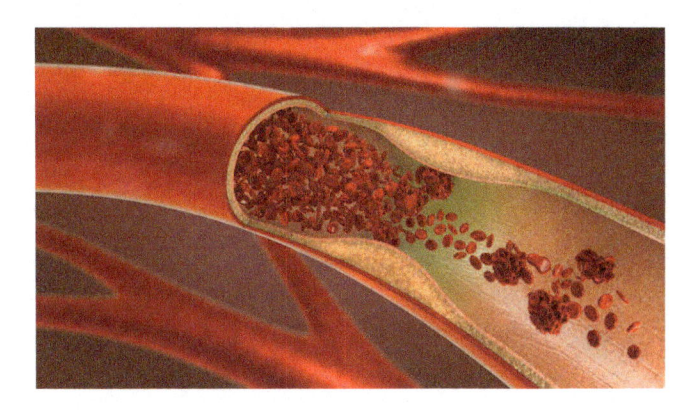

FIGURE 6.2 Arterial inflammation with plaque formation and early narrowing of the coronary artery.

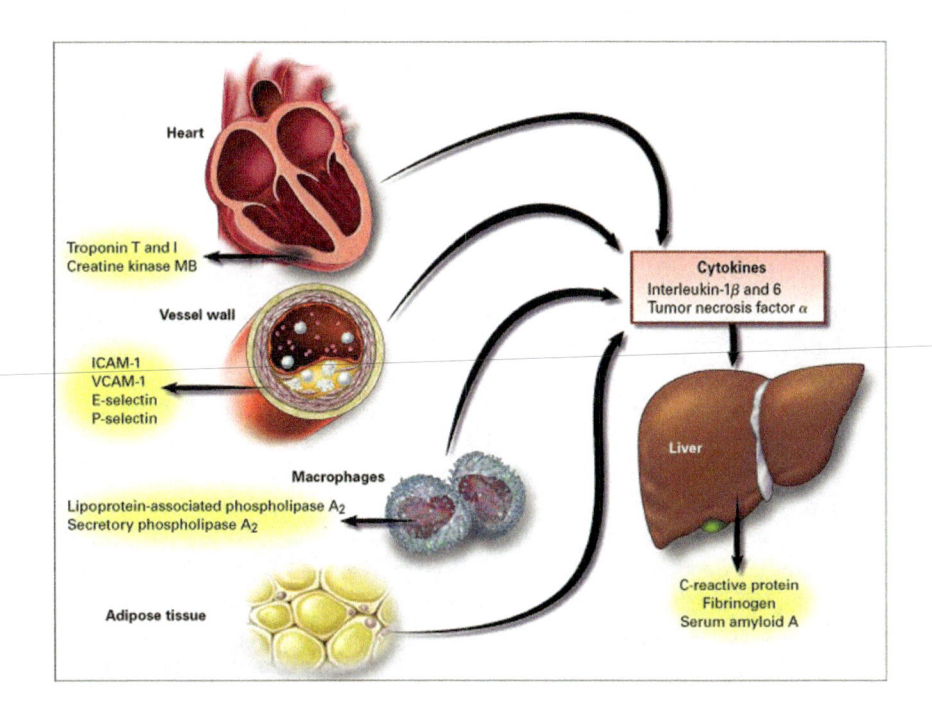

FIGURE 6.3 Formation of HS-CRP in the liver from the proinflammatory cytokines, interleukin 1b, interleukin 6, and TNF alpha that are from the heart, blood vessels, macrophages, and adipose tissue.

glucose, obesity with a high percentage of belly fat (visceral fat), smoking, high homocysteine, and many others. In addition, lack of sleep, obstructive sleep apnea, lack of exercise, infections of any type (viral, bacterial, parasites, fungus, tuberculosis), arthritis, autoimmune diseases, heavy metals, high uric acid, cancer, and other diseases will elevate HS-CRP. If the HS-CRP is over 1 mg/L, then it is both a risk marker and a risk factor for CHD. This means that the HS-CRP can actually cause artery damage by itself. Nitric oxide is inhibited by HS-CRP, which increases the risk of hypertension, CHD, and MI. The HS-CRP should be lowered to normal, the underlying cause(s) for the elevation identified, and the causes removed and treated.

How to Treat Vascular Inflammation

There are many nutritional recommendations, natural supplements, and lifestyle changes that will reduce vascular inflammation, lower HS-CRP, interleukins, TNF alpha, and other cytokines. Here are some of the best recommendations for you to consider (see Chapters 12 and 13).

1. Start an anti-inflammatory diet. Reduce refined carbohydrates and starches to less than 25 grams per day. For example, avoid all breads, pasta, white rice, white potatoes, sweets, sugars, sodas, and desserts. Avoid all artificial sweeteners. Include extra virgin olive oil (EVOO) four to five tablespoons per day on your salads and other foods.
2. Eat 12 servings of fresh fruits and vegetables daily. Include more vegetables of variable colors (8 servings) and many types of fruits, especially berries (4 servings).
3. Practice fasting. Intermittent fasting, overnight fasting, or the fasting mimicking diet (FMD) will reduce weight, body fat, and belly fat and lower inflammation. The FMD has also been shown to increase the production of stem cells, and to reduce CHD, and to slow aging in primates and other animal species. Stem cells are cells from the bone marrow or mesenchymal fat tissue that have the potential to form younger cells that are healthy in many tissues.
4. Avoid all trans fats and dramatically reduce long-chain saturated fats (C12 and higher) and processed foods (Chapters 12 and 13).
5. Maintain an ideal body weight, body mass index (BMI), total body fat, and belly fat (visceral fat) of 22% or less in women and 16% or less in men. Fat contains substances called "adipokines" that increase inflammation. Maintain good lean muscle mass by doing resistance exercises and increasing organic high-quality proteins from various sources.
6. Exercise for one hour per day, 4 days per week or more doing both resistance and aerobic training. This will be discussed later in the book with the ABCT exercise program (Chapter 21).
7. Stop using all tobacco products and smokeless tobacco.
8. Get eight hours of restful sleep per night.
9. Reduce stress, anxiety, and depression.

SUPPLEMENTS

All supplements should be of high quality and obtained from reputable nutrition companies. I have listed some of the best supplement companies below next to the supplement indicated. Other supplement companies are listed in the Sources section at the back of this book. When you call to order, please tell them you read about them in this book and give them my name and the Hypertension Institute.

Biotics Research Corporation: 800-231-5777
Designs for Health: 860-623-6314
Ortho Molecular Lab: 800-332-2351

I personally use these companies and recommend their products as well to patients. If a company is not listed below, then the Sources section will have other suggestions.

1. Omega-3 fatty acids: 1–4 grams per day; EFA Sirt Supreme (Biotics).
2. Curcumin: 1–2 grams per day; Curcumin Rx (Biotics) or Curcumin Evail (Designs for Health).
3. Boswellia: 500 mg twice a day.
4. Bromelin: 200 mg twice a day.
5. Cacao and dark chocolate: 30 grams per day.
6. Quercetin with nettles: 500 mg twice a day (Designs for Health).
7. Carnosine: 500 mg twice a day (Designs for Health).
8. Coenzyme Q 10 (CoQNOL): 200 mg per day (Designs for Health).
9. Vitamin D liquid emulsion: 2000 IU per day (Designs for Health).
10. Vitamin C: 500 mg twice a day.
11. Vitamin E: as gamma tocopherol 100 mg per day (Designs for Health or AC Grace).
12. Vitamin K2 MK 7: 360 micrograms per day (Ortho Molecular).
13. Lycopene: 20 mg per day.
14. Selenium: 200 micrograms per day.
15. GLA (gamma linolenic acid): 1 gram per day.
16. R lipoic acid: 100 mg per twice a day (Designs for Health).
17. Lutein: 5 mg per day.
18. Zinc: 50 mg per day (Designs for Health).
19. Ginseng: 200 mg per day.
20. Pomegranate seeds 1/4 cup per day or pomegranate juice 6 ounces per day.
21. Pycnogenol: 200 mg per day.
22. Green tea extract (EGCG): 500 mg twice per day (Designs for Health).
23. Trans Resveratrol: 250 mg per day; Resveratrol HP (Biotics).
24. Grape seed extract: 300 mg twice per day (Designs for Health).

In conclusion, inflammation is due to white blood cells, called T cells and B cells, that attack invaders such as bacteria, viruses, and other insults to contain, remove, and kill them. The redness, swelling, and pain that you would see on your skin after a burn, a cut, or an infection with a bacteria or other pathogen is inflammation. This same response occurs in the coronary arteries and causes CHD. Acute and chronic inflammation with abnormal vascular immune responses and involvement of pattern recognition receptors (PRR) and toll-like receptors (TLR) are definitely involved in

CHD (Figure 6.7). The PRR and TLR are inflammation receptors that are on the surface of all the arteries. There are numerous inflammatory compounds that can be measured in the blood that are excellent markers for CHD.

OXIDATION AND OXIDATIVE STRESS

Almost everyone has heard of antioxidants such as vitamin C, vitamin E, and vitamin A. But what does an antioxidant actually do? They quench or neutralize oxidants, also called free radicals. A free radical is produced as part of the normal energy process in all of our cells which require oxygen. The "atomic power plants" of our cells are called mitochondria, and they make adenosine triphosphate (ATP), which is like gasoline for your car engine. In the process of making ATP, a lot of energy is made. This process is very efficient with about 99% of the oxygen being used to make ATP energy. However, about one percent of the oxygen leaks out of the mitochondria and forms these highly reactive oxygen molecules (free radicals) that damage our cells, proteins, cell membranes, and genes. These free radicals steal electrons from other molecules and start a chain reaction of damage. Oxidation, like inflammation, is a normal biological process that is required for life. However, if it goes unchecked and becomes chronic, then the arteries become damaged, and this leads to glycocalyx and endothelial dysfunction, low levels of nitric oxide, and eventually CHD. In addition to CHD, oxidative stress can cause many other diseases and accelerate aging (Box 6.1).

BOX 6.1   Some Diseases Caused by Oxidative Stress

1. Aging
2. Coronary heart disease
3. Myocardial infarction
4. Heart failure
5. Stroke
6. Kidney disease
7. Diabetes mellitus
8. Alzheimer's disease
9. Arthritis
10. Inflammatory bowel disease
11. Parkinson's disease
12. Neurodegenerative diseases
13. Autoimmune diseases
14. Multiple sclerosis
15. Psoriasis
16. Cancer
17. Lung disease

Oxidative stress is an imbalance of the oxidative stress molecules coupled with a decrease in substances that provide an oxidative defense. The oxidative stress molecules include reactive oxygen species (ROS) and reactive nitrogen species (RNS) with a decrease in antioxidant defenses that contribute to CHD. In CHD, ROS and RNS are increased in the arteries, especially in the heart.

The predominant ROS produced is superoxide anion, which is generated by numerous cellular sources. The interaction of superoxide anion with nitric oxide (NO) will partially or completely eliminate NO. In addition, there may be production of downstream ROS such as peroxynitrite, hydroxyl ion, and hydrogen peroxide, which induce more vascular damage. All of these events lead to an oxidation of LDL cholesterol, reduction in NO bioavailability, glycocalyx, and endothelial dysfunction, loss of vascular elasticity with stiffness of the arteries, vascular and cardiac enlargement (thickening of the muscle), hypertension, vascular oxidative stress, vascular inflammation, vascular immune dysfunction, CHD, and MI.

Oxidative stress is caused by poor nutrition, an excess of refined carbohydrates, and high trans fat intake. The long-chain saturated fats (C12 or higher) may increase oxidative stress (see Chapters 12 and 13). In addition, obesity, lack of sleep, obstructive sleep apnea, lack of exercise, smoking, pollution, elevated blood iron, increased levels of myeloperoxidase (MPO, an enzyme that is made by white blood cells to kill bacteria), infections of any type (viral, bacterial, parasites, fungus, tuberculosis), arthritis, autoimmune diseases, heavy metals, high uric acid, high homocysteine, low levels of antioxidants, radiation therapy, chemotherapy, excessive sun exposure, cancer, and other diseases will increase oxidative stress.

How Is Oxidative Stress Measured in the Body?

There are numerous blood and urine tests to measure the presence and severity of oxidative stress. These tests measure the by-products of oxidative stress and vascular and cell damage.

Oxidative stress can be measured indirectly by measuring the levels of DNA/RNA genetic damage, lipid damage or peroxidation, and protein damage or oxidation and nitration, rather than a direct measurement of reactive oxygen species (ROS). These oxidative stress markers are more enduring than reactive oxygen species. All of these can be measured by your doctor.

DNA/RNA Damage

Deoxyribonucleic acid (DNA) and ribonucleic acid (RNA) are your genetic material. Several types of DNA/RNA damage occur that can be measured as oxidative stress markers. 8-Hydroxydeoxyguanosine (8-OHdG) in the blood or urine is probably the most commonly used DNA damage marker for oxidative stress. Comet assays are also used which is a microscopic examination of what the DNA looks like in the lab.

Lipid Peroxidation

Lipid Peroxidation is damage to the lipids, fats, and cell membranes, and Malondialdehyde (MDA) is the most commonly used lipid marker of oxidative stress. It is formed via peroxidation or damage of polyunsaturated fatty acids, which are part of all cell membranes. The MDA is released into the blood and will damage cells. It is measured in the lab using a TBARS assay test. There are other lipid peroxidation markers that can be measured in the blood and urine.

Protein Oxidation and Nitration

Oxidative damage to proteins can take the form of protein oxidation (oxygen damage) and protein nitration (nitrogen damage) Reactive oxygen species can also cause the formation of advanced glycation end products (AGE) that are related to high blood glucose and advanced oxidation protein products (AOPP), which indicates damage to proteins. All of these markers can be measured by standard assays.

ANTIOXIDANTS FIGHT OXIDATIVE STRESS

The body has many antioxidants to fight oxidation, ROS, and RNS. These are called endogenous (those made in the body) and exogenous (those that are obtained from an outside source) antioxidants. These are shown in Box 6.2. It is known that high levels of antioxidants in our body are associated with a reduced risk of CHD and MI and that low levels of antioxidants in our body are associated with a higher risk of CHD and MI.

BOX 6.2   Exogenous and Endogenous Antioxidants

Exogenous antioxidants

- Vitamin A and carotenoids, such as alpha and beta carotene and lycopene
- Vitamin C (ascorbic acid)
- Vitamin E (tocopherols and tocotrienols)
- Cysteine (whey protein, N-acetyl cysteine)
- Coenzyme Q 10
- Uric acid
- Lutein
- Flavonoids and polyphenols (teas, herbs, grape skin, and other foods)
- Sulfhydryl group (broccoli, lipoic acid, garlic)
- Melatonin
- Fruits and vegetables
- Mediterranean diet

Endogenous afcntioxidants

- Superoxide dismutase (SOD) breaks down the superoxide anion to water and oxygen
- Catalase (CAT) breaks down hydrogen peroxide (H2O2) into water and oxygen
- Glutathione peroxidase (GSH-Px), an important enzyme that breaks down dangerous peroxides in the blood and in the cells, such as hydrogen peroxide (H_2O_2). This enzyme requires vitamin C and selenium to be effective

- Glutathione is the most important intracellular antioxidant. It can be increased by consuming NAC (N-acetyl cysteine, lipoic acid, and whey protein.

For example, glutathione peroxidase (GSH-Px) lowers blood pressure and reduces CHD and MI. GSH-Px confers more cell, tissue, and organ protection than does SOD or catalase, or the combination of both. Patients with higher levels of GSH-Px had a 71% lower risk of MI compared to patients with the lowest levels. A low level of GSH-Px is a major CHD risk factor (*NEJM* 2003; 349:1605–13). Abnormal genes for GSH-Px also exist which increase the risk for CHD and MI. (*Coronary Artery Disease* 2003; 14:149–153).

Antioxidant Mechanisms to Reduce CHD and MI

1. Quench or neutralize many of the free electrons to decrease ROS and RNS.
2. Inhibit WBC, macrophages, and oxidized LDL from sticking to the artery wall by inhibiting molecules called cell adhesion molecules (CAMs), which are "glue-like".
3. Block oxidation of LDL, which is the form of LDL that starts the atherosclerosis process in the artery leading to CHD.
4. Reduce clotting in the arteries.
5. Modify the expression of our genes that produce the three finite responses and thus reduce inflammation, oxidative stress, and immune dysfunction in the arteries.
6. Lower blood pressure.
7. Lower LDL and increase HDL cholesterol.
8. Lower blood glucose.
9. Lower homocysteine.
10. Improve endothelial and glycocalyx function.
11. Improve arterial elasticity and expansion.
12. Improve blood flow.

How is Antioxidant Defense Measured?

Antioxidant enzymes and other oxidative defense molecules counteract the ROS and RNS that cause oxidative damage. There are three classes of antioxidants used as oxidative stress markers: small molecules, enzymes, and proteins (such as albumin).

A number of tests exist to measure the total antioxidant capacity of the blood. One of the most common total antioxidant capacity assays is the Trolox equivalent antioxidant capacity assay (TEAC). The oxygen radical antioxidant capacity (ORAC) assay and ferric reducing antioxidant power (FRAP) are other common oxidative stress tests.

Antioxidant activity can also be measured at the level of specific enzymes and molecules, such as glutathione, catalase, SOD, and GSH-Px and others shown in Box 6.3. Your physician can order these for you.

> **BOX 6.3   Tests to Measure Oxidative Defense**
>
> TEAC
> ORAC
> FRAP
> Ascorbic acid
> NAD/NADH
> NADP/NADPH
> GSH/GSSG ratio
> Thiol
>
> Intracellular glutathione (GSH)
>
> Superoxide dismutase
> Glutathione reductase
> Xanthine oxidase
> Glutathione peroxidase
> Aconitase
> Catalase
> Thioredoxin reductase

How to Treat Oxidative Stress

There are many foods and supplements that will treat oxidative stress. Eating 12 servings of fresh fruits and vegetables per day is the foundation of oxidative defense to reduce oxidative stress. The Mediterranean diet is one of the best-proven diets to decrease oxidative stress and reduce CHD and MI (Box 6.4 and Table 6.1).

> **BOX 6.4   Top Antioxidant Foods and Supplements with Their Antioxidant Capacity**
>
> **Foods**
>
> 1. Small red beans
> 2. Wild blueberries
> 3. Red kidney beans
> 4. Pinto beans
> 5. Cultivated blueberries
> 6. Cranberries
> 7. Artichokes
> 8. Blackberries
> 9. Prunes
> 10. Raspberries

11. Strawberries (Organic to avoid pesticides)
12. Red delicious apples (Organic to avoid pesticides)
13. Granny Smith apples (Organic to avoid pesticides)
14. Pecans
15. Sweet cherries
16. Black plums
17. Russet potatoes
18. Black beans
19. Black plums
20. Gala apples
21. Dark leafy greens
22. Dark chocolate
23. Kale
24. Spinach
25. Red cabbage
26. Beets

TABLE 6.1
Best Sources of Food Antioxidants: Top 20 Fruits, Vegetables, and Nuts (as Measured by Total Antioxidant Capacity Per Serving Size)

Rank	Food item	Serving size	Total antioxidant capacity per serving size
1	Small red beans (dried)	Half cup	13,727
2	Wild blueberry	1 cup	13,427
3	Red kidney bean (dried)	Half cup	13,259
4	Pinto bean	Half cup	11,864
5	Blueberry (cultivated)	1 cup	9019
6	Cranberry	1 cup (whole)	8983
7	Artichoke (cooked)	1 cup (hearts)	7904
8	Blackberry	1 cup	7701
9	Prune	Half cup	7291
10	Raspberry	1 cup	6058
11	Strawberry	1 cup	5938
12	Red delicious apple	One	5900
13	Granny Smith apple	One	5381
14	Pecan	1 ounce	5095
15	Sweet cherry	1 cup	4873
16	Black plum	One	4844
17	Russet potato (cooked)	One	4649
18	Black bean (dried)	Half cup	4181
19	Plum	One	4118
20	Gala apple	One	3903

Top Antioxidant Supplements

All supplements should be high quality and obtained from reputable companies. I have listed two of the best supplement companies below next to the supplement if indicated. Also see the Sources section at the end of this book. When you call to order, please tell them you read about them in this book and give them my name and the Hypertension Institute.

Biotics Research Corporation: 800-231-5777
Designs for Health: 860-623-6314
1. Omega-3 fatty acids: 1–4 grams per day; EFA Sirt Supreme (Biotics).
2. Curcumin: 1–2 grams per day; Curcumin Rx (Biotics) or Curcumin Evail (Designs for Health).
3. Kyolic garlic: one twice per day.
4. Lutein: 5 mg per day.
5. Lycopene: 20 mg per day.
6. Pycnogenol: 200 mg per day.
7. Green tea extract (EGCG): 500 mg twice per day (Designs for Health).
8. Trans Resveratrol: 250 mg per day Resveratrol HP (Biotics).
9. Grape seed extract: 300 mg twice per day (Designs for Health).
10. Selenium: 200 micrograms per day.
11. Vitamin C: 500 mg twice per day.
12. Vitamin E as gamma tocopherol: 100 mg per day (Designs for Health or AC Grace).
13. Cacao and dark chocolate: 30 grams per day.
14. R lipoic acid: 100 mg twice per day (Designs for Health).
15. Zinc: 50 mg per day (Designs for Health).
16. Co enzyme Q 10 (COQNOL): 200 mg per day (Designs for Health).

In conclusion, oxidative stress is an imbalance of radical oxygen species (ROS) and radical nitrogen species (RNS) (the bad guys) with a decrease in antioxidant defenses (the good guys) that contributes to CHD and MI in humans based on genetics and environment. Oxidative stress is like having too much fire in the arteries but no fire extinguishers to put out the fire, i.e., not having enough antioxidants. The oxidative stress will damage cells to the point that they do not function or they die. If this happens in the coronary arteries, the result is CHD or MI. It is important to balance the oxidative stress and the oxidative defense. The predominant ROS produced by cells is called superoxide anion and is part of our normal metabolism with oxygen and the breakdown of our food to make energy or ATP. In addition, the superoxide anion reduces nitric oxide and produces other downstream ROS and RNS, which leads to endothelial and glycocalyx dysfunction and CHD. Our antioxidant defense is supplied by various compounds (enzymes) that will break down the ROS and also by the intake of vitamins, minerals, and antioxidants in our diet or in supplements.

Vascular Immune Dysfunction

The immune system is like a continuous "security system" that is monitoring your blood for invaders, such as bacteria, viruses and other organisms, cancer cells, and modified types of LDL cholesterol. Vascular immune system dysfunction is one of the primary causes of hypertension, high cholesterol, atherosclerosis, CHD, and MI in the general population. Immune dysfunction of the arteries occurs with an elevated WBC count and involvement of T cells, also called T-helper cells, and cytotoxic T cells to induce CHD. Monocytes are a type of WBC that crosses the damaged glycocalyx and the endothelial lining, invades into the underlying subendothelial layer, and transforms into macrophages and various T-cell subtypes, which promote vascular damage.

All three finite vascular responses occur at the same time. (Figures 6.4–6.6). Five different sizes of native LDL cholesterol (also called LDL lipoprotein) and other cholesterol particles called remnant lipoproteins (RLP) in the blood can penetrate a damaged endothelium, especially the small, dense LDL particles. Once they invade the subendothelial layer, they become trapped and are then changed into a different type of LDL that is oxidized (oxLDL) or modified LDL. This is recognized by the immune cells as a foreign invader, and it is taken up by monocytes which leave the blood and cross the glycocalyx and endothelium to become macrophages

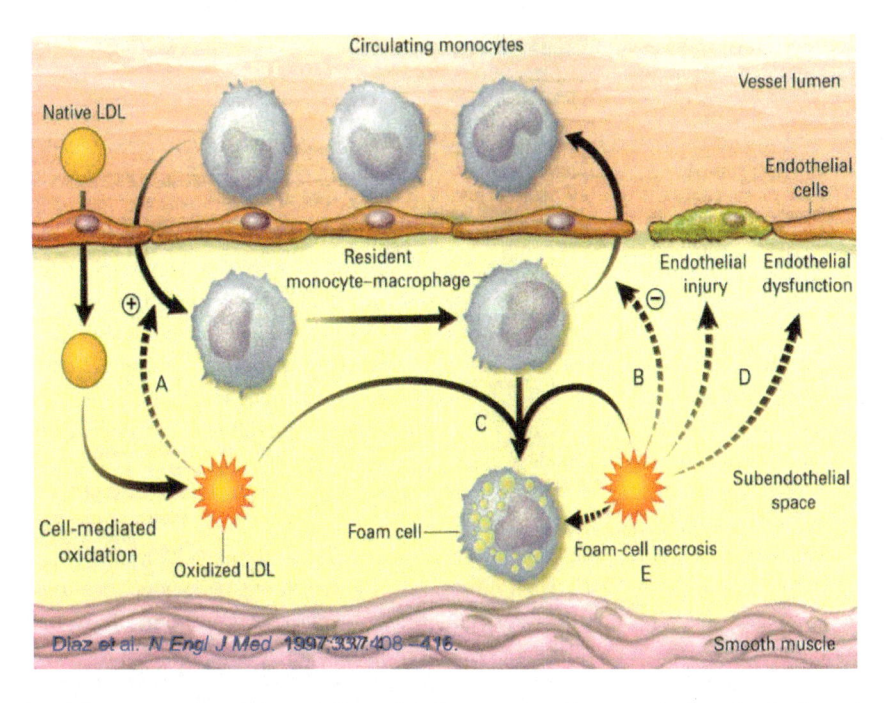

FIGURE 6.4 Vascular inflammation, oxidative stress, and immune dysfunction in the endothelium and artery.

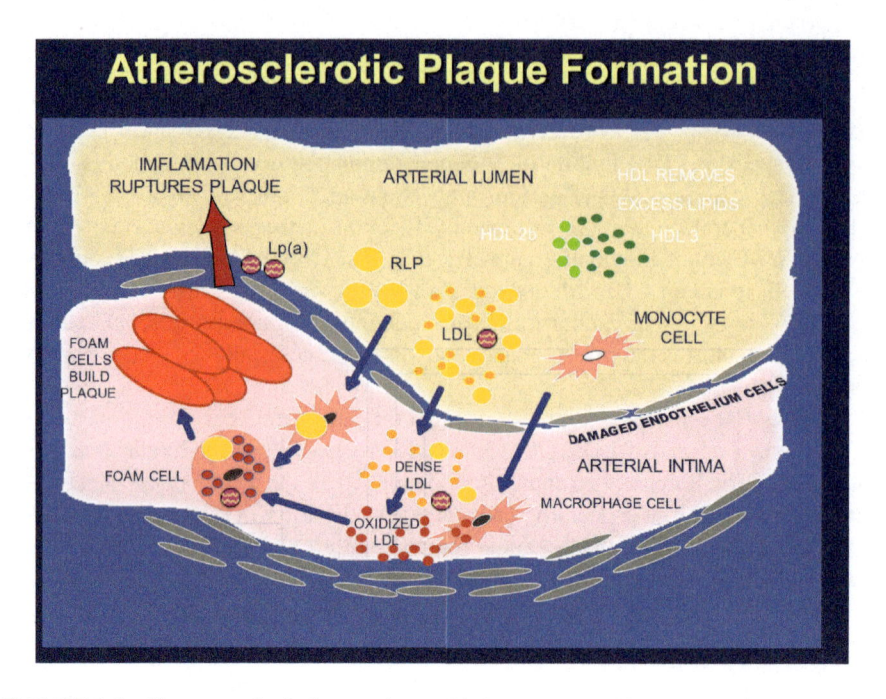

FIGURE 6.5 How vascular inflammation, oxidative stress, and immune dysfunction cause arterial plaque to form.

in an attempt to control the invasion. The macrophages literally eat these oxidized LDL particles with no "appetite control" and become full of LDL cholesterol and form larger macrophages called "foam cells". These accumulate and get larger and spread out to form "fatty streaks". During this time, the macrophages send out messages to other cells in the blood to assist in this battle (Figure 6.6). These cells have many names and produce many inflammatory, oxidative stress and immune substances such as PAI-1, tissue factor, adhesion molecules, colony-stimulating factors, TLRs, NODs, heat shock proteins (HSPs), monocyte chemotactic protein (MCPs) that try to contain the damage done by oxidized and modified LDL cholesterol. (Figures 6.4–6.6) It is not important to remember these complicated names but simply to recognize that these responses are designed to protect the artery from damage. This is a normal and appropriate response in the short term. However, if the insults continue to damage the coronary arteries, then it becomes a chronic process that is now out of control and will produce more arterial damage. The more damage that occurs in the artery, which is now like an "innocent bystander", the faster and more severe this process progresses. Eventually, the accumulation of very bad substances forms an atherosclerotic plaque that grows and grows, invades the lumen of the artery, and then ruptures and spews all of the nasty material into the blood. A clot (thrombus) then forms that blocks the artery, stops blood flow, and causes an MI.

In conclusion, the immune system is closely involved with vascular inflammation and vascular oxidative stress with T cells, B cells, and other immune mediators

What Do Arteries Do When They Are Damaged?

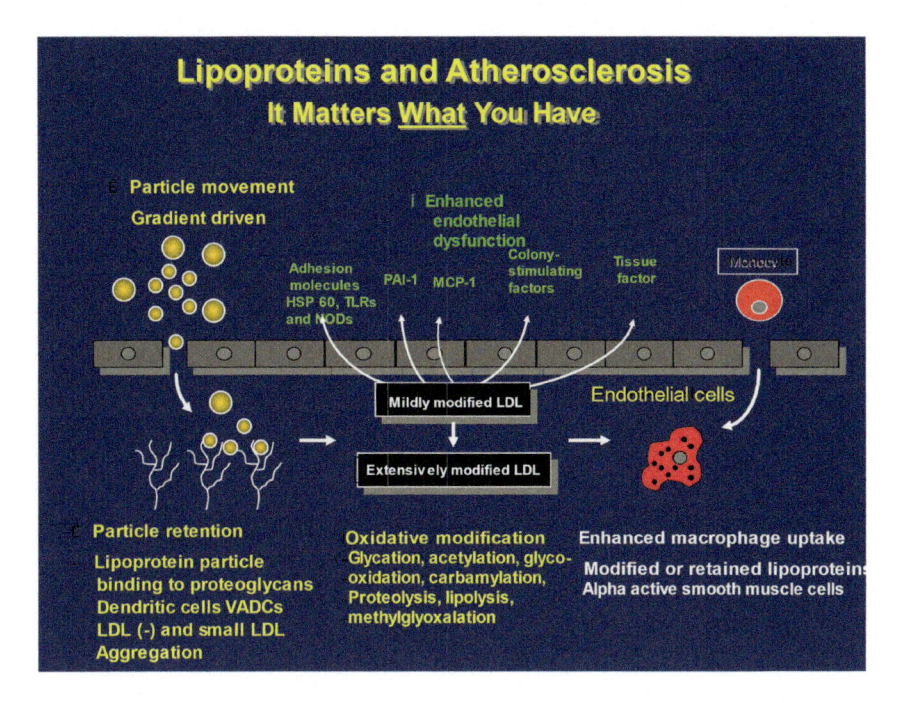

FIGURE 6.6 LDL cholesterol, LDL lipoprotein, vascular inflammation, and oxidative stress in the endothelium and in the underlying layer of the artery (subendothelial layer). Monocytes invade the glycocalyx and endothelial linings and form new and larger cells, called macrophages. . However, if those immune responses become chronic or imbalanced, then the coronary arteries and the heart will be damaged and result in CHD and MI.

to initially provide vascular protection from foreign invaders. This acute initial response is appropriate and correct. However, with chronic insults to the artery of many types, this process becomes chronic, dysregulated, and counterproductive, resulting in atherosclerosis and CHD.

FINAL CONCLUSION

CHD and MI are the result of the many insults, both biomechanical and biochemical, that will damage the coronary arteries and the heart (Figure 6.7). The receptors in the coronary arteries are like a "lock and key system". The insults are the keys and the arterial receptors are the locks. Blood pressure is the primary biomechanical insult, and the other insults are called "biochemical" and include high cholesterol, high blood sugar, diabetes mellitus, toxins, metabolic issues, and infections that can also stimulate the same receptors or cause CHD. As more and more insults are added, then CHD is more common and more severe. For example, hypertension with high cholesterol or diabetes mellitus is much worse than any of these alone in causing CHD. Some of the insults are considered to be like a "foreign invader" and are called antigens or neoantigens (new antigen or "like" an antigen). This simply

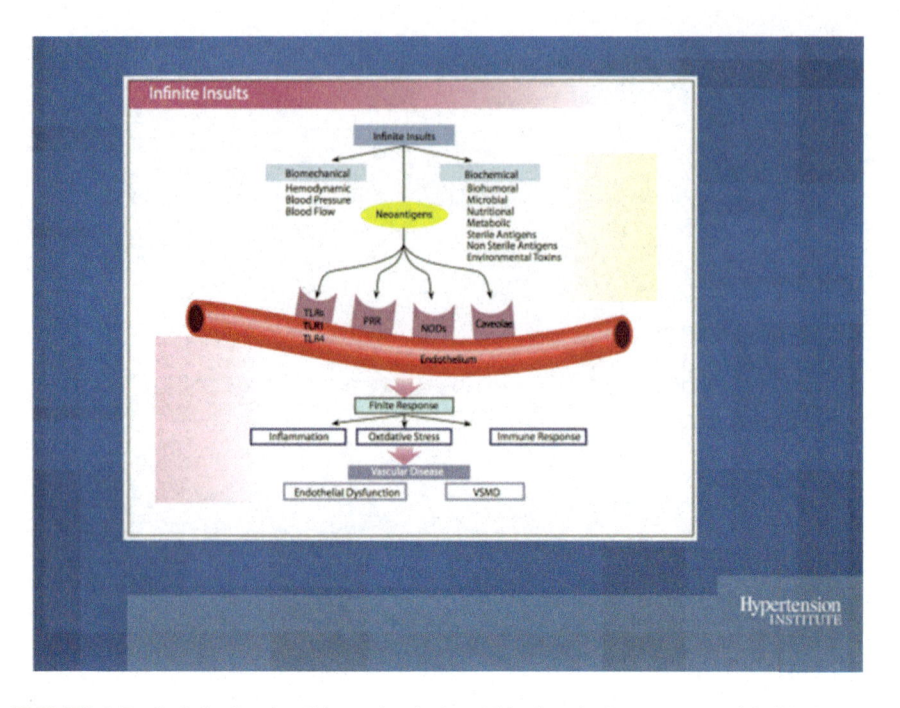

FIGURE 6.7 Infinite insults (biomechanical and biochemical groups noted in blue) to the blood vessel (noted in red) and the resultant three finite responses of inflammation, oxidative stress, and vascular immune dysfunction. These insults attach to receptors (in purple) on the artery to induce damage and disease in the coronary arteries and the heart. The receptors have various names, like pattern recognition receptors (PRR) and toll-like receptors (TLR) (noted in purple), and others that control what happens in the glycocalyx endothelium and the artery (called vascular smooth muscle dysfunction or VSMD, as noted).

means that the body does not recognize them as friendly and will send out the troops and military to fight. The finite processes are designed to protect the coronary arteries from damage in the acute setting. However, if they become chronic, then the processes to cause CHD are abnormal, dysregulated, and severe as they continue.

SUMMARY AND KEY TAKE AWAY POINTS

1. The coronary arteries have three finite or final responses: vascular inflammation, oxidative stress, and immune dysfunction.
2. These responses are generated as the result of a large number of environmental and internal insults, coupled with our genetics and nutrition.
3. Inflammation in the coronary arteries is part of the injury process that repairs and heals the acute damage to the glycocalyx and endothelium. If it is chronic, then the arteries become innocent bystanders to more damage leading to CHD.
4. Inflammation can be measured in the blood using HS-CRP, interleukins, TNF alpha, and other inflammatory markers. There are many foods and supplements that reduce inflammation.

5. Oxidation is a normal process in the body that uses oxygen to make energy that is needed for life. But if there is an imbalance with more oxidative stress than oxidative defense, then CHD is more common.
6. Oxidative stress can be measured in the blood with many tests. There are many effective foods and supplements to provide oxidative defense.
7. Vascular immune dysfunction involves various types of WBCs, such as leukocytes, lymphocytes, monocytes, T cells, and macrophages that attempt to control the invasion of the glycocalyx and endothelium of the artery by LDL cholesterol.
8. All three finite vascular responses occur at the same time and lead to foam cells, fatty streaks, and atherosclerotic plaque that may rupture and form a clot (thrombus) in the artery, leading to MI.

BIBLIOGRAPHY

1. Houston MC. *What Your Doctor May Not Tell You About Heart Disease. The Revolutionary Book that Reveals the Truth Behind Coronary Illnesses and How You Can Fight Them.* New York: Grand Central Life and Style, Hachette Book Group; 2012.
2. O'Donnell CJ, Nabel EG. Genomics of cardiovascular disease. *N Engl J Med* 2011;365:2098–109.
3. Houston MC. Nutrition and nutraceutical supplements in the treatment of hypertension. *Expert Rev Cardiovasc Ther.* 2010;8:821–33.
4. Epelman S, Liu PP, Mann DL. Role of innate and adaptive immune mechanisms in cardiac injury and repair. *Nat Rev Immunol* 2015;15:117–129.
5. Wen Y, Crowley SD. Renal effects of cytokines in hypertension. *Curr Opin Nephrol Hypertens* 2018; 27:70–76.
6. Van Laecke S, Malfait T, Schepers E, Van Biesen W. Cardiovascular disease after transplantation: An emerging role of the immune system.*Transpl Int* 2018;31: 689–699.
7. Petrie JR, Guzik TJ, Touyz RM. Diabetes, hypertension, and cardiovascular disease: Clinical insights and vascular mechanisms. *Can J Cardiol* 2018;34: 575–584. _ VOL. 25 NO. 4 191
8. Caillon A, Mian MO, Fraulob-Aquino JC, et al. Gamma delta t cells mediate angiotensin II-induced hypertension and vascular injury. *Circulation* 2017;135:2155–62.
9. Solanki A, Bhatt LK, Johnston TP. Evolving targets for the treatment of atherosclerosis. *Pharmacol Ther* 2018;187:1–12.
10. De Ciuceis C, Agabiti-Rosei C, Rossini C, et al. Relationship between different subpopulations of circulating CD41 T lymphocytes and microvascular or systemic oxidative stress in humans. *Blood Press* 2017;26: 237–45.
11. Justin Rucker A, Crowley SD. The role of macrophages in hypertension and its complications. *Pflugers Arch* 2017;469: 419–30.
12. Ritchie RH, Drummond GR, Sobey CG, et al. The opposing roles of NO and oxidative stress in cardiovascular disease. *Pharmacol Res* 2017;116:57–69.
13. Lundberg AM, Yan ZQ. Innate immune recognition receptors and damage-associated molecular patterns in plaque inflammation. *Curr Opin Lipidol.* 2011;22: 343–9.
14. Eaton SB, Eaton SB III, Konner MJ. Paleolithic nutrition revisited: a twelve-year retrospective on its nature and implications. *Eur J Clin Nutr* 1997; **51**:207–16.
15. Layne J, Majkova Z, Smart EJ, Toborek M and Hennig B. Caveolae: A regulatory platform for nutritional modulation of inflammatory diseases. *Journal of Nutritional Biochemistry* 2011;22:807–11.

16. Dandona P, Ghanim H, Chaudhuri A, Dhindsa S Kim SS. Macronutrient intake induces oxidative and inflammatory stress: Potential relevance to atherosclerosis and insulin resistance. *Exp Mol Med* 2010;42(4):245–53.
17. Kizhakekuttu TJ and Widlansky ME. Natural antioxidants and hypertension: Promise and challenges. *Cardiovasc Ther.* 2010;28(4): e20–e32.
18. Nayak DU, Karmen C, Frishman WH, Vakili BA. Antioxidant vitamins and enzymatic and synthetic oxygen-derived free radical scavengers in the prevention and treatment of cardiovascular disease. *Heart Disease* 2001;3:28–45.
19. Ritchie RH, Drummond GR Sobey CG, De Silva TM, Kemp-Harper BK The opposing roles of NO and oxidative stress in cardiovascular disease. *Pharmacol Res.* 2017;116:57–69.
20. Dhalla NS, Temsah RM, Netticadam T. The role of oxidative stress in cardiovascular diseases. *J Hypertens* 2000;18:655–73.
21. Amer MS, Elawam AE, Khater MS, Omar OH, Mabrouk RA and Taha HM. Association of high-sensitivity C reactive protein with carotid artery intima media thickness in hypertensive older adults. *J Am Soc Hypertension.* Sep-Oct 2011;5(5):395–400.
22. Rodriquez-Iturbe B, Franco M, Tapia E, Quiroz Y and Johnson RJ. Renal inflammation, autoimmunity and salt-sensitive hypertension. *Clin Exp Pharmacol Physiol.* 2012 Jan;39(1):96–103.
23. Razzouk, Munter P, Bansilal S, Kini AS, Aneja A, Mozes J, Ivan O, Jakkula M, Sharma S and Farkouh ME. C reactive protein predicts long-term mortality independently of low-density lipoprotein cholesterol in patients undergoing percutaneous coronary intervention. *Am Heart J* 2009;158(2):277–83.

7 New Concepts in Coronary Heart Disease (CHD)

Over the past few years there have been many new scientific breakthroughs and published medical articles on coronary heart disease (CHD). These new concepts include discussions on what causes CHD, how to diagnose it correctly, and many new treatments for the prevention of CHD. In this chapter, I will discuss many of these and explain the implications for you and how to prevent CHD and MI.

CONCEPT 1

As reviewed in the last chapter, the blood vessel has only three finite responses to an infinite number of insults:

Vascular inflammation.
Vascular oxidative stress.
Vascular immune dysfunction and imbalance.

A large number of insults stimulate various receptors on the blood vessel leading to these three finite responses, which can be measured in the blood or urine. Once these are identified, they can be treated to prevent or slow the progression to CHD and MI. Finding the genesis or beginning of this process and identification of these insult(s), then removing them and treating any downstream problems, is the new approach to CHD and MI.

CONCEPT 2

The blood vessel responds acutely to internal and external insults that are "correct and normal", but if these insults become chronic, the result is a long-term severely abnormal and unregulated arterial system that does not perform its normal functions. This leads to preclinical abnormal labs and other tests, then preclinical CHD or MI, and then clinical CHD and MI. The subsequent environmental–gene interactions cause specific expression patterns by our genes that produce downstream substances which damage the arteries. Proper assessment, comprehension, and treatment of the top five CHD risk factors, top 25 modifiable CHD risk factors, and the other 375 CHD risk factors and the downstream substances are required to prevent CHD. These risk factors will be reviewed in detail in Chapter 18.

CONCEPT 3

All cells are surrounded by a protective layer called a "cell membrane". This keeps unwanted things outside the cell and allows for all the parts inside the cell to function normally and not become damaged. All cell membranes are composed of "fatty acids". These include omega-3 fatty acids and monounsaturated fatty acids—good fatty acids that promote a healthy membrane which is elastic, soft, and "fluid like". However, other fatty acids, such as some long-chain saturated fats (over C12 carbon length) and all trans fats are bad and damage the cell membrane, making it hard, nonelastic, and stiff. When this happens, the cell membrane leaks and the normal receptors of the membrane do not work properly. The cell then becomes damaged inside, dysfunctional, and eventually dies. This is the same process that happens in the coronary arteries leading to CHD.

It is important to make sure that the cell membranes are mostly omega-3 and monounsaturated fatty acids. These are contained in coldwater fish, nuts, olives and olive oil, and supplements. Trans fats, such as those found in many processed foods, hydrogenated oils, and saturated fats should be reduced to less than 10% of total daily fat intake.

CONCEPT 4

There is a continuum of risks with the CHD risk factors that affect the blood vessel. For example, a blood pressure of 120/80 mm Hg is normal. For each increase in blood pressure by only 1 mm Hg the risk for CHD increases. If you have high cholesterol, for each one mg increase in LDL cholesterol the risk for CHD increases. If you have high blood sugar or diabetes mellitus, each increase in your blood sugar by one mg or your HbA1c by even one-tenth of a unit will increase CHD. This leads initially to functional abnormalities such as glycocalyx and endothelial dysfunction, and then to structural abnormalities such as stiff arteries, plaque formation, decreased blood flow, CHD, and MI.

CONCEPT 5

You have known friends or family members who appear to do everything right. They eat healthy foods, exercise, are not overweight, do not smoke, get plenty of sleep, and do not have anxiety, stress, or depression. However, one day they are diagnosed with CHD or have an MI and possibly die. You have seen others who do not eat healthy, do not exercise, are overweight, have CHD risk factors, or even smoke, who do not get CHD or MI. You ask the important question, "Why? How can this happen?"

Do the risk factors translate into CHD?

Does the absence of risk factors translate into no CHD?

There are many possible reasons for these findings:

1. Genetics. Genetics are important in predicting the risk for CHD. These are often not measured by your doctor (Chapters 9 and 10).

2. Proper testing. We must measure more sensitive indicators for CHD and measure them earlier. For example, it is important to measure endothelial dysfunction and conduct other noninvasive tests on the vascular system and heart to determine if there is early CHD. If so, then an aggressive, early prevention, and treatment program is warranted.
3. CHD risk factors. The standard or top five CHD risk factors, as they are often touted, do not adequately identify individuals at risk for CHD. In addition, there are over 400 CHD risk factors. These are not measured, not known by many doctors, or they are ignored. For example, blood pressure should be measured in the office, at home, and by using a 24-hour ambulatory blood pressure monitor. The blood cholesterol and other lipids should be measured using an advanced blood lipid testing method that measures not just total cholesterol, LDL, HDL, and triglycerides but also measures each of these related to particle numbers and size for LDL, HDL, and triglycerides. In addition, the HDL function should be measured.

A combination of targeted and integrated, personalized, and precision treatments with genetics, nutrition, nutraceutical supplements, vitamins, minerals, anti-inflammatory, antioxidant, anti-immunological agents, and drugs will be needed to reduce CHD.

CONCEPT 6

The cardiovascular disease starts very early in life (in utero) and is subclinical for 10 to 30 years or more prior to any CHD event. Endothelial and glycocalyx dysfunction is the earliest finding in CHD. This is followed by changes in arterial stiffness, elasticity, and cardiac changes. Functional changes in the cardiovascular system precede structural changes. All of these changes must be detected early in life, prevented, and treated.

SUMMARY AND KEY TAKE AWAY POINTS

CHD starts early in life, but it can be identified with new cardiovascular testing that will allow early and aggressive prevention and treatment. Assessment of the three finite vascular responses, removing the vascular insult(s), and measuring correctly all the CHD risk factors coupled with a personalized and precision cardiovascular and genetics program will help to reduce CHD and MI.

BIBLIOGRAPHY

1. Houston M. Three finite vascular responses that cause coronary heart disease and cardiovascular disease: Implications for diagnosis and treatment. *Alternative and Complimentary Medicine.* 2019;25(4):181.
2. *Nutrition and Integrative Strategies in Cardiovascular Medicine.* Houston and Sinatra, Editors. Boca Raton, FL: CRC Press; 2015.

3. *Personalized and Precision Integrative Cardiovascular Medicine.* Houston M, Editor and Contributor. Philadelphia and Chicago: Wolters Kluwer Publishers; November 2019.
4. Houston MC. Saturated fats and coronary heart disease. *Annals of Nutritional Disorders and Therapy.* 2017;4(1):1038–40.
5. Houston MC. The role of saturated fats in coronary heart disease. *Journal of Heart and Stroke.* 2017;2(3):1025–26.
6. Houston M. The relationship of saturated fats and coronary heart disease: Fact or fiction: A commentary. *Ther Adv Cardiovascular Disease.* 2017; 1:5.
7. Houston M, Minich D, Sinatra ST, Kahn JK, Guarneri M. Recent science and clinical application of nutrition to coronary heart disease. *J Am Coll Nutr.* 2018 Jan 9:1–19.
8. Houston M. The role of noninvasive cardiovascular testing, applied clinical nutrition and nutritional supplements in the prevention and treatment of coronary heart disease. *Ther Adv Cardiovasc Dis.* 2018 Mar;12(3):85–108. Epub 2018 Jan 10.
9. Houston M. Cardiovascular disease: Good fats and bad fats. *EC Cardiology.* 2018;5(5):239–46.

8 The Causes of Chest Pain
Is It a Heart Attack or Something Else?

You should never ignore any type of chest pain. When it occurs, go to the nearest emergency department or the hospital for immediate evaluation. Remember that CHD and MI are the number-one causes of death in the US. There are many causes of chest pain that with variable presentations with or without other symptoms. Many of these are emergencies just like a heart attack and need immediate medical care as they are life threatening. It is not your job to try to determine the cause of the chest pain but to seek medical attention as soon as possible.

REVIEW OF CHD SYMPTOMS

1. Pain with exercise that is diffuse or substernal chest pressure, tightness, heaviness, chest pain, ache, cramping, and stabbing or discomfort with or without radiation to the arms, neck, shoulders, or back.
2. Shortness of breath (dyspnea).
3. Sweating (diaphoresis).
4. Nausea and vomiting.
5. Dizziness.
6. Pale color of face and skin.
7. Palpitations (fast or irregular heart rate).
8. Weakness and fatigue.
9. Stomach and abdominal pain with a "heart burn sensation".

CAUSES OF CHEST PAIN IN ADDITION TO CHD OR MI

1. Blood clot in the lung called a pulmonary embolus (PE).
2. Lung disease, such as chronic obstructive pulmonary disease (COPD); chronic bronchitis; pleurisy (pain that occurs with deep breathing from irritation of the lining around the lung (pleura)); asthma; infections of all types (viral, bacteria, fungal, tuberculosis); pneumonia; collapsed lung; sarcoidosis; pneumothorax (air around the lung or air in the middle chest cavity); pleural effusions.
3. Chest wall pain in the ribs, tendon, and ligaments of the chest; muscle strain and inflammation of the muscles in the chest; injured ribs (costochondritis or Tietze's syndrome).

4. Fibromyalgia (a disease of unknown cause with generalized muscle pain or aching).
5. Slipping rib syndrome.
6. Aortic aneurysm or dissection of the thoracic aorta that is in the chest. This is dilatation of the aorta or an actual tear in the aortic wall that can result in rapid blood loss, shock, and death.
7. Pericarditis (inflammation of the lining of the heart).
8. Myocarditis (inflammation of the heart muscle).
9. Hypertrophic cardiomyopathy (thickened heart muscle that can block the flow of blood out of the heart (left ventricle)).
10. Abnormal heart valves, such as mitral valve prolapse. However, any valve can be involved with stenosis of leakage such as the aortic, tricuspid, or pulmonary valve, as well as the mitral valve.
11. Pulmonary hypertension (high blood pressure in the arteries of the lung).
12. Coronary artery dissection (the wall of the artery separates with blood in the wall of the artery).
13. Gastroesophageal reflux (GERD) (acid from the stomach goes into the esophagus).
14. Hiatal hernia (part of the stomach moves into the chest cavity through a hole in the diaphragm).
15. Peptic ulcer (stomach ulcer).
16. Gastritis (lining of the stomach has inflammation).
17. Esophageal spasms and contraction disorders.
18. Esophagus is missing contractions caused by nerve loss (achalasia).
19. Esophageal rupture.
20. Biliary tree diseases, spasm, dysfunction, gallstones.
21. Pancreatitis and pancreatic cancer.
22. Abnormalities in the deep structures in the chest cavity called the mediastinum, such as infections, abscess, bleeding, tumors, or enlarged lymph nodes.
23. Lung cancer.
24. Esophageal cancer.
25. Stomach cancer.
26. Herpes zoster (shingles).
27. Sickle cell crisis.
28. Stress, anxiety, or depression.

SUMMARY AND KEY TAKE AWAY POINTS

There are many causes of chest pain. Some of these are serious and life threatening. Angina chest pain or MI can have variable types of chest pain depending on age, gender, and other issues. If you have chest pain with or without other symptoms you need to get an immediate medical evaluation by your doctor, the emergency department, or hospital. Do not wait and assume that the chest pain is not serious. Your life depends on rapid action and evaluation.

BIBLIOGRAPHY

1. Chang AM, Fischman DL, Hollander JE. Evaluation of chest pain and acute coronary syndromes. *Cardiol Clin.* 2018 Feb;36(1):1–12.
2. Sharron Rushton MJC. Chest pain: If it is not the heart, what is it? *Nurs Clin North Am.* 2018 Sep;53(3):421–31.

9 CHD Genetics Part I
What Is in the Genes That Place You at Risk

THE HUMAN GENOME

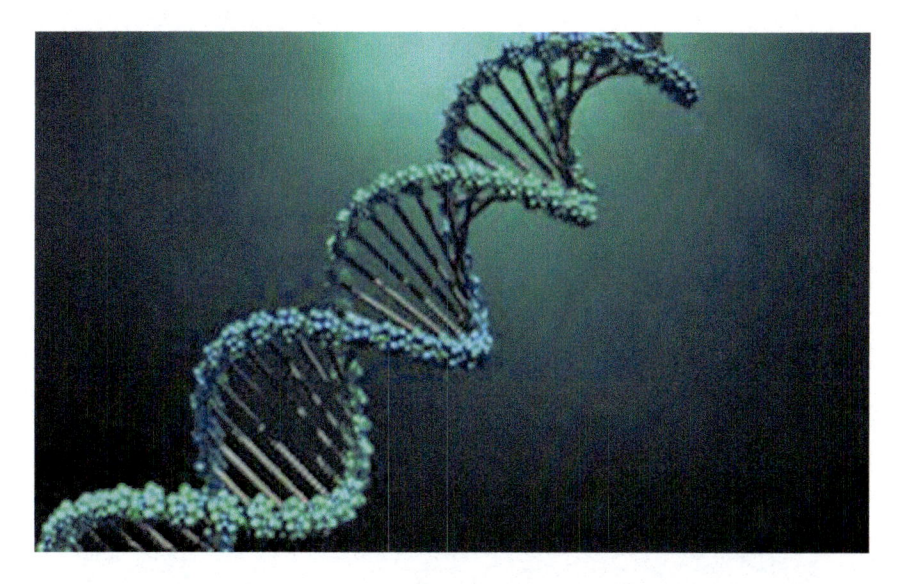

The human genome (all of your genes and DNA) is 99.9% identical to your Paleolithic ancestors, but the changes in our nutrition, the macronutrients (fats, proteins, and carbohydrates), and the micronutrients (vitamins, antioxidants, minerals, and electrolytes) that we consume have reduced our ability to prevent CHD and MI. (Table 9.1I) There are many genes that have been identified that cause CHD and MI. However, in some cases, a specific gene—which could have indicated a positive family history—is not found. Vibrant America Labs has the best genetic testing for CHD, MI, the CHD risk factors, and overall risk for cardiovascular disease. The test is called CARDIA-X and can be done on blood or saliva and ordered by your doctor (see the Sources section). Vascular biology (the way that the arteries respond to our environment, nutrition, supplements, drugs, and overall lifestyle), assumes a pivotal role in CHD. Radical oxygen species (ROS) and radical nitrogen species (RNS), coupled with impaired oxidative defense in the arteries, vascular inflammation, vascular

TABLE 9.1I

Contrasting the Paleolithic and Modern Intakes of Nutrients Involved in Vascular Biology. Evolutionary Nutritional Imposition

Nutrient	Paleolithic intake	Modern intake
Potassium	> 256 grams	2 grams
Sodium	< 1.2 grams	4–10 grams
Sodium/potassium ratio	< 0.13/day	> 0.67/day
Fiber	> 100 grams/day	9 grams/day
Protein	37%	20%
Carbohydrate	41%	40–50%
Fat	22%	30–40%
Polyunsaturated/saturated fats ratio	1.4	0.4

Note the marked changes from our early ancestors diets that were very healthy to our present diets that are very unhealthy and increase the risk of CHD.

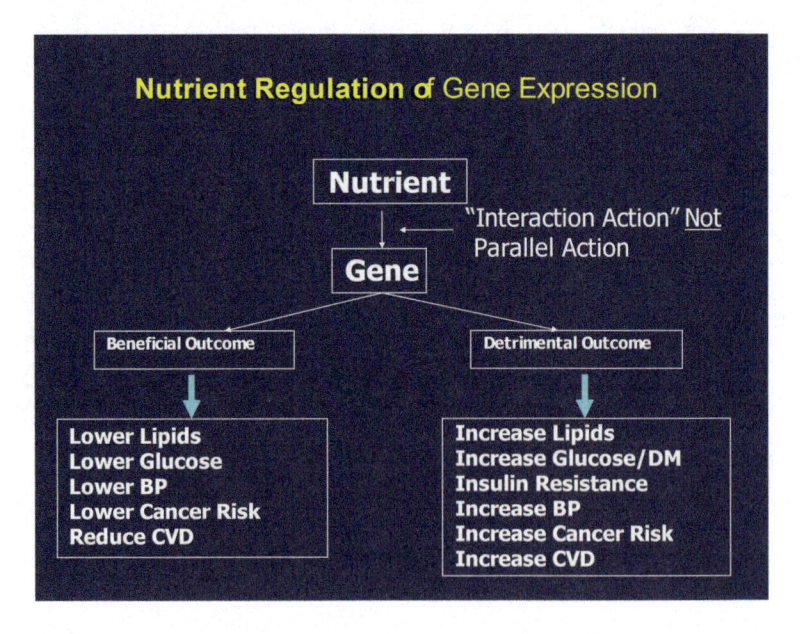

FIGURE 9.1I Nutrient gene interactions and gene expression. When the gene expresses itself, the result may be a good or bad outcome depending on the information it receives.

immune dysfunction, glycocalyx and endothelial dysfunction, loss of nitric oxide bioavailability, and our genes contribute to CHD (Figure 9.1I). Our genes cannot be changed, but the way they express themselves can be altered. This is called "gene expression". Gene expression means that the gene may respond in a good way or a bad way depending on the message it receives. That response could be an increase or decrease in CHD depending on what the gene does related to a specific stimulus such as nutrition, a supplement, lifestyle, or a drug (Figure 9.1I). Food is like the input into your computer. Food communicates with and directs your genes to perform. Food is the information that directs gene expression. This performance may be improved with good nutrition, but it could be made worse with poor nutrition.

The modern diet consists of high sodium intake and low potassium and magnesium intake. In addition, there is a lower dietary intake of fiber and complex carbohydrates (vegetables), decreased consumption of fruits and vegetables, and lower quality and quantity of protein, omega-3 fatty acids (FA) (fish, nuts), and monounsaturated fats (MUFA) (nuts, olive products, avocado). At the same time, there is an increased consumption of inflammatory omega-6 FA (hydrogenated fats, fried foods, processed foods), saturated fat (SFA) and trans fatty acids (TFA), sugar, and simple or refined carbohydrate-containing foods. In addition, obesity and our sedentary lifestyle with inadequate exercise, stress, excess alcohol, and smoking increases the risk of CHD. The implications of the genetic–nutrition/environmental connections mentioned above are enormous, in relation to the prevention and treatment of CHD.

SUMMARY AND KEY TAKE AWAY POINTS

1. Our modern diet is not healthy. We consume too much sodium, refined sugars, bad fats (trans fats and some saturated fats), and omega-6 fatty acids, but not enough potassium, magnesium, fruit, fiber, vegetables, high-quality protein, omega-3 fatty acids, and monounsaturated fats. These nutritional habits contribute to CHD and MI.
2. Your genes interact with your nutrition and your environment (gene expression) to determine your risk for CHD and MI. You can test many CHD genes with CARDIA-X from Vibrant America Labs (see Sources section).

BIBLIOGRAPHY

1. Houston MC. *What Your Doctor May Not Tell You About Heart Disease.* New York: Grand Central Press; February 2012.
2. Sinatra S and Houston M, Editors. *Nutrition and Integrative Strategies in Cardiovascular Medicine.* Boca Raton: CRC Press; 2015.
3. Eaton SB, Eaton SB III, Konner MJ. Paleolithic nutrition revisited: A twelve-year retrospective on its nature and implications. *Eur J Clin Nutr.* 1997; 51:207–16.
4. Layne J, Majkova Z, Smart EJ, Toborek M Hennig B. Caveolae: A regulatory platform for nutritional modulation of inflammatory diseases. *Journal of Nutritional Biochemistry.* 2011; 22:807–811.
5. Dandona P, Ghanim H, Chaudhuri A, Dhindsa S and Kim SS. Macronutrient intake induces oxidative and inflammatory stress: potential relevance to atherosclerosis and insulin resistance. *Exp Mol Med.* 2010;42(4):245–253.
6. Nayak DU, Karmen C, Frishman WH, Vakili BA. Antioxidant vitamins and enzymatic and synthetic oxygen-derived free radical scavengers in the prevention and treatment of cardiovascular disease. *Heart Disease* 2001;3:28–45.

7. Ritchie RH, Drummond GR Sobey CG, De Silva TM, Kemp-Harper BK The opposing roles of NO and oxidative stress in cardiovascular disease. *Pharmacol Res.* 2017; 116:57–69.
8. Dhalla NS, Temsah RM, Netticadam T. The role of oxidative stress in cardiovascular diseases. *J Hypertens* 2000;18:655–673.
9. Houston MC. The role of cellular micronutrient analysis and minerals in the prevention and treatment of hypertension and cardiovascular disease. *Therapeutic Advances in Cardiovascular Disease.* 2010;4:165–83.

9 CHD Genetics Part II
What Is in the Genes That Place You at Risk

The risk for CHD and MI is higher for individuals with a first-degree relative who developed premature CHD (with a threshold at the age of 55 years for a male or 65 years for a female). This Kaplan–Meier curve (Figure 9.1II) shows the risk of developing premature CHD with percentages from zero to 100 on the vertical line. The horizontal line shows the age of a parent that has CHD or MI. As you can see, before the age of 55 for a male or age 65 for a female in your parents, your risk for CHD and MI is quite high. However, the age of onset in the parent is after that age, your CHD risk drops with each year.

Numerous genes that cause CHD or some of the risk factors, such as hypertension, high cholesterol, or diabetes mellitus can be modified, that is change the gene expression with treatment with a drug, a supplement, an electrolyte, exercise, nutrition, or other lifestyle changes. There are numerous genes for CHD that have now been discovered that help us to understand the previously unknown details and causes underlying the concept of "family history of CHD". This genetic testing will allow your doctor to measure many of the genes that cause CHD or the other risk factors, such as hypertension, high cholesterol, or diabetes mellitus and provide a more direct, integrated, personalized, and precision treatment program without any of the guesswork. This means that your risk for CHD will be reduced. The Cardia-X genetic profile from Vibrant Labs America measures over 20 different genes related to CHD and CHD risk factors. The "family history" is your primary risk factor to develop CHD, and now you can find out what the specific genes are in many cases.

Genetics and nutrigenomics (the effect of nutrition and supplements on genes) provide us with an expanded perspective on the prevention and treatment of CHD. In CHD management, nutrigenomics encompasses genetic testing and the identification of single nucleotide polymorphisms (SNPs) that are part of your genes, nutrient–genetic interactions, and how the genes express themselves. The genes may be "turned on" or "turned off" by nutrition, supplements, drugs, and other environmental or lifestyle factors. This is referred to as "gene expression".

The majority of the CHD genes, once turned on, promote vascular inflammation, oxidative stress, and immune dysfunction. Regardless of the type of insult, coronary arteries respond to insults via the same three mechanisms.

Consequently, the inflammatory pathways have become the primary focus in the management of genetic expression and genetic risk for CHD. The prevention and the reduction of CHD are not likely to improve without using genetic testing with

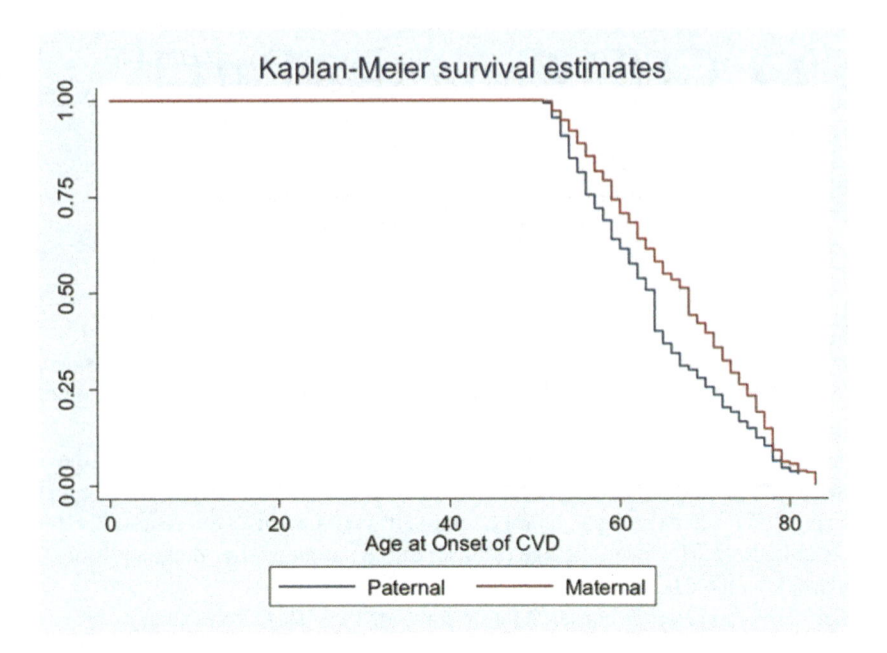

FIGURE 9.1II The risk of CHD and MI in first-degree relatives with a parent having CHD or MI at various ages is presented in this Kaplan–Meier curve. The blue line represents paternal, and the red line represents maternal.

integrated, precision, and personalized cardiovascular medicine. Let us look at some of these influences on your genes, such as nutrients and nutrition, electrolytes, supplements, lifestyle, and drugs.

NUTRIENTS

Nutritional factors provide information that determines whether our genes are turned on or turned off, with a corresponding beneficial or detrimental outcome. One change in a single nutrient, such as magnesium, may cause 300 different changes in downstream mediators and cell metabolism related to cardiovascular function and health. This is just one example of environmental influences and the importance of genetic expression. When there is interference with a metabolic pathway, a single area of abnormality can result in a large number of defects and a spoke-like effect in a wheel, resulting in a ripple of downstream changes in many metabolic pathways.

EPIGENETICS

There are several issues we want to define in patients. One is their genetic profile, the genes they were dealt. There are also epigenetic influences that are not genetic that alter the function of deoxyribonucleic acid (DNA) in many different ways that can affect future generations. These are called methylation, histone modifications, and noncoding messenger ribonucleic acids (RNA). These influences are not in the genetic code but can be passed on from mother to fetus and from generation to generation. For example, a mother that is malnourished during pregnancy is more likely to have a child that develops hypertension, diabetes mellitus, or obesity later in life. The risk for these diseases can then be passed on "epigenetically" to future generations. The final aspect is gene expression, as genes express themselves in response to nourishment or insults from different types of information coming in from the environment. Genetics has become important in determining not only the best nutritional program but also medication use in many patients, based on their genetic profiles.

NUTRITION

MEDITERRANEAN DIET

We know that the Mediterranean diet (MedDiet) turns on numerous beneficial genetic pathways that can reduce the risk for CHD, MI, hypertension, type II diabetes, and other cardiovascular diseases. If you consume a Western diet, this will result in totally different outcomes in terms of gene expression, since most of the foods included in a Western dietary pattern have been shown to express 30 to 40 different inflammatory, oxidative stress, and immune pathways.

The MedDiet has an advantageous effect on many genes. In a clinical trial of this diet, prevalent beneficial effects were related to CHD, MI, atherosclerosis, diabetes, and hypertension. The Mediterranean diet, in combination with CoEnyzme Q 10 (CoQ10), has been shown to be the most beneficial intervention for healthy aging and preventing diseases related to chronic oxidative stress, hypertension, and CHD. Changes in genetic expression toward a protective mode were often associated with improvement in systemic markers for inflammation, immune function, oxidative stress, hypertension, blood glucose, and CHD.

PRITIKIN AND DASH DIETS

The Pritikin diet is one of the most effective ways to turn off the gene expression that increases the risk for CHD, hypertension, and cardiovascular disease. The Pritikin diet can reduce the risk of CHD by as much as 30–35%. That benefit is directly correlated with the diet itself but is also enhanced when supplemented with nutrients such as CoQ10. The DASH-1 and DASH-2 diets (Dietary Approaches to Stop Hypertension) have also been found beneficial in relation to changes in inflammatory genes, reducing blood pressure, as well as improved blood glucose, weight, and response to the types of medications prescribed for hypertension.

SPECIFIC NUTRIENTS

ELECTROLYTES

Electrolytes, particularly sodium, potassium, and magnesium, can change genetic expression, salt sensitivity, intravascular volume, blood pressure, and risk for CHD, MI, cardiac arrhythmias, congestive heart failure, and stroke.

OMEGA-3 FATTY ACIDS

Omega-3 fatty acids affect a large number of genes that reverse changes in our metabolic profile and in our genes that can improve mitochondrial health. As a result, adenosine triphosphate (ATP) production goes up, cells are healthier, and patients live longer. ATP is the energy produced by the mitochondria in our cells from the metabolism of food. The mitochondria are like small "nuclear power plants". We know that omega-3 fatty acids by themselves have dramatic effects on many receptors that can have enormous influences, reversing inflammation, oxidative stress, blood pressure, and risk for CHD and MI. In specific studies, omega-3 fats changed the expression of 610 genes in men and 250 genes in women.

MONOUNSATURATED FATS

Olive oil, olives, and nuts contain monounsaturated fat, which has a positive impact on different SNPs and receptors, improving hypertension, CHD, and diabetes mellitus. Even without the MedDiet, the regular ingestion of olive oil, olives, and nuts can have dramatic and highly beneficial influences on genetic expression related to the three finite vascular responses for reducing blood pressure and CHD.

GENES RELEVANT TO CARDIOVASCULAR RISK

Every patient should have their cardiovascular genetics tested by their physician. This allows for an integrated, precise, and personalized evaluation of CHD risk, prevention, and treatment program with nutrition, supplements, or drugs (see Figure 9.1II).

CHD AND CHD RISK FACTOR GENES

Gene 9p21 is one of the primary genes we are now measuring that increases the risk of atherosclerosis, CHD, and MI. Patients who have a heterozygote SNP (one-half of the gene) for 9p21 have a risk of CHD and MI that is increased by 50%. When a patient has a homozygote SNP (both halves of the gene), the risk goes up to approximately 100%. However, there are many other genes that should also be evaluated not just for CHD but also for hypertension and dyslipidemia (abnormal cholesterol). 9p21 also causes inflammation, plaque rupture, thrombosis, aortic aneurysms, atherosclerosis, and diabetes mellitus.

GLU 1q25 increases the risk of CHD and MI in diabetes mellitus.

APOE4 genotype increases the risk for CHD and MI. Management of risk factors for patients with the APOE4 allele, especially with the homozygote E4/E4 type addresses issues such as:

- Increased cholesterol absorption and delayed clearance, resulting in higher serum LDL cholesterol (the bad form of cholesterol).

CHD Genetics Part II

- Increased CHD with smoking and alcohol intake and overall increased incidence of MI, Alzheimer's disease, and dementia.
- Inability to repair the vascular endothelium to produce nitric oxide, resulting in an increase in blood pressure.
- Less response to statins for lowering cholesterol.
- The best reduction of LDL occurs through dietary restriction of carbohydrates, with low-fat diets and omega-3 fatty acids.

COMT Polymorphisms. One of the latest genes to be recommended for evaluation is catechol-O-methyltransferase (COMT), which provides instructions for the breakdown of norepinephrine and epinephrine (adrenaline). If this genetic SNP is present, the patient will have higher levels of norepinephrine and epinephrine in the blood and urine, with an increased risk of hypertension, CHD, and MI. There is a variation in response depending on which of the specific COMT SNPs the patient carries. For example, aspirin or vitamin E may be beneficial for patients with one type of COMT SNP, but detrimental if one of the other SNPs is present.

Glutathione-Related SNPs. The risk of MI can be increased by 71% if an SNP affecting glutathione metabolism (glutathione peroxidase, GSH-Px) is present. This selenium-dependent enzyme expresses different capacities to neutralize oxidative molecules related to increases in oxidative stress and cardiovascular disease. For these patients, glutathione peroxidase and selenium levels would be key measurements to track for the risk of MI:

- Low GSH-Px is a major CHD and MI risk factor.
- Higher levels of glutathione peroxidase support more rapid recycling of glutathione, resulting in higher availability of glutathione which is one of the most important antioxidants inside the cell.
- Increased glutathione peroxidase (GSH-Px) decreases blood pressure, CHD, and MI.

6p24.1 increases the risk for CHD and MI.

ACE I/D increases the risk for CHD, MI, hypertension, left ventricular hypertrophy (heart enlargement), carotid artery disease, and kidney failure.

MTHFR (A1298C and C677T) is associated with CHD, MI, methylation, endothelial dysfunction, hypertension, thrombosis, stroke, and high homocysteine.

CYP1A2. Caffeine intake in slow metabolizers increases the risk for fast heart rate, hypertension, stiff arteries, CHD, and MI.

CYP 11 B2. Related hypertension, high aldosterone levels (a hormone made in the adrenal gland that causes salt and water retention, hypertension, and CHD).

NOS 3. Patients have low levels of nitric oxide and a higher risk for hypertension, CHD, and MI.

MMP-2, MMP-9, and TIMP-1. CHD, MI, and hypertension are more common.

GENES RELEVANT TO HYPERTENSION

There are a whole host of genetic influences on blood pressure. In fact, probably over 30 different genes have been recognized to date, all of which are helpful in determining both risk for hypertension and risk for CHD, MI, and stroke. These genes are also helpful to determine the response to diet and nutrition, various nutrients, supplements, electrolytes, caffeine, and medications (Figure 9.1II and Box 9.1).

> **BOX 9.1 Recommended Genetic Testing (Vibrant America Labs)**
>
> 1. 9p21 causes inflammation, plaque rupture, thrombosis, aortic aneurysms atherosclerosis, CHD, MI, and diabetes mellitus
> 2. 6p24.1 causes CHD
> 3. 4q25 causes atrial fibrillation
> 4. ACE I/D: hypertension, left ventricular hypertrophy, CHD, MI, carotid artery disease, and kidney failure
> 5. COMT: CHD, MI, and hypertension
> 6. 1q25 (GLUL): CHD in diabetes
> 7. APO E (E4/E4): CHD, lipids, and dietary response to fats.
> 8. MTHFR (A1298C and C677T) for methylation, endothelial dysfunction, hypertension, thrombosis, CHD, MI stroke, and high homocysteine levels in the blood
> 9. CYP 1A2: caffeine hypertension, CHD, and MI
> 10. Corin: hypertension
> 11. CYP 11 B2: hypertension and aldosterone and CHD
> 12. GSH-Px (glutathione peroxidase): CHD and MI
> 13. NOS 3: Nitric oxide, hypertension, and CHD
> 14. ADR B2: hypertension and DASH diet.
> 15. APO A1 and A2: cholesterol and other lipids
> 16. CYP4AII and CYP4F2: hypertension
> 17. MMP-2, MMP-9, and TIMP-1: CHD, MI, and hypertension
> 18. AGTR1, NR3C2, HSD11B1, and B2: hypertension
> 19. AT1R-AA (AT1R autoantibodies): hypertension
> 20. Blood group type A, B, and AB: thrombosis and CHD

CYP-450-1A2. We know, for example, that someone who consumes caffeine, in the form of caffeinated coffee and tea, has the SNP cytochrome P-450-1A2, and is a slow metabolizer of caffeine, will have an increased risk of tachycardia (fast heart rate), hypertension, aortic stiffness, CHD, and MI. Of course, one could have the fast metabolizer type of SNP for caffeine detoxification and that will *reduce* their risk of all of these problems. The risk of having this gene is about 50% of the US population. The 50% of the population with this gene who are slow metabolizers of caffeine actually increase their risk directly based on the amount of caffeine consumption and their age. Before you drink caffeine (coffee, tea) you need to check the gene for cytochrome P-450 1A2 function. If you are a slow metabolizer of caffeine then you should stop taking caffeine from all sources.

CYP 11 B2. The CYP 11 B2 is related to resistant hypertension, salt and water retention, increased blood volume, high aldosterone levels (a hormone produced by the adrenal gland). and low blood potassium. Genetic hypertension is best treated with the drugs spironolactone or eplerenone. High aldosterone levels and hypertension both increase the risk for CHD.

CYP4A11. In terms of salt sensitivity and resistant hypertension, one of the most important is CYP4A11 which relates to sodium and water reabsorption and the role of the epithelial sodium channel (ENaC) function in the kidney. These patients have avid reabsorption of sodium in the kidney tubules from the ENaC which increases the blood volume, blood pressure, and risk for CHD. This type of hypertension is best treated with the drug Amiloride, which blocks the ENaC and results in a diuresis of salt and water to lower the blood pressure. Some of these patients may need a short- or long-term treatment with another type of diuretic, Indapamide.

ACE I/D. The ACE I/D (DD allele) is associated with hypertension, left ventricular hypertrophy (enlargement of the heart), CHD, MI, carotid artery disease, kidney failure, microalbuminuria (loss of the protein, albumin, in the urine due to kidney damage), and kidney disease. This type of hypertension is best treated with the antihypertensive drug class called "angiotensin-converting enzyme inhibitors" (ACEIs).

MTHFR gene is related to methylation and folic acid and other B vitamins and, if defective, will cause hypertension and CHD. If you have one of these genes then you will need to take a complex of B vitamins with methylated folic acid.

NOS 3 is an important enzyme in the production of nitric oxide, which improves heart health and lowers blood pressure. If the NOS 3 is defective, then blood pressure increases as will the risk for CHD and MI. A high nitrate/nitrite diet with dark green leafy vegetables, like kale and spinach, and beets or beetroot extracts, like NEO 40, will improve nitric oxide levels.

ADR B2. The ADR B2 gene is related to how effective the DASH diet will be in lowering blood pressure. If you have a defect in this gene, then the DASH 2 diet results in a reflex increase in the enzyme renin, which increases the formation of the potent vasoconstrictor angiotensin II that increases blood pressure, CHD, and MI. If this occurs, then you should take one of the antihypertensive drug classes that will block the effects of renin and angiotensin II. The drug classes are called "angiotensin receptor blockers" (ARBs) or angiotensin-converting enzyme inhibitors (ACEIs).

AGTR1(AT1R-AA) is related to autoantibodies and hypertension. An "autoantibody" is an abnormal antibody that is part of your immune system and which attacks your own organs (i.e., attacks "the self"). This autoantibody stimulates a receptor called angiotensin receptor I (AT1R) that causes hypertension and increases the risk for CHD and MI. This form of hypertension is best treated with an ARB.

GENES RELATED TO CHOLESTEROL AND DIABETES MELLITUS THAT INCREASE RISK FOR CHD AND MI

Cardiovascular SNPs. Obviously, there are large numbers of cardiovascular SNPs that we could check. At this point, I recommend testing for those that have the best validation and highest correlation with risk prediction, are easily attainable, and have implications for specific treatment of CHD and CHD risk factors. The genetic tests listed in the (Box 9.1) define risk for CHD and CHD risk factors. The best genetic lab test for these is Cardia-X from Vibrant Labs America (see Sources section).

SUMMARY AND KEY TAKE AWAY POINTS

1. If one of your parents had an MI at an early age, you have a higher chance of having an MI.
2. The genetic expression of numerous genes that cause CHD and hypertension can be altered through nutrition, supplements, exercise, drugs, and lifestyle.
3. Evaluation of specific genetic SNPs for CHD and CHD risk factors using the Cardia-X genetic profile from Vibrant Labs America is recommended.
4. The traditional Mediterranean diet with five tablespoons EVOO/day (50 grams), nuts, and CoQ10 reduces the risk for CHD.
5. Omega-3 fatty acids should be given to all patients, dose dependent (1–5 grams per day to lower blood pressure and reduce CHD and MI risk).
6. Avoid caffeine in CYP 1A2 SNP if a slow metabolizer of caffeine to reduce the risk of CHD and MI.
7. Selective use of ASA and vitamin E depending on COMT phenotype.
8. Methyl folate and B vitamins depending on MTHFR genotype for methylation.
9. Selenium should be given with the GSH-Px gene if defective.

BIBLIOGRAPHY

1. Allport SA Kikah N, Saif NA, Ekokobe F and Folefac D. Atem parental age of onset of cardiovascular disease as a predictor for offspring age of onset of cardiovascular disease. *PLoS One*. 2016;11(12): e0163334.
2. Sinatra ST and Houston MC (editors). *Nutritional and Integrative Strategies in Cardiovascular Medicine*. Boca Raton, FL: CRC Press; 2015.
3. Price PT, Nelson CM, Clarke SD Omega-3 polyunsaturated fatty acid regulation of gene expression. *Curr Opin Lipidol*. 2000;11(1):3–7.
4. McNiven EM, German JB, Slupsky CM Analytical metabolomics: Nutritional opportunities for personalized health. *J Nutr Biochem*. 2011;22(11):995–1002.
5. O'Donnell CJ, Nabel EG Genomics of cardiovascular disease. *N Engl J Med*. 2011;365(22):2098–109.
6. Nuno NB, Heuberger R Nutrigenetic associations with cardiovascular disease. *Rev Cardiovasc Med*. 2014;15(3):217–25.
7. Holdt LM, Teupser D. From genotype to phenotype in human atherosclerosis--recent findings. *Curr Opin Lipidol*. 2013;24(5):410–8.
8. Roberts R, Stewart AF. Genetics of coronary artery disease in the 21st century. *Clin Cardiol*. 2012;35(9):536–40.
9. Houston, MC. *What Your Doctor May Not Tell You About Heart Disease*. New York: Grand Central Press; February 2012.
10. Houston M. The role of noninvasive cardiovascular testing, applied clinical nutrition and nutritional supplements in the prevention and treatment of coronary heart disease. *Ther Adv Cardiovasc Dis*. 2018;12(3):85–108.

11. Houston MC. New concepts in cardiovascular disease. *J of Restorative Medicine* 2013;2:30–44.
12. Webster AL, Yan MS, Marsden PA. Epigenetics and cardiovascular disease. *Can J Cardiol.* 2013;29(1):46–57.
13. Castañer O, Corella D, Covas MI, Sorlí JV, Subirana I, Flores-Mateo G, Nonell L, Bulló M, de la Torre R, Portolés O, Fitó M; PREDIMED study investigators In vivo transcriptomic profile after a Mediterranean diet in high-cardiovascular risk patients: a randomized controlled trial. *Am J Clin Nutr.* 2013;98(3):845–53.
14. Konstantinidou V, Covas MI, Sola R, Fitó M Up-to date knowledge on the *in vivo* transcriptomic effect of the Mediterranean diet in humans. *Mol Nutr Food Res.* 2013;57(5):772–83
15. Corella D, Ordovás JM. How does the Mediterranean diet promote cardiovascular health? Current progress toward molecular mechanisms: gene-diet interactions at the genomic, transcriptomic, and epigenomic levels provide novel insights into new mechanisms. *Bioessays.* 2014;36(5):526–37.
16. Estruch R, Ros E, Salas-Salvadó J, Covas MI, Corella D, Arós F, Gómez-Gracia E, Ruiz-Gutiérrez V, Fiol M, Lapetra J, Lamuela-Raventos RM, Serra-Majem L, Pintó X, Basora J, Muñoz MA, Sorlí JV, Martínez JA, Martínez-González MA; PREDIMED study investigators. Primary prevention of cardiovascular disease with a Mediterranean diet. *N Engl J Med.* 2013;368(14):1279–90.
17. González-Guardia L, Yubero-Serrano EM2, Delgado-Lista J, Perez-Martinez P, Garcia-Rios A, Marin C, Camargo A, Delgado-Casado N Roche HM, Perez-Jimenez F, Brennan L, López-Miranda J. Effects of the Mediterranean diet supplemented with coenzyme q10 on metabolomic profiles in elderly men and women. *J Gerontol A Biol Sci Med Sci.* 2015;70(1):78–84.
18. Appel LJ, Moore TJ, Obarzanek E, Vollmer WM, Svetkey LP, Sacks FM, Bray GA, Vogt TM, Cutler JA, Windhauser MM, Lin PH, Karanja N. A clinical trial of the effects of dietary patterns on blood pressure. DASH Collaborative Research Group. *N Engl J Med.* 1997;336(16):1117–24.
19. Sacks FM, Svetkey LP, Vollmer WM, Appel LJ, Bray GA, Harsha D, Obarzanek E, Conlin PR, Miller ER 3rd, Simons-Morton DG, Karanja N, Lin PH; DASH-Sodium Collaborative Research Group Effects on blood pressure of reduced dietary sodium and the dietary approaches to stop hypertension (DASH) diet. DASH-Sodium Collaborative Research Group. *N Engl J Med.* 2001;344(1):3–10.
20. Ornish D, Scherwitz LW, Billings JH, Brown SE, Gould KL, Merritt TA, Sparler S, Armstrong WT, Ports TA, Kirkeeide RL, Hogeboom C, Brand RJ. Intensive lifestyle changes for reversal of coronary heart disease. *JAMA.* 1998 Dec 16;280(23):2001–7.
21. Laffer CL, Elijovich F, Eckert GJ, Tu W Pratt JH, Brown NJ. Genetic variation in CYP4A11 and blood pressure response to mineralocorticoid receptor antagonism or ENaC inhibition: an exploratory pilot study in African Americans. *J Am Soc Hypertens.* 2014 1;8(7):475–80.
22. Vanden Heuvel JP. Nutrigenomics and nutrigenetics of ω3 polyunsaturated fatty acids. *Prog Mol Biol Transl Sci.*2012;108:75–112.
23. Varela LM, Ortega-Gomez A, Lopez S, Abia R, Muriana FJ, Bermudez B. The effects of dietary fatty acids on the postprandial triglyceride-rich lipoprotein/apoB48 receptor axis in human monocyte/macrophage cells. *J Nutr Biochem.* 2013;24(12):2031–9.
24. Silva S, Bronze MR, Figueira ME, Siwy J, Mischak H, Combet E, Mullen W. Impact of a 6-wk olive oil supplementation in healthy adults on urinary proteomic biomarkers of coronary artery disease, chronic kidney disease, and diabetes (types 1 and 2): A randomized, parallel, controlled, double-blind study. *Am J Clin Nutr.* 2015;101(1):44–54.

25. Costanza AC, Moscavitch SD, Faria Neto HC, Mesquita ET. Probiotic therapy with *Saccharomyces boulardii* for heart failure patients: a randomized, double-blind, placebo-controlled pilot trial. *Int J Cardiol.* 2015;179:348–50.
26. Khalesi S, Sun J, Buys N, Jayasinghe R Effect of probiotics on blood pressure: A systematic review and meta-analysis of randomized, controlled trials. Hypertension. 2014 Oct;64(4):897–903.
27. Tuohy KM, Fava F, Viola R. The way to a man's heart is through his gut microbiota'--dietary pro- and prebiotics for the management of cardiovascular risk. *Proc Nutr Soc.* 2014 May;73(2):172–85.
28. Khanna S, Tosh PK. A clinician's primer on the role of the microbiome in human health and disease. *Mayo Clin Proc.* 2014 Jan;89(1):107–14.
29. O'Donnell CJ, Nabel EG Genomics of cardiovascular disease. *N Engl J Med.* 2011;365(22):2098–109.
30. Paloaki GE, Melillo S, Bradley LA. Association between 9p21 genomic markers and heart disease: a meta-analysis. *JAMA.* 2010;303:648–56.
31. Qi L, Qi Q, Prudente S, Mendonca C, Andreozzi F, di Pietro N Sturma M, Novelli V. Association between a genetic variant related to glutamic acid metabolism and coronary heart disease in individuals with type 2 diabetes. *JAMA.* 2013 Aug 28;310(8):821–8.
32. Schaefer EJ Lipoproteins, nutrition, and heart disease. *Am J Clin Nutr.* 2002;75(2):191–212.
33. Skulas-Ray AC, Kris-Etherton PM, Harris WS, Vanden Heuvel JP, Wagner PR, West SG Dose-response effects of omega-3 fatty acids on triglycerides, inflammation, and endothelial function in healthy persons with moderate hypertriglyceridemia. *Am J Clin Nutr.* 2011;93(2):243–52.
34. Hall KT, Nelson CP, Davis RB, Buring JE, Kirsch I Mittleman MA, Loscalzo J, Samani NJ, Ridker PM, Kaptchuk TJ, Chasman DI Polymorphisms in catechol-*O*-methyltransferase modify treatment effects of aspirin on risk of cardiovascular disease. *Arterioscler Thromb Vasc Biol.* 2014;34(9):2160–7.
35. Winter JP, Gong Y, Grant PJ, Wild CP Glutathione peroxidase 1 genotype is associated with an increased risk of coronary artery disease. *Coron Artery Dis.* 2003;14(2):149–53.
36. Blankenberg S, Rupprecht HJ, Bickel C, Torzewski M, Hafner G, Tiret L, Smieja M, Cambien F, Meyer J, Lackner KJ; AtheroGene investigators glutathione peroxidase 1 activity and cardiovascular events in patients with coronary artery disease. *N Engl J Med.* 2003;349(17):1605–13.
37. Korkor MT, Meng FB, Xing SY, Zhang MC, Guo JR, Zhu XX, Yang P Microarray analysis of differential gene expression profile in peripheral blood cells of patients with human essential hypertension. *Int J Med Sci.* 2011;8(2):168–79.
38. Harrap SB Blood pressure genetics: time to focus. *J Am Soc Hypertens.* 2009;3(4):231–7.
39. Zhou L, Xi B, Wei Y, Shen W, Li Y. Meta-analysis of the association between the insertion/deletion polymorphism in ACE gene and coronary heart disease among the Chinese population. *J Renin Angiotensin Aldosterone Syst.* 2012;13(2):296–304.
40. Fernández-Llama P Poch E, Oriola J, Botey A, Coll E, Darnell A, Rivera F, Revert L. Angiotensin converting enzyme gene I/D polymorphism in essential hypertension and nephroangiosclerosis. *Kidney Int.* 1998;53(6):1743–7.
41. Gardemann A, Fink M, Stricker J, Nguyen QD, Humme J, Katz N, Tillmanns H, Hehrlein FW, Rau M, Haberbosch W. ACE I/D gene polymorphism: presence of the ACE D allele increases the risk of coronary artery disease in younger individuals .*Atherosclerosis.* 1998;139(1):153–9.
42. Castellano M, Muiesan ML, Rizzoni D, Beschi M, Pasini G, Cinelli A, Salvetti M, Porteri E, Bettoni G, Kreutz R, Angiotensin-converting enzyme I/D polymorphism and arterial wall thickness in a general population. The Vobarno Study. *Circulation.* 1995;91(11):2721–4.

43. Mesas AE, Leon-Muñoz LM, Rodriguez-Artalejo F, Lopez-Garcia E.The effect of coffee on blood pressure and cardiovascular disease in hypertensive individuals: A systematic review and meta-analysis. *Am J Clin Nutr.* 2011;94(4):1113–26.
44. Palatini P, Ceolotto G, Ragazzo F, Dorigatti F, Saladini F, Papparella I, Mos L, Zanata G, Santonastaso M.CYP1A2 genotype modifies the association between coffee intake and the risk of hypertension. *J Hypertens.* 2009;27(8):1594–601.
45. Armaly Z, Assady S, Abassi Z. Corin: a new player in the regulation of salt-water balance and blood pressure. *Curr Opin Nephrol Hypertens.* 2013;22(6):713–20.
46. Zhou Y, Wu Q. Corin in natriuretic peptide processing and hypertension. *Curr Hypertens Rep.* 2014;16(2):415.
47. Peng H, Zhang Q, Cai X, Liu Y, Ding J, Tian H, Chao X, Shen H, Jiang L, Jin J, Zhang Y. Association between high serum soluble corin and hypertension: A cross-sectional study in a general population of China. *Am J Hypertens.* 2015;28(9):1141–9.
48. Fontana V, de Faria AP, Barbaro NR, Sabbatini AR, Modolo R, Lacchini R, Moreno H. Modulation of aldosterone levels by −344 C/T CYP11B2 polymorphism and spironolactone use in resistant hypertension. *J Am Soc Hypertens.* 2014;8(3):146–51.
49. Levinsson A Olin AC, Björck L, Rosengren A, Nyberg F. Nitric oxide synthase (NOS) single nucleotide polymorphisms are associated with coronary heart disease and hypertension in the INTERGENE study. *Nitric Oxide.* 2014;39:1–7.
50. Sun B, Williams JS, Svetkey LP, Kolatkar NS, Conlin PR. Beta2-adrenergic receptor genotype affects the renin-angiotensin-aldosterone system response to the Dietary Approaches to Stop Hypertension (DASH) dietary pattern. *Am J Clin Nutr.* 2010;92(2):444–9.
51. Chen Q, Turban S, Miller ER, Appel LJ. The effects of dietary patterns on plasma renin activity: results from the dietary approaches to stop hypertension trial. *J Hum Hypertens.* 2012;26(11):664–9.
52. Laffer CL, Elijovich F, Eckert GJ, Tu W, Pratt JH, Brown NJ. Genetic variation in CYP4A11 and blood pressure response to mineralocorticoid receptor antagonism or ENaC inhibition: An exploratory pilot study in African Americans. *J Am Soc Hypertens.* 2014;8(7):475–80.
53. Ward NC, Tsai IJ, Barden A, van Bockxmeer FM, Puddey IB, Hodgson JM, Croft KD. A single nucleotide polymorphism in the CYP4F2 but not CYP4A11 gene is associated with increased 20-HETE excretion and blood pressure. *Hypertension.* 2008;51(5):1393–8.
54. Sun Y, Liao Y, Yuan Y, Feng L, Ma S, Wei F, Wang M, Zhu F Influence of autoantibodies against AT1 receptor and AGTR1 polymorphisms on candesartan-based antihypertensive regimen: results from the study of optimal treatment in hypertensive patients with anti-AT1-receptor autoantibodies trial. *J Am Soc Hypertens.* 2014;8(1):21–7.
55. Yang X, Sethi A, Yanek LR, Knapper C, Nordestgaard BG, Tybjærg-Hansen A, Becker DM, Mathias RA, Remaley AT, Becker LC. SCARB1 gene variants are associated with the phenotype of combined high high-density lipoprotein cholesterol and high lipoprotein (a). *Circ Cardiovasc Genet.* 2016;9(5):408–418.
56. Mendoza S, Trenchevska O King SM, Nelson RW, Nedelkov D, Krauss RM, Yassine HN. Changes in low-density lipoprotein size phenotypes associate with changes in apolipoprotein C-III glycoforms after dietary interventions. *J Clin Lipidol.* 2017 Jan;11(1):224–233.
57. Wyler von Ballmoos MC Haring B, Sacks FM. The risk of cardiovascular events with increased apolipoprotein CIII: A systematic review and meta-analysis. *J Clin Lipidol.* 2015;9(4):498–510.
58. Zheng C. Updates on apolipoprotein CIII: Fulfilling promise as a therapeutic target for hypertriglyceridemia and cardiovascular disease. *Curr Opin Lipidol.* 2014;25(1):35–9.

59. Thomas GS, Voros S, McPherson JA, Lansky AJ, Winn ME, Bateman TM, Elashoff MR, Lieu HD, Johnson AM, Daniels SE, Ladapo JA, Phelps CE, Douglas PS, Rosenberg S A blood-based gene expression test for obstructive coronary artery disease tested in symptomatic nondiabetic patients referred for myocardial perfusion imaging the COMPASS study. *Circ Cardiovasc Genet.* 2013;6(2):154–62.
60. McPherson JA Davis K, Yau M, Beineke P, Rosenberg S, Monane M, Fredi JL. The clinical utility of gene expression testing on the diagnostic evaluation of patients presenting to the cardiologist with symptoms of suspected obstructive coronary artery disease: results from the IMPACT (Investigation of a Molecular Personalized Coronary Gene Expression Test on Cardiology Practice Pattern) trial. *Pathw Cardiol.* 2013;12(2):37–42.
61. Wingrove JA, Daniels SE, Sehnert AJ, Tingley W, Elashoff MR, Rosenberg S, Buellesfeld L, Grube E, Newby LK, Ginsburg GS. Kraus WECorrelation of peripheral-blood gene expression with the extent of coronary artery stenosis. *Circulation: Cardiovascular Genetics.* 2008;1(1):31–8.
62. Rosenberg S, Elashoff MR, Lieu HD, Brown BO, Kraus WE, Schwartz RS, Voros S, Ellis SG, Waksman JR, McPherson JA, Lansky AJ, Topol EJ. PREDICT Investigators whole blood gene expression testing for coronary artery disease in nondiabetic patients: Major adverse cardiovascular events and interventions in the PREDICT trial. *Cardiovasc Transl Res.* 2012;5(3):366–74.
63. Ganna A, Salihovic S, Sundström J Broeckling CD, Hedman AK, Magnusson PK, Pedersen NL, Larsson A, Siegbahn A, Zilmer M, Prenni J, Arnlöv J, Lind L, Fall T, Ingelsson E. Large-scale metabolomic profiling identifies novel biomarkers for incident coronary heart disease. *PLoS Genet.* 2014:11;10–15.
64. Granger CB, Newgard CB, Califf RM, Newby LK, Shah SH, Sun JL, Stevens RD, Bain JR, Muehlbauer MJ, Pieper KS, Haynes C, Hauser ER, Kraus WE. Baseline metabolomic profiles predict cardiovascular events in patients at risk for coronary artery disease. Am *Heart* J. 2012;163(5):844–50.
65. Rizza S, Copetti M, Rossi C, Cianfarani MA Zucchelli M, Luzi A, Pecchioli C, Porzio O, Di Cola G, Urbani A, Pellegrini F, Federici M. Metabolomics signature improves the prediction of cardiovascular events in elderly subjects. *Atherosclerosis.* 2014;232(2):260–4.

10 What Are Your Genes Doing? Gene Expression in Coronary Heart Disease (CHD) and How to Treat

TREATMENT OF YOUR GENES AND GENE EXPRESSION

Following is a list of genes and the treatments to "turn off" or modify the gene expression, thereby reducing your risk of CHD, MI, and other cardiovascular events and diseases. Please note that if a specific supplement is recommended and the name of the company is provided, then go to the Sources section at the end of the book for the contact information and phone number.

1. **9p21**. Start an anti-inflammatory diet. Reduce intake of refined carbohydrates and starches to less than 25 grams per day. Avoid white rice, white potatoes, all breads, wheat, pasta of all types, sugars, sweets, soft drinks, and artificial sweeteners. Limit fruit juices. Start the Mediterranean diet. Consume 12 servings of vegetables and fruits each day. Maintain an ideal body weight, body fat, visceral fat, and exercise at least four times per week or more for one hour each time. Take one low-dose coated 81 mg aspirin per day with food.
2. **1q25**. Start an anti-inflammatory diet. Reduce intake of refined carbohydrates and starches to less than 25 grams per day. Avoid white rice, white potatoes, all breads, wheat, pasta of all types, sugars, sweets, soft drinks, and artificial sweeteners. Limit fruit juices. Start the Mediterranean diet. Consume 12 servings of vegetables and fruits each day. Maintain an ideal body weight, body fat, visceral fat, and exercise at least four times per week or more for one hour each time. Take one low-dose coated 81 mg aspirin per day with food.
3. **CYP 11 B2**. If you have high blood pressure, a specific drug will be recommended such as spironolactone or eplerenone. Reduce sodium intake to 1500 mg per day or less. Increase dietary potassium to 6000 mg per day and magnesium to 1000 mg per day.

4. **GSH PX**. Start NAC (N acetylcysteine) one, twice per day; and Whey Cool Protein, one scoop per day in a liquid of your choice (Designs for Health). Also, start Selenium at 200 micrograms per day. Get this from a health food store. Reduce sodium intake to less than 1500 mg per day. Increase dietary potassium to 6000 mg per day and magnesium to 1000 mg per day. Consider taking a liposomal glutathione supplement orally each day.
5. **APO C3**. Start **CardioLux,** two capsules twice a day with food (Metagenics). Eat pomegranate seeds, ¼ cup per day either fresh or frozen (Whole Foods, Sprouts, Publix, Kroger, or order from Amazon.)
6. **6p 24.1**. Start an anti-inflammatory diet. Reduce intake of refined carbohydrates and starches to less than 25 grams per day. Avoid white rice, white potatoes, all breads, wheat, pasta of all types, sugars, sweets, soft drinks, and artificial sweeteners. Limit fruit juices. Start the Mediterranean diet. Consume 12 servings of vegetables and fruits each day. Maintain an ideal body weight, body fat, visceral fat, and exercise at least four times per week or more for one hour each time. Take one low-dose coated 81 mg aspirin per day with food.
7. **ACE I/D**. If you have high blood pressure you will be given an angiotensin-converting enzyme inhibitor (ACEI) drug. Reduce sodium intake to less than 1500 mg per day. Increase dietary potassium to 6000 mg per day and magnesium to 1000 mg per day.
8. **MTHFR**. Start Homocysteine Supreme, one, twice per day (Designs for Health). This is a complex of folate and B vitamins.
9. **NOS 3**. Start NEO 40, one, twice per day (Human N) and Arterosil (Calroy Labs) one, twice per day. Eat dark green leafy vegetables and beets.
10. **APO A 1**. Start CardioLux (Metagenics), two capsules, twice a day with food.
11. **APO A 2**. Start CardioLux (Metagenics), two capsules, twice per day with food.
12. **CYP 4 F2**. Stop all caffeine from all sources, such as coffee, tea, chocolate, or other caffeine-containing beverages, and supplements or medications (either prescription or over the counter). You metabolize caffeine slowly. Take one low-dose coated 81 mg aspirin per day with food. If you have high blood pressure or kidney disease, you should be started on a specific medication called Amiloride. Reduce sodium intake to 1500 mg per day or less.
13. **4q25**. Start an anti-inflammatory diet. Reduce intake of refined carbohydrates and starches to less than 25 grams per day. Avoid white rice, white potatoes, all breads, wheat, pasta of all types, sugars, sweets, soft drinks, and artificial sweeteners. Limit fruit juices. Start the Mediterranean diet. Consume 12 servings of vegetables and fruits each day. Maintain an ideal body weight, body fat, visceral fat, and exercise at least four times per week or more for one hour each time. Take one low-dose coated 81 mg aspirin per day with food. Avoid caffeine, alcohol, trans fats, and long-chain saturated fats. Take Magnesium Malate (Designs for Health), two, twice per

day. Consume four tablespoons of extra virgin olive oil per day with your salads and other foods.
14. **COMT**. If you have abnormal genes that are heterozygote or homozygote for COMT, stop all caffeine from all sources such as coffee, tea, chocolate, or other caffeine-containing beverages, and supplements or medications (either prescription or over the counter). You metabolize caffeine slowly. Start one low-dose coated 81 mg aspirin per day. Start vitamin E in the form of gamma tocopherol one per day from Designs for Health.
15. **APO E: APO E 2 and 4 Genotypes.** Start an anti-inflammatory diet. Reduce intake of refined carbohydrates and starches to less than 25 grams per day. Avoid white rice, white potatoes, all breads, wheat, pasta of all types, sugars, sweets, soft drinks, and artificial sweeteners. Limit fruit juices. Start the Mediterranean diet. Consume 12 servings of vegetables and fruits each day. Maintain an ideal body weight, body fat, visceral fat, and exercise at least four times per week or more for one hour each time. Take one low-dose coated 81 mg aspirin per day with food. Avoid caffeine, alcohol, trans fats, and long-chain saturated fats. Start Omega-3 fatty acids in the supplement EFA Sirt Supreme (Biotics Research), take three capsules twice per day with food. Start Neo 30 (Human N Labs), one, twice per day.
16. **CYP 1A2**. Stop all caffeine from all sources such as coffee, tea, chocolate, or other caffeine-containing beverages, and supplements or medications (either prescription or over the counter). You metabolize caffeine slowly. Take one low-dose coated 81 mg aspirin per day with food.
17. **SCARB1**. Start CardioLux (Metagenics), two capsules, twice a day with food. Eat pomegranate seeds, one-quarter cup per day, either fresh or frozen (Whole Foods, Sprouts, Publix, Kroger, or order from Amazon).
18. **CORIN**. If you have high blood pressure or kidney disease, you will be started on specific antihypertensive medications. Reduce sodium intake to 1500 mg per day or less. Increase dietary potassium to 6000 mg per day and magnesium to 1000 mg per day.
19. **ADRB2**. If you have high blood pressure or kidney disease, you will be started on specific medications for blood pressure. Reduce sodium intake to 1500 mg per day or less.
20. **CYP 4A 11**. If you have high blood pressure or kidney disease, you will be started on a specific medication called Amiloride. Reduce sodium intake to 1500 mg per day or less. Increase dietary potassium to 6000 mg per day and magnesium to 1000 mg per day.
21. **AGTR1**. If you have high blood pressure, you will be started on an angiotensin receptor blocker. Reduce sodium intake to 1500 mg per day or less. Increase dietary potassium to 6000 mg per day and magnesium to 1000 mg per day.

SUMMARY AND KEY TAKE AWAY POINTS

All of the genes listed in this chapter have specific treatments that will modify the gene expression to reduce CHD risk and CHD risk factors, such as hypertension, dyslipidemia, and blood glucose, and avoid the three finite responses. These include treatments with nutrition, nutritional supplements, lifestyle changes, and medications.

BIBLIOGRAPHY

1. Sinatra ST, Houston MC (Editors). *Nutritional and Integrative Strategies in Cardiovascular Medicine*. Boca Raton, FL: CRC Press; 2015.
2. Price PT, Nelson CM, Clarke SD Omega-3 polyunsaturated fatty acid regulation of gene expression. *Curr Opin Lipidol.* 2000;11(1):3–7.
3. McNiven EM, German JB, Slupsky CM Analytical metabolomics: Nutritional opportunities for personalized health. *J Nutr Biochem.* 2011;22(11):995–1002.
4. O'Donnell CJ, Nabel EG Genomics of cardiovascular disease. *N Engl J Med.* 2011;365(22):2098–109.
5. Nuno NB, Heuberger R Nutrigenetic associations with cardiovascular disease. *Rev Cardiovasc Med.* 2014;15(3):217–25.
6. Holdt LM, Teupser D. From genotype to phenotype in human atherosclerosis--recent findings. *Curr Opin Lipidol.* 2013;24(5):410–8.
7. Roberts R, Stewart AF. Genetics of coronary artery disease in the 21st century. *Clin Cardiol.* 2012;35(9):536–40.
8. Houston, MC. *What Your Doctor May Not Tell You About Heart Disease*. New York: Grand Central Press; February 2012.
9. Webster AL, Yan MS, Marsden PA. Epigenetics and cardiovascular disease. *Can J Cardiol.* 2013;29(1):46–57.
10. Castañer O, Corella D, Covas MI, Sorlí JV, Subirana I, Flores-Mateo G, Nonell L, Bulló M, de la Torre R, Portolés O, Fitó M; PREDIMED study investigators In vivo transcriptomic profile after a Mediterranean diet in high-cardiovascular risk patients: A randomized controlled trial. *Am J Clin Nutr.* 2013;98(3):845–53.
11. Konstantinidou V, Covas MI, Sola R, Fitó M Up-to date knowledge on the in vivo transcriptomic effect of the Mediterranean diet in humans. *Mol Nutr Food Res.* 2013;57(5):772–83.
12. Corella D, Ordovás JM. How does the Mediterranean diet promote cardiovascular health? Current progress toward molecular mechanisms: Gene-diet interactions at the genomic, transcriptomic, and epigenomic levels provide novel insights into new mechanisms. *Bioessays.* 2014;36(5):526–37.
13. Estruch R, Ros E, Salas-Salvadó J, Covas MI, Corella D, Arós F, Gómez-Gracia E, Ruiz-Gutiérrez V, Fiol M, Lapetra J, Lamuela-Raventos RM, Serra-Majem L, Pintó X, Basora J, Muñoz MA, Sorlí JV, Martínez JA, Martínez-González MA; PREDIMED Study Investigators. Primary prevention of cardiovascular disease with a Mediterranean diet. *N Engl J Med.* 2013;368(14):1279–90.
14. González-Guardia L, Yubero-Serrano EM2, Delgado-Lista J, Perez-Martinez P, Garcia-Rios A, Marin C, Camargo A, Delgado-Casado N Roche HM, Perez-Jimenez F, Brennan L, López-Miranda J. Effects of the Mediterranean diet supplemented with coenzyme q10 on metabolomic profiles in elderly men and women. *J Gerontol A Biol Sci Med Sci.* 2015;70(1):78–84.
15. Appel LJ, Moore TJ, Obarzanek E, Vollmer WM, Svetkey LP, Sacks FM, Bray GA, Vogt TM, Cutler JA, Windhauser MM, Lin PH, Karanja N. A clinical trial of the effects of dietary patterns on blood pressure. DASH Collaborative Research Group. *N Engl J Med.* 1997;336(16):1117–24.
16. Sacks FM, Svetkey LP, Vollmer WM, Appel LJ, Bray GA, Harsha D, Obarzanek E, Conlin PR, Miller ER 3rd, Simons-Morton DG, Karanja N, Lin PH; DASH-sodium collaborative research group effects on blood pressure of reduced dietary sodium and the dietary approaches to stop hypertension (DASH) diet. DASH-sodium collaborative research group. *N Engl J Med.* 2001;344(1):3–10.

17. Ornish D, Scherwitz LW, Billings JH, Brown SE, Gould KL, Merritt TA, Sparler S, Armstrong WT, Ports TA, Kirkeeide RL, Hogeboom C, Brand RJ. Intensive lifestyle changes for reversal of coronary heart disease. *JAMA*. 1998 Dec 16;280(23):2001–7.
18. Laffer CL, Elijovich F, Eckert GJ, Tu W Pratt JH, Brown NJ. Genetic variation in CYP4A11 and blood pressure response to mineralocorticoid receptor antagonism or ENaC inhibition: An exploratory pilot study in African Americans. *J Am Soc Hypertens*. 2014 1;8(7):475–80.
19. Vanden Heuvel JP Nutrigenomics and nutrigenetics of ω3 polyunsaturated fatty acids. *Prog Mol Biol Transl Sci*.2012;108:75–112.
20. Varela LM, Ortega-Gomez A, Lopez S, Abia R, Muriana FJ, Bermudez B The effects of dietary fatty acids on the postprandial triglyceride-rich lipoprotein/apoB48 receptor axis in human monocyte/macrophage cells. *J Nutr Biochem*. 2013;24(12):2031–9.
21. Silva S, Bronze MR, Figueira ME, Siwy J, Mischak H, Combet E, Mullen W Impact of a 6-wk olive oil supplementation in healthy adults on urinary proteomic biomarkers of coronary artery disease, chronic kidney disease, and diabetes (types 1 and 2): A randomized, parallel, controlled, double-blind study. *Am J Clin Nutr*. 2015;101(1):44–54.
22. O'Donnell CJ, Nabel EG Genomics of cardiovascular disease .*N Engl J Med*. 2011;365(22):2098–109.
23. Paloaki GE, Melillo S, Bradley LA. Association between 9p21 genomic markers and heart disease: A meta-analysis. *JAMA*. 2010;303:648–56.
24. Qi L, Qi Q, Prudente S, Mendonca C, Andreozzi F, di Pietro N Sturma M, Novelli V. Association between a genetic variant related to glutamic acid metabolism and coronary heart disease in individuals with type 2 diabetes. *JAMA*. 2013 Aug 28;310(8):821–8.
25. Skulas-Ray AC1, Kris-Etherton PM, Harris WS, Vanden Heuvel JP, Wagner PR, West SG Dose-response effects of omega-3 fatty acids on triglycerides, inflammation, and endothelial function in healthy persons with moderate hypertriglyceridemia. *Am J Clin Nutr*. 2011;93(2):243–52.
26. Hall KT, Nelson CP, Davis RB, Buring JE, Kirsch I Mittleman MA, Loscalzo J, Samani NJ, Ridker PM, Kaptchuk TJ, Chasman DI Polymorphisms in catechol-O-methyltransferase modify treatment effects of aspirin on risk of cardiovascular disease. *Arterioscler Thromb Vasc Biol*. 2014;34(9):2160–7.
27. Winter JP, Gong Y, Grant PJ, Wild CP Glutathione peroxidase 1 genotype is associated with an increased risk of coronary artery disease. *Coron Artery Dis*. 2003;14(2):149–53.
28. Blankenberg S, Rupprecht HJ, Bickel C, Torzewski M, Hafner G, Tiret L, Smieja M, Cambien F, Meyer J, Lackner KJ; AtheroGene Investigators Glutathione peroxidase 1 activity and cardiovascular events in patients with coronary artery disease. *N Engl J Med*. 2003;349(17):1605–13.
29. Korkor MT, Meng FB, Xing SY, Zhang MC, Guo JR, Zhu XX, Yang P Microarray analysis of differential gene expression profile in peripheral blood cells of patients with human essential hypertension. *Int J Med Sci*. 2011;8(2):168–79.
30. Harrap SB Blood pressure genetics: Time to focus. *J Am Soc Hypertens*. 2009;3(4):231–7.
31. Zhou L, Xi B, Wei Y, Shen W, Li Y. Meta-analysis of the association between the insertion/deletion polymorphism in ACE gene and coronary heart disease among the Chinese population. *J Renin Angiotensin Aldosterone Syst*. 2012;13(2):296–304.
32. Fernández-Llama P Poch E, Oriola J, Botey A, Coll E, Darnell A, Rivera F, Revert L. Angiotensin converting enzyme gene I/D polymorphism in essential hypertension and nephroangiosclerosis. *Kidney Int*. 1998;53(6):1743–7.
33. Gardemann A, Fink M, Stricker J, Nguyen QD, Humme J, Katz N, Tillmanns H, Hehrlein FW, Rau M, Haberbosch W. ACE I/D gene polymorphism: Presence of the ACE D allele increases the risk of coronary artery disease in younger individuals. *Atherosclerosis*. 1998 1;139(1):153–9.

34. Castellano M, Muiesan ML, Rizzoni D, Beschi M, Pasini G, Cinelli A, Salvetti M, Porteri E, Bettoni G, Kreutz R, Angiotensin-converting enzyme I/D polymorphism and arterial wall thickness in a general population. The Vobarno Study. *Circulation.* 1995;91(11):2721–4.
35. Mesas AE, Leon-Muñoz LM, Rodriguez-Artalejo F, Lopez-Garcia E.The effect of coffee on blood pressure and cardiovascular disease in hypertensive individuals: A systematic review and meta-analysis. *Am J Clin Nutr.* 2011;94(4):1113–26.
36. Palatini P, Ceolotto G, Ragazzo F, Dorigatti F, Saladini F, Papparella I, Mos L, Zanata G, Santonastaso M.CYP1A2 genotype modifies the association between coffee intake and the risk of hypertension. *J Hypertens.* 2009;27(8):1594–601.
37. Armaly Z, Assady S, Abassi Z. Corin: A new player in the regulation of salt-water balance and blood pressure. *Curr Opin Nephrol Hypertens.* 2013;22(6):713–20.
38. Zhou Y, Wu Q. Corin in natriuretic peptide processing and hypertension. *Curr Hypertens Rep.* 2014;16(2):415.
39. Peng H, Zhang Q, Cai X, Liu Y, Ding J, Tian H, Chao X, Shen H, Jiang L, Jin J, Zhang Y. Association between high serum soluble corin and hypertension: A cross-sectional study in a general population of China. *Am J Hypertens.* 2015;28(9):1141–9.
40. Fontana V, de Faria AP, Barbaro NR, Sabbatini AR, Modolo R, Lacchini R, Moreno H. Modulation of aldosterone levels by −344 C/T CYP11B2 polymorphism and spironolactone use in resistant hypertension. *J Am Soc Hypertens.* 2014;8(3):146–51.
41. Levinsson A Olin AC, Björck L, Rosengren A, Nyberg F. Nitric oxide synthase (NOS) single nucleotide polymorphisms are associated with coronary heart disease and hypertension in the INTERGENE study. *Nitric Oxide.* 2014;39:1–7.
42. Sun B, Williams JS, Svetkey LP, Kolatkar NS, Conlin PR. Beta2-adrenergic receptor genotype affects the renin-angiotensin-aldosterone system response to the dietary approaches to stop hypertension (DASH) dietary pattern. *Am J Clin Nutr.* 2010;92(2):444–9.
43. Chen Q, Turban S, Miller ER, Appel LJ. The effects of dietary patterns on plasma renin activity: Results from the Dietary Approaches to Stop Hypertension trial. *J Hum Hypertens.* 2012;26(11):664–9.
44. Laffer CL, Elijovich F, Eckert GJ, Tu W, Pratt JH, Brown NJ. Genetic variation in CYP4A11 and blood pressure response to mineralocorticoid receptor antagonism or ENaC inhibition: An exploratory pilot study in African Americans. *J Am Soc Hypertens.* 2014;8(7):475–80.
45. Ward NC, Tsai IJ, Barden A, van Bockxmeer FM, Puddey IB, Hodgson JM, Croft KD. A single nucleotide polymorphism in the CYP4F2 but not CYP4A11 gene is associated with increased 20-HETE excretion and blood pressure. *Hypertension.* 2008;51(5):1393–8.
46. Sun Y, Liao Y, Yuan Y, Feng L, Ma S, Wei F, Wang M, Zhu F Influence of autoantibodies against AT1 receptor and AGTR1 polymorphisms on candesartan-based antihypertensive regimen: Results from the study of optimal treatment in hypertensive patients with anti-AT1-receptor autoantibodies trial. *J Am Soc Hypertens.* 2014;8(1):21–7.
47. Yang X, Sethi A, Yanek LR, Knapper C, Nordestgaard BG, Tybjærg-Hansen A, Becker DM, Mathias RA, Remaley AT, Becker LC SCARB1 gene variants are associated with the phenotype of combined high high-density lipoprotein cholesterol and high lipoprotein (a). *Circ Cardiovasc Genet.* 2016;9(5):408–418.
48. Mendoza S, Trenchevska O King SM, Nelson RW, Nedelkov D, Krauss RM, Yassine HN. Changes in low-density lipoprotein size phenotypes associate with changes in apolipoprotein C-III glycoforms after dietary interventions. *J Clin Lipidol.* 2017 Jan;11(1):224–233.

49. Wyler von Ballmoos MC Haring B, Sacks FM. The risk of cardiovascular events with increased apolipoprotein CIII: A systematic review and meta-analysis. *J Clin Lipidol.* 2015;9(4):498–510.
50. Zheng C. Updates on apolipoprotein CIII: Fulfilling promise as a therapeutic target for hypertriglyceridemia and cardiovascular disease. *Curr Opin Lipidol.* 2014;25(1):35–9.
51. Thomas GS, Voros S, McPherson JA, Lansky AJ, Winn ME, Bateman TM, Elashoff MR, Lieu HD, Johnson AM, Daniels SE, Ladapo JA, Phelps CE, Douglas PS, Rosenberg S. A blood-based gene expression test for obstructive coronary artery disease tested in symptomatic nondiabetic patients referred for myocardial perfusion imaging the COMPASS study. *Circ Cardiovasc Genet.* 2013;6(2):154–62
52. McPherson JA Davis K, Yau M, Beineke P, Rosenberg S, Monane M, Fredi JL The clinical utility of gene expression testing on the diagnostic evaluation of patients presenting to the cardiologist with symptoms of suspected obstructive coronary artery disease: Results from the IMPACT (Investigation of a Molecular Personalized Coronary Gene Expression Test on Cardiology Practice Pattern) trial. Pathw Cardiol. 2013 12(2):37–42.
53. Wingrove JA, Daniels SE, Sehnert AJ, Tingley W, Elashoff MR, Rosenberg S, Buellesfeld L, Grube E, Newby LK, Ginsburg GS, Kraus WE. Correlation of peripheral-blood gene expression with the extent of coronary artery stenosis. *Circ Cardiovasc Genet.* 2008;1(1):31–8.
54. Rosenberg S, Elashoff MR, Lieu HD, Brown BO, Kraus WE, Schwartz RS, Voros S, Ellis SG, Waksman JR, McPherson JA, Lansky AJ, Topol EJ. PREDICT investigators whole blood gene expression testing for coronary artery disease in nondiabetic patients: Major adverse cardiovascular events and interventions in the PREDICT trial. *Cardiovasc Transl Res.* 2012;5(3):366–74.
55. Ganna A, Salihovic S, Sundström J Broeckling CD, Hedman AK, Magnusson PK, Pedersen NL, Larsson A, Siegbahn A, Zilmer M, Prenni J, Arnlöv J, Lind L, Fall T, Ingelsson E. Large-scale metabolomic profiling identifies novel biomarkers for incident coronary heart disease. *PLoS Genet.* 2014:11;10–15.
56. Granger CB, Newgard CB, Califf RM, Newby LK Shah SH, Sun JL, Stevens RD, Bain JR, Muehlbauer MJ, Pieper KS, Haynes C, Hauser ER, Kraus WE. Baseline metabolomic profiles predict cardiovascular events in patients at risk for coronary artery disease. Am *Heart* J. 2012;163(5):844–850.
57. Rizza S, Copetti M, Rossi C, Cianfarani MA Zucchelli M, Luzi A, Pecchioli C, Porzio O, Di Cola G, Urbani A, Pellegrini F, Federici M. Metabolomics signature improves the prediction of cardiovascular events in elderly subjects. *Atherosclerosis.* 2014;232(2):260–4.

11 The Gut and Heart Connection
Gut Microbiome

There is now a proven connection between your gut (or gastrointestinal tract, GI) and the risk of coronary heart disease (CHD), myocardial infarction (MI), heart failure, hypertension, high cholesterol, diabetes mellitus, obesity, and many other medical problems and diseases. The medical term that is used to describe all the organisms in your gut is the "gut microbiome" (GM).

The gut microbiome is a collection of bacteria, fungi, viruses, and other organisms that are often referred to as "microbes". They inhabit the human intestine and play an essential role in human health and disease (Figure 11.1). There exists a very close information exchange between your body and the intestinal microbes, as they perform a vital role in digestion, immune defense, nervous system regulation, and metabolism. Studies have shown that the composition of the GM and the substances that they make are undeniably related to the occurrence of various diseases—especially CHD and MI. Many researchers have demonstrated that the intestinal microbiome is a "virtual organ" with endocrine function. The active substances produced by it can affect your physiology, health, functionality, and risk of cardiovascular diseases.

This knowledge allows us to now identify, test, measure, and treat GM to prevent CHD and other diseases. In this chapter, I will discuss how the gut microbiome is related to CHD and MI. In addition, you will begin to understand that until you have good gut health, you will not likely have good heart health. The treatment program that I will discuss allows you to improve your gut health, the microbiome, and your risk for many of the CHD risk factors and for CHD. I will review recent science and the clinical applications that will allow you to alter your nutrition, supplements, and lifestyle in such a way that will improve the GM and thus reduce your risk of CHD and MI.

The human gastrointestinal (GI) tract is predominantly bacteria with a separate and unique environment called an "ecosystem" (microbiome) that contains over 100 trillion microbial cells, with the highest microbe densities found in the colon. Gut microbes are, for the most part, codependent, meaning that they require one another, from additional members of the community and from humans, for metabolic support and for survival in a mutually beneficial or helpful (symbiotic) relationship. For example, gut microbes help with the digestion of nutrients, prevent significant colonization of bad bacteria called pathogens, and promote gut immunity, while the host provides a favorable environment for microbial survival.

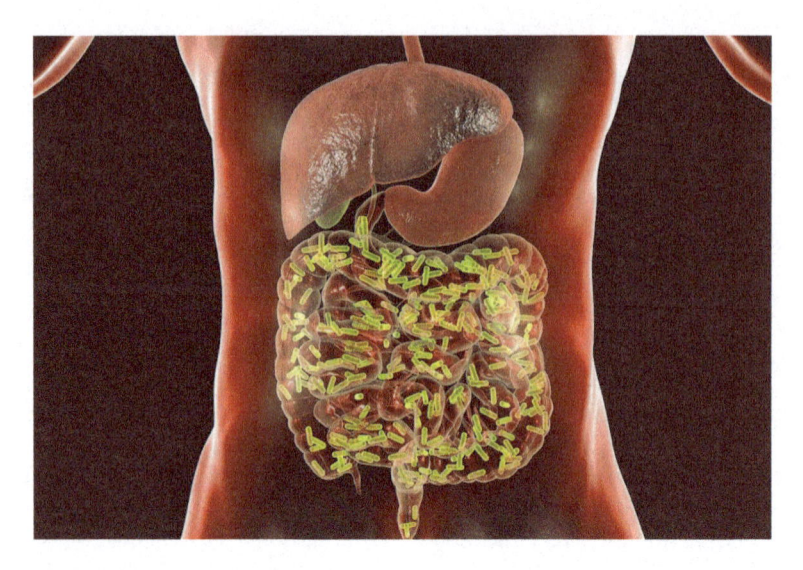

FIGURE 11.1 The human intestine and the gut microbiome with the esophagus, stomach, small intestine and large intestine (colon), and rectum.

If the gut microbiome changes into an abnormal or adverse condition (dysbiosis) it leads to long-term increases in many diseases that can originate early in life, similar to traditional CHD risk factors. The microbial inhabitants within the host often contribute to chemical changes, and this can increase the risk for many diseases, sometimes emerging decades later.

Alterations in the gut or intestinal barrier or lining (called the "enterocytes") lead to dysfunction and increased intestinal leakage Certain types of bacteria called gram-negative bacteria may enter the blood to produce severe inflammation and damage to the arteries. The medical term for this is *endotoxemia* (internal toxins), which means that the bacteria, the components of the bacteria, the types of food that we eat, and the health of the gut may allow toxins to enter the blood and cause arterial damage. The chronic dietary intake of long-chain saturated fatty acids (SFAs) with refined carbohydrates and starches increases the gram-negative bacterial concentration by 70%. On the other hand, a vegetable intake, fiber, and probiotics decrease gram-negative gut concentration.

There are many products and chemicals produced by the bad bacteria in the gut. One of these is called trimethylamine N-oxide (TMAO) which comes from compounds (such as meat and eggs) that contain choline, phosphatidyl choline, and carnitine and is associated with a higher risk for CHD and MI. However, a type of fatty acid called short-chain fatty acid (SCFA) such as butyrate, propionate, and acetate, improve gut and colonic health, insulin resistance, diabetes mellitus, high cholesterol, hypertension, and CHD. In addition, bile acids are important in glucose metabolism, cholesterol and fat metabolism, and in obesity regulation.

Increases in the abundance of certain gut bacteria can increase or decrease the risk for CHD and MI. You can measure these bacteria with a stool test and determine

The Gut and Heart Connection

which bacteria are present and also the numbers of the bacteria in the gut. CHD plaque also has a higher ratio of some bad bacteria and their respective cell walls or their genetic material that have traveled from the gut to the coronary arteries. It has been shown that a high-fiber diet reduces some of the bad bacteria, both in the gut and in the plaque in the heart arteries, and thus reduces the risk of CHD.

A large number of bacteria in the gut have over 100× as many genes compared to the number of human genes. It is important to have a large variety of bacteria with many functions in the gut to improve health and reduce CHD. For example, a lower variety of bacteria is observed in obesity, meat-based diets, diabetes mellitus, and CHD. A higher variety of bacteria is found in vegetarian-based diets with high dietary fiber, nondigestible carbohydrates, resistant starches, SCFA, prebiotics, and probiotics. These patients have a lower incidence of diabetes mellitus and CHD. An example of one vegetable group is the basilica family such as cabbages and broccoli, which contain high amounts of important chemicals that have favorable actions on the gut to reduce gut leakage, improve gut repair, reduce endotoxemia, increase SCFA, and increase production of vitamins and other good compounds by the bacteria and the gut lining.

BAD FOOD CHOICES EQUAL BAD HEART ARTERIES

Atherosclerosis and CHD are primarily related to what happens after you eat your meals. This is referred to in medicine as the "CHD postprandial phenomena". Inflammatory foods, coupled with high blood sugar (hyperglycemia), high fructose (soft drinks and other foods), high sucrose, and high fats in the blood called triglycerides (TG) (hypertriglyceridemia) cause vascular oxidative stress, vascular inflammation, vascular immune dysfunction, endotoxemia, and endothelial dysfunction with a reduction in nitric oxide. (Figure 11.2) The reduction in nitric oxide levels may be as high as 20–42%. Fructose is particularly bad, as it contributes to fatty liver, damages to the gut lining, increased inflammation, elevated triglycerides, and CHD.

These effects are particularly common with sugars, such as glucose, sucrose, and fructose, refined carbohydrates and starches, all breads (white, brown or wheat); white potatoes, white rice, and all wheat pastas; and with long-chain SFA and trans fats (TFA). Microbial products and the actual bacteria also cross into the gut and the blood to induce even more endotoxemia. There are increases in inflammatory chemicals that activate the various receptors on the arteries.

In addition, the body has what is called "metabolic memory". This means that these vascular responses are remembered each time that these types of foods are eaten, they are more severe and last longer each time compared to the first consumption. This chronic ingestion of sugars, such as glucose and fructose, refined carbohydrates and starches, and TFA and long-chain SFA may activate fat or adipose tissue pathways leading to inflammation that comes from the actual fat cells all over the body, especially those in the belly (visceral fat). A type of vitamin E called gamma tocopherol reduces many of these changes as well as vitamin C and a nutritional supplement called "lipoic acid". The implications for this are very real

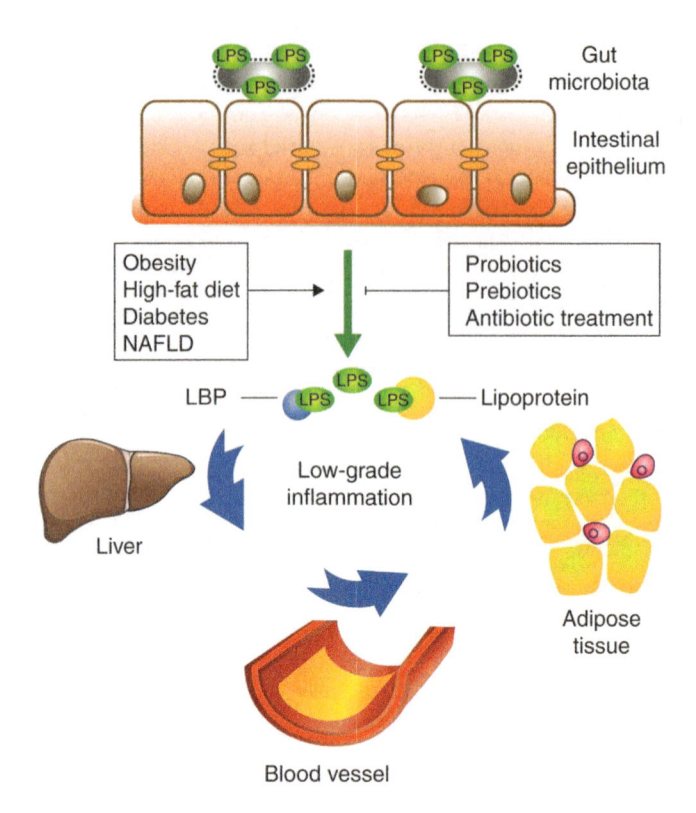

FIGURE 11.2 The enterocyte responses to various dietary intake (foods) and the association with inflammation and diseases.

and important for your eating habits in order to reduce CHD. Eating good, healthy food and taking about 200 mg of gamma tocopherol, 250 mg of vitamin C and 100 mg of lipoic acid before eating could be important to reduce the adverse effects of these types of food on your arteries. Other supplements and selected foods may also be of benefit (Box 11.4).

Postprandial (post-meal) blood glucose predicts CHD better than fasting blood glucose in both diabetic and nondiabetic patients. Postprandial blood glucose (hyperglycemia) and a high intake of fructose (common in soft drinks) also induce higher triglycerides, inflammation, and high sensitivity C reactive protein (HS-CRP). These abnormalities in blood sugar and triglycerides may last for 6–8 hours depending on the amount, type, and content of bad food eaten. This means that the arterial abnormalities with inflammation, oxidative stress, immune and endothelial dysfunction, and lower nitric oxide levels also continue. You can thus imagine and understand that if you eat some of the bad food types every 6–8 hours you will never have normal arterial health. The idea of fasting for 12–14 hours, intermittent fasting, the fasting mimicking diet, and eating good, healthy food is extremely important in your daily life to reduce the incidence of CHD.

Lipoprotein saccharide (LPS) is the main component of the bacteria cell wall. LBP is a lipoprotein binding protein, and NAFLD is nonalcoholic fatty liver disease. Obesity, a high-fat diet, diabetes, and NAFLD increase inflammation. Probiotics, prebiotics, and some antibiotics decrease inflammation (see the next section.

PROBLEMS WITH MEDICATIONS THAT REDUCE STOMACH ACID AND CAUSE DECREASED ABSORPTION OF VITAMINS AND MINERALS

Avoid long-term use of medications that block acid production in your stomach such as proton pump inhibitors (PPIs) (Prilosec, Protonix, Nexium, Prevacid, AcipHex, and others) and H2 blockers (Zantac, Pepcid, Tagamet, Axid, and others), even the over-the-counter brands. All of these should be used for only a few months unless you have a severe medical condition that requires them and your doctor approves them. All of the drugs that reduce acid, also reduce protein absorption and vitamin and mineral absorption, such as vitamin B 12, vitamin C, and also iron, calcium, and magnesium. They will lower nitric oxide and cause endothelial dysfunction by reducing nitrites in the blood. They will increase the risk of CHD, MI, and kidney disease. The reductions can result in muscle loss, memory loss, anemia, fatigue, bone loss, palpitations, and more.

THE TREATMENT FOR THE GUT MICROBIOME

Treatment with good nutrition, such as a plant-based diet, increased dietary fiber, prebiotics, probiotics, extra virgin olive oil (EVOO) and omega-3 fatty acids will improve gut health and reduce the risk for CHD and MI (Figure 11.2). In addition, daily dietary intake of long-chain saturated fatty acids (SFAs) should be reduced to less than ten percent of your total fat intake (Box 11.1). Decrease the daily intake of TFA to zero (Box 11.2). Reduce the intake of all refined carbohydrates and sugars to less than 25 grams per day (Box 11.3). Increase dietary intake of fruits, vegetables, and legumes to 12 servings per day. Reduce alcohol intake to less than ten grams per day. Reduce or stop alcohol and reduce caffeine, sodas, and artificial sweeteners. Take a good multivitamin and mineral complex with zinc at 50 mg per day and the recommended daily intake of vitamins such as A, D, C, and magnesium. Consume glutamine at 2 grams per day to improve gut healing and reduce leakage. Get plenty of multicolored polyphenol-rich plants and plant extracts and fermented foods.

BOX 11.1   Saturated Fats Are Found in These Foods

- Butter
- Ghee
- Lard
- Coconut oil and palm oil

- Biscuits
- Fatty cuts of meat
- Sausages
- Bacon
- Cured meats like salami, chorizo, and pancetta
- Cheese
- Pastries, such as pies, quiches, sausage rolls, and croissants
- Cream
- Sour cream
- Ice cream
- Milkshakes

Chocolate and chocolate spreads

BOX 11.2 Trans Fats in Your Food

- Baked goods, such as cakes, cookies, and pies
- Shortening
- Microwave popcorn
- Frozen pizza
- Refrigerated dough, such as biscuits and rolls
- Fried foods, including French fries, doughnuts, and fried chicken
- Nondairy coffee creamer
- Stick margarine

BOX 11.3 Sugars and Refined Carbohydrates Are in These Foods

White bread
Pizza dough
Wheat pasta, pastries
White flour
White rice
White potatoes
Sweet desserts
Many breakfast cereals
Soft drinks
Sports drinks, energy drinks
Vitamin water
Some fruit beverages
Granola, breakfast
Cereal bars

The Gut and Heart Connection

High-quality prebiotics and probiotics should be consumed on a daily basis. Prebiotics are a source of food for your gut's healthy bacteria. They are carbohydrates your body can't digest. So they go to your lower digestive tract, where they act like food to help the healthy bacteria grow. Probiotics are live yeasts and good bacteria that live in your body and are good for your digestive system. They are found in foods such as yogurt and sauerkraut. You should take a high-quality probiotic with a variety of bacterial strains and at least 100 billion colony-forming units (CFU) per day. Prebiotics are in foods such as whole grains, bananas, greens, onions, garlic, soybeans, and artichokes. In addition, probiotics and prebiotics are added to some foods and available as dietary supplements (see Box 11.4).

> **BOX 11.4 Supplements and Foods That Can Be Taken before Meals to Improve the Post-Meal Adverse Effects on Enterocytes and Arteries**
>
> - Gamma tocopherol
> - Vitamin C
> - Lipoic acid
> - Neo 40
> - Probiotics with at least 100 billion colony-forming units (CFU) per day
> - Prebiotics
> - Curcumin
> - Quercetin
> - Omega-3 fatty acids
> - Luteolin
> - Resveratrol
> - EGCG and green tea
> - Cocoa
> - Basilica family of plants (cabbage and broccoli)
> - Fiber
> - Avocado
> - Olive oil and olive products

SUMMARY AND KEY TAKE AWAY POINTS

1. The gut microbiome has more genes than the human genome.
2. The gut microbes help with the digestion of nutrients, prevent significant colonization of pathogens, and promote gut immunity, while the host provides a favorable environment for microbial survival.
3. The microbiome has direct effects on the risks for CHD, MI, hypertension, high cholesterol, diabetes mellitus, and obesity.
4. A damaged or leaky gut allows bacteria and other products into the blood and causes endotoxemia, inflammation, and arterial damage.

5. CHD is a post-meal disease. Eating healthy foods and taking specific supplements before eating will reduce the arterial damage, higher blood sugar, and triglycerides.
6. Treatments to repair the gut include proper nutrition with low intake of long-chain SFA, TFA, and sugars, such as glucose and fructose. You should consume more fruits, vegetables, fiber, prebiotics, probiotics, omega-3 fatty acids, olive products, specific vitamins, minerals, and glutamine.
7. Avoid long-term use of medications that block acid production in your stomach, such a proton pump inhibitors (PPIs) and H2 blockers, as they will increase risk for endothelial dysfunction, lower nitric oxide, CHD, and MI.

BIBLIOGRAPHY

1. Jing X and Yuejin Y. Implications of gut microbiome on coronary artery disease. *Cardiovasc Diagn Ther.* 2020 Aug;10(4):869–80.
2. Naofumi Y, Yamashita T, Hirata K. Gut microbiome and cardiovascular diseases. *Diseases.* 2018;6(3):56.
3. Jie Z, Xia H, Kristiansen K. The gut microbiome in atherosclerotic cardiovascular disease. *Nature Communications.* 2017;8:845.
4. Trøseid M, Andersen GØ, Roksund Hov KBJ. The gut microbiome in coronary artery disease and heart failure: Current knowledge and future directions. *EBioMedicine* 2020 Feb;52:102649.
5. Lin L, Xuyu H, Yingqing F. Coronary heart disease and intestinal microbiota. *Coronary Artery Disease.* 2019 Aug;30(5):384–9.

12 Nutrition Part I

The Importance of Nutrition in Coronary Heart Disease Prevention and Treatment

INTRODUCTION

One of the greatest causes of mortality in industrialized societies continues to be coronary heart disease (CHD). Moreover, the ability to decrease the incidence of CHD has reached a limit utilizing traditional diagnostic evaluations, prevention, and treatment strategies for the top five cardiovascular risk factors that include hypertension, diabetes mellitus, dyslipidemia (abnormal cholesterol), obesity, and smoking. It is well known that about 80% of CHD can be prevented with optimal nutrition, coupled with exercise, weight management, and smoking cessation. Among all of these factors, optimal nutrition provides the basic foundation for prevention and treatment of CHD. Numerous nutrition studies have shown dramatic reductions in the incidence of CHD. As nutritional science, genetics, and nutrient gene interaction research continues, our ability to adjust the best nutrition with an individualized, precise, and integrated approach is emerging. This will be a new era in personalized and precision nutrition for the prevention and treatment of CHD. Coronary heart disease remains the number one cause of morbidity and mortality in the United States (1, 2). The annual cost (direct and indirect) of treating CHD is approximately $320 billion (2). One in every three deaths is due to CHD with over 2200 US citizens dying from CHD and myocardial infarction (MI) daily (2–5). The top five CHD risk factors include hypertension, dyslipidemia (abnormal cholesterol), diabetes mellitus, obesity, and smoking (1). However, more than 400 CHD risk factors have been defined (2). There are numerous insults to the cardiovascular system, but there are only three finite vascular responses which include vascular inflammation, vascular oxidative stress, and vascular immune dysfunction, which lead to plaque formation, CHD, and MI (2–12).

NUTRITION AND CHD

Targeted nutrition in combination with other lifestyle changes is a foundational recommendation for the reduction of CHD. National and international nutritional guidelines are still evolving as new science, nutrition, and genetic studies (nutrigenomics)

DOI: 10.1201/b22808-13

are published. There are many recent clinical trials that provide new information in this quest to improve CHD outcomes related to nutrition (13–15) (Table 12.1).

MEDITERRANEAN DIET

In the 4.8-year primary prevention (PREDIMED diet), the rate of major cardiovascular events from CHD and MI, cerebrovascular accidents (CVA), or total cardiovascular deaths were reduced by 28% with the consumption of nuts and reduced by 30% with extra-virgin olive oil (EVOO) (16). The reduction in MI was significant at 23% overall with a 20% reduction with EVOO and a 26% reduction from mixed nuts. Total cardiovascular deaths were reduced by 17% (1–15). New onset type 2 diabetes mellitus (T2DM) was decreased by 40% with EVOO and 18% with mixed nuts (17). These reductions were associated with decreases in high-sensitivity C reactive protein (HS-CRP) and interleukin (IL-6). The high content of nitrate (NO3) that is converted to nitrite (NO2) (average of 400 mg per day), the increased amount of omega-3 fatty acids, good omega-6 fatty acids, polyphenols (such as quercetin, resveratrol, and catechins, in grapes and wine), provide many of the beneficial outcomes in CHD (18). In large secondary prevention in patients that had already had an MI, the Lyon Heart study (19) demonstrated significant reductions in all events including cardiac death, nonfatal MI, unstable angina, CVA, congestive heart failure (CHF), and hospitalization at four years using the traditional Mediterranean diet (TMD) supplemented with alpha-linolenic acid (ALA) compared to a prudent Western diet. Compared to the control, the Mediterranean style diet with ALA demonstrated a 73% lower risk of cardiac death and nonfatal MI during the study period (19). Olive oil was associated with a decreased risk of overall mortality and a significant reduction in cardiovascular disease (CVD) mortality in a large Mediterranean study of 40,622 subjects (20). For each increase in olive oil by 10 grams there was a 13% decrease in CV mortality. In the highest intake of olive oil, there was a 44% decrease in CV mortality (20).

One of the mechanisms by which the TMD, particularly if supplemented with extra-virgin olive oil at 50 grams per day, can exert CV health benefits is through changes in the expression of cardiovascular genes related to cardiovascular risk that include genes for atherosclerosis, inflammation, oxidative stress, vascular immune dysfunction, T2DM, and hypertension(16, 17, 21–23). This includes genes such as ADRB2 (adrenergic beta 2 receptor), IL7R (interleukin 7 receptor), IFN gamma (interferon gamma), MCP1 (monocyte chemotactic protein), TNFα (tumor necrosis factor alpha), interleukin 6 (IL-6) and HS-CRP (high-sensitivity C reactive protein) (16, 17, 20–23). In summary, the TMD has been shown to have the following effects (15–17, 21–23):

- Lowers blood pressure.
- Improves serum lipids: lowers total cholesterol (TC), low-density lipoprotein (LDL), triglycerides (TG); increases high-density lipoprotein (HDL); lowers oxidized LDL (oxLDL) and lipoprotein a (Lp(a)). In addition, the TMD increases LDL size and decreases the LDL particle number (LDL P) to a less atherogenic profile that reduces the risk for CHD and MI.

TABLE 12.1
Summary of Nutrition, Nutrients, and Daily Intake

Nutrient	Daily Intake
Diets that benefit cardiovascular health:	
Mediterranean Diet and PREDIMED+ALA	
DASH 1 and 2 Modified	
Vegetarian Diet	
• Potential for nutrient deficiencies, including vitamin B12, vitamin D, omega-3 fatty acids, iron, calcium, carnitine, zinc, and protein	
Paleolithic Diet	
Caloric Restriction and Intermittent Fasting and FMD	
Low AGEs	
Alkaline Diet	
• No definitive results but appears in line with DASH and TMD	
Fats:	Less than 35% Total Caloric Intake
SFA	<7–9% of total diet
• LCFA and MCFA have variable effects, with LCFA having a higher risk; SCFA are neutral	Replace with PUFA or MUFA
Coconut Oil	
• No recommendation for prevention or treatment of CHD or CVD, but it is a possible substitute for high glycemic carbohydrates in low amounts	
Trans Fat	Avoid trans fat
PUFA:	Omega-3 to Omega-6 ratio at 4:1
Omega-3 Fatty Acids	>1 g of EPA+DHA per day
• Opt for balanced formulation with DHA, EPA, GLA, and gamma-delta tocopherols	1.1 g/day for women
	1.6 g/day for men
	~2% total daily calories
MUFA (164):	50 g/day
Extra-virgin olive oil (EV00)	
Diet Elements:	
Animal Protein	
• Avoid processed red meat	
• Aim for lean, organic cuts	
Fish	1–2 servings/week
• Choose fish with high omega-3 content and low mercury levels; coldwater fish is best	20 g/day
Nuts	>5 servings/week; 28 g/day
Vegetables and Fruits	200–800 g/day
• Dark leafy greens have the strongest effect on CHD risk	
Milk and Milk Products (whole milk)	
• Intake has an inverse association with CVD	
Eggs	6–10 eggs per week as part of a
• No association with increased risk, except possibly for diabetics	healthy cardiovascular diet
Special recommendation for diabetics	

(Continued)

TABLE 12.1 (CONTINUED)
Summary of Nutrition, Nutrients, and Daily Intake

Nutrient	Daily Intake
Refined Carbs, Sugar, and Sugar Substitutes	Reduce or eliminate from diet
Alcohol	1–2 drinks/day for women
	2–4 drinks/day for men
Isolated Nutrients and Neutraceutical Supplements	
Curcumin	
Cinnamaldehyde (cinnamon)	
Sulforaphane (broccoli)	
Resveratrol	
Luteolin	
Quercetin	
Caffeine:	Avoid caffeine from all sources
• Different effect on fast metabolizers compared to slow metabolizers	in slow metabolizers
Soy Protein (fermented)	15–30 g/day
Whey Protein	20–30 g/day
Gluten	
• Possible but weak link even in those with celiac disease	
• Choose 100% whole grains	
Sodium	Less than 1500 mg per day. Low sodium to potassium ratio
Potassium	4.7 g/day, preferably from food
Magnesium	1000 mg per day

- Improves T2DM and dysglycemia.
- Improves oxidative defense and reduces oxidative stress.
- Reduces inflammation.
- Reduces the risks of clots or thrombosis.
- Decreases the risk of congestive heart failure.
- Increases nitrates/nitrites that increase nitric oxide in the arteries.
- Improves membrane health of all the cells.
- Reduces MI, CHD, and CVA.
- Reduces homocysteine, which is a risk factor related to CHD and folic acid metabolism.

DIETARY APPROACHES TO STOP HYPERTENSION: DASH DIETS 1 AND 2

The Dietary Approaches to Stop Hypertension (DASH) diets reduce blood pressure (BP) and CHD. Both DASH 1 and DASH 2 diets emphasize the increased daily intake of fruits, vegetables, whole grains, beans, fiber, low-fat dairy products, poultry, fish, seeds and nuts, but limited red meat, sweets, and sugar-containing beverages.

Nutrition Part I

The intake of potassium, magnesium, and calcium are increased but with a variable restriction in dietary sodium (24, 25). The DASH diets evaluated borderline or stage 1 hypertension (< 160/80–95 mm Hg) in 379 subjects who were drug-free over eight weeks. A control diet was prescribed for three weeks, and then the study subjects were randomized to the control diet—a fruit and vegetable diet with 8.5 servings or a combined fruit and vegetable diet with 10 servings and low-fat dairy. The content of sodium, potassium, magnesium, calcium, and fiber were the same in the two diets. The control diet had less potassium, magnesium, and calcium by 50%, less fiber by 22 grams, and only four servings of fruit and vegetables but was otherwise the same as the other. Both DASH diets reduced blood pressure within four weeks by approximately 10/5 mm Hg. The blood pressure remained stable as long as there was good adherence to the diets. The results on blood pressure of the various types of DASH-I and DASH-II diets are shown below:

1. DASH-I overall combination diet vs control diet: 5 /3 mm Hg.
2. DASH-I hypertensive patients. Combination diet vs control diet: 10.7/5.2 mm Hg.
3. DASH-II overall combination vs low-sodium DASH diet: 8.9/4.5 mm Hg vs control high-sodium diet.
4. DASH-II hypertensive patients. Combination of low-sodium DASH 11.5 /6.8 mm Hg in the diet vs high-sodium control diet.

Limiting refined carbohydrates, despite an increase in dietary saturated fatty acids (SFA), improves the lipid profile with both of the DASH diets. The DASH diets are as effective in BP reduction as one antihypertensive medication and also decrease HS-CRP and serum lipids. In the Nurses' Health Study (NHS), adherence to the DASH dietary pattern was associated with a lower risk of CHD by 14% in those with the highest adherence to the diet (26).

The DASH diets provide various mechanisms for the improvement of all the cardiovascular risk factors and CHD risks including the following:

1. Increased nitric oxide and increased plasma nitrate.
2. Increased kidney excretion of salt and water, like a diuretic.
3. Decreased oxidative stress and increased oxidative defense.
4. Reduced oxidative stress.
5. Improved endothelial function.
6. Decreased arterial stiffness.

DIETARY FATS

OMEGA-3 FATTY ACIDS

The role of fats in CHD has been evaluated in numerous clinical trials (14, 27–71). A large study of over 830,000 patients taking omega-3 FA (30) examined EPA+DHA (eicosapentaenoic acid and docosahexaenoic acid) from foods or supplements and the relationship to CHD, MI, sudden cardiac death, coronary death, and angina in primary and secondary prevention. There was a 6% reduction in CHD risk with EPA+DHA. However, in those with high triglycerides over 150 mg/dL, and elevated low-density lipoprotein cholesterol above 130 mg/dL, there was a statistically significant 14–16% CHD risk reduction with EPA+DHA. Other studies resulted a 18% significant reduction of CHD for higher intakes of EPA+DHA over one gram per day. The sudden cardiac death (SCD) rate was reduced 47%. The greatest reduction in CHD (25%) occurred in those with high TG over 150 mg/dL and doses of omega-3 FA over one gram per day. These results and others indicate that EPA+DHA may be associated with a reduction in CHD risk, with the greatest benefit observed among higher-risk populations and those taking higher doses of EPA+DHA. Omega-3 FAs reduce ventricular arrhythmias (57) and decrease cardiovascular and total mortality (58). Omega-3 FAs are typically found in coldwater fish, such as salmon, mackerel, and others as well as plant-based products, like algae, flax, chia, and hemp seeds; as fatty fish eat algae, they serve as a supply for these essential fats. Omega-3 fatty acids decrease MI and CHD 18% more with concomitant use of statins (61), reduce stent restenosis, (59), reduce post MI mortality (62), coronary artery bypass graft occlusion (CABG occlusion) (63, 64), plaque formation, coronary artery calcification, and atherosclerosis (65, 66); improve the lipid profile (10, 67); lower glucose and improve insulin resistance (68–70); and reduce blood pressure (2, 4, 11, 71). The dose prescribed will depend on the condition being treated as well as age, body weight, and use of concomitant medications and other nutritional supplements. It is best to use a balanced formulation with DHA, EPA, gamma-linolenic acid (GLA) and gamma-delta tocopherols. This will prevent oxidation in the cell membranes and reduce depletions of the EPA and DHA by GLA or vice versa (10, 67, 71).

MONOUNSATURATED FATS

The effects of monounsaturated fatty acids (MUFA) on the risk of CHD, CHD mortality, serum lipids, and endothelial vascular function are favorable (49, 72, 73). Partial replacement of SFA with MUFA improves the blood lipids and reduces the risk of CHD (49, 73, 74). The Nurses' Health Study and the Health Professionals Follow-up Study (HPFS) followed over 84,000 patients for 24–30 years (51). Replacing 5% of energy from SFA with equivalent energy intake from MUFA was associated with a 15% lower risk of CHD. In the 4.8-year primary prevention (PREDIMED diet), the rate of major cardiovascular events from myocardial infarction, cerebrovascular accidents, and total CV deaths were reduced by 28% with mixed nuts and 30% with extra-virgin olive oil (16).

SATURATED FATTY ACIDS

Clinical studies offer conflicting conclusions regarding the role of SFA in the risk of CHD. This has led to confusion in the lay public that is exacerbated by recently published national best sellers and different nutrition recommendations by national and international committees (14, 18, 27, 28, 31–56). The source of the confusion lies within the complexity, accuracy, and coordination of the results and conclusions in basic science, clinical epidemiology, and clinical studies. Some of the misconceptions and improper interpretations are related to the source of the SFA, its chemical structure (such as the carbon length, absorption, and the replacement nutrient(s)), how our genes respond to dietary SFAs, and the composition of the gut microbiome (31–35).

SFAs also have variable effects on blood lipids, fatty liver disease, thrombosis, inflammation, oxidative stress, and immune vascular function (31–41). Stearate is a long-chain fatty acid that has a carbon length of 18 (C-18) and has minimal effect on CHD risk or serum lipids due to its rapid change to MUFA by an enzyme (31–33). Long-chain fatty acids (LCFA) enhance gastrointestinal growth of gram-negative bacteria and their cell walls, inducing inflammation, increasing gastrointestinal permeability and the risk of endotoxemia (31, 36–40).

Published clinical trials and reviews have provided more accurate insights into the relationship of SFA and CHD (27, 28, 42–49). The largest meta-analysis of three large studies (Health Professionals Follow-Up Study (HPFS), the Nurses' Health Study (NHS 1), and the NHS-2), utilizing a 5% equal energy replacement in calories (isocaloric energy replacement) of SFA with polyunsaturated fatty acids (PUFA, or vegetable fat), was associated with a 24% and 10% reduction in CHD risk, respectively (43). Replacement of 1% of energy from SFAs with PUFAs lowers LDL cholesterol, which predicts a 2–8% reduction in CHD (43).

SFA intake and CHD were positively associated in the prospective, longitudinal cohort studies of over 115,000 men and women in the HPFS and the NHS over a 34–38-year follow-up (44). SFAs were mostly lauric acid (12:0), myristic acid (14:0), palmitic acid (16:0), and stearic acid (18:0) at 9–11.3% of energy intake. Comparing the highest versus the lowest groups of individual SFA intakes, CHD increased 7% for 12:0, 13% for 14:0, 18% for 16:0, 18% for 18:0, and 18% for all four SFAs combined. The reduction in CHD after 1% energy isocaloric replacement of SFA 12:0–18:0 was 8% for PUFA, 5% for MUFA, 6% for whole grains and 7% for plant proteins (45).

The PREvención con DIeta MEDiterránea (PREDIMED) was a six-year prospective study of 7038 subjects with a high CVD risk that included CHD, MI, CVA, or death from CV causes (28). The dietary consumption of saturated fatty acid (SFA) and trans fatty acid (TFA) from the highest to the lowest daily intake increased overall CVD by 81% and 67%, respectively. The intake of PUFAs and MUFAs reduced the risk of CVD and death. The isocaloric replacement of SFAs or TFA with MUFAs and PUFAs reduced CHD (28). SFA from processed foods increased CHD (28).

CONCLUSIONS AND SUMMARY ON SFAS

SFA are diverse compounds and cannot be "lumped" into a single category, as they have variable effects on CHD. It is prudent to replace long-chain fatty acids (LCFA) with PUFA, MUFA, short chain fatty acids (SCFA), whole grains, plant proteins, and, perhaps, medium chain fatty acids (MCFA). The daily recommended grams per day or percent of SFA relative to total fat or total calories cannot be accurately determined at this time. Some studies suggest that the SFA dietary intake should be well below 9% of the total caloric intake. The overall relationship of the human diet to CHD should include the totality of our nutrition and avoid reductionist evaluations of single macronutrients. New nutritional guidelines should promote dietary patterns that improve CHD based on validated science. Refined carbohydrates, high fructose corn syrup, starches, and TFA increase the risk of CHD. Omega-6 FAs appear to be neutral or improve CHD risk, whereas omega-3 Fas, PUFAs, MUFAs, fermented foods, fiber, fruits and vegetables, and the PREDIMED diet reduce CHD and CVD (14, 31–56).

SATURATED FATTY ACID KEY TAKE AWAY POINTS AND CONCLUSIONS

1. Dietary SFA intake is associated with an increased CHD risk, and reducing dietary SFA through isocaloric (ISC) replacement with PUFA, MUFA, omega-6 FA, whole grains, and plant proteins decreases CHD risk.
2. The source of the SFA is associated with the risk for CHD. Dietary intake of meat and animal fat have the greatest risk with a range of 6–48%.
3. LCFA are the most likely SFA associated with CHD risk. SCFA are not associated with CHD risk, but additional studies are needed to confirm this.
4. The carbon chain number of the SFA may be associated with CHD risk.
5. Replacement of SFA with PUFA reduces CHD risk.
6. Replacement of SFA with MUFA reduces CHD risk
7. Replacement of SFA with omega-6 FAs decreases CHD risk.
8. Replacement of SFA with refined carbohydrates (CHO) increases CHD risk.

TRANS FATTY ACIDS

A study of 126,233 participants from the NHS and the HPFS analyzed the relationship between choices of dietary fats and overall mortality (75). During the follow-up, 33,304 deaths were documented. Dietary TFA had the most significant adverse impact on health. Every 2% higher intake of TFA was associated with a 16% higher chance of premature death and a 25% increase in CHD death and nonfatal MI during the study period (75). The overall recommendations were to reduce omega-6 fatty acids, increase omega-3 fatty acids and the ratio of omega-3 to omega-6 fatty acids and reduce SFA in addition to the elimination of TFA. The cardiovascular adverse effects of industrially produced TFAs are shown below (14, 76, 77,78).

1. Dyslipidemia.
    A. Increases total cholesterol— 8%.
    B. Increases LDL-C— 9%.
    C. Increases TG— 9%.
    D. Lowers HDL-C 2—3%.

E. Increases apolipoprotein B— 8%.
   F. Increases lipoprotein (a) (Lp (a))— 4%.
2. Increases adipose tissue TFA levels.
3. Increases TG.
4. Increases insulin resistance, glucose intolerance, and T2DM risk.
5. Increases thrombogenic risk and plaque vulnerability.
6. Increases risk of CHD and MI.
7. Increases risk of primary cardiac arrhythmias and sudden death.
8. Increases all-cause mortality by 25% from lowest to high intake.
9. Every 2% energy increase from TFA intake results in a 25% increase in CHD (CHD death and nonfatal MI).
10. Hypertension.
11. Endothelial Dysfunction.
12. Obesity.
13. Increases inflammation.

COCONUT OIL

Coconut oil has been inappropriately promoted for a reduction in CHD and other CV events with no evidence to support it in human clinical studies. In a meta-analysis of 21 studies with 8 clinical trials and 13 observational studies, coconut oil increased TC and LDL more than PUFA but less than butter, and it increased HDL and TG with no change in TC/HDL ratio. There was no change in any cardiovascular or CHD events (79, 80). Coconut oil is 92% SFA, mostly lauric acid C 12:0 up to 51% (MCFA), then myristic acid (C14:0), and palmitic (C-16). SFA with a carbon length over 12 are considered LCFA, which may increase LDL and CHD risk. Coconut oil should not be recommended at this time for prevention or treatment of CHD or CVD due to the lack of prospective studies on CV outcomes and the mixed effects on serum lipids. It would seem based on the percentage of SFA types that it may be neutral in the effects on CHD and lipids. See below for the types of fats in coconut oil.

MILK, MILK PRODUCTS, AND PEPTIDES

Recent clinical studies indicate that milk, milk peptides, and milk products reduce blood pressure, CHD, DM, CVA, and atherosclerosis (81–83). In a recent meta-analysis of 27 studies there was an inverse association between total dairy intake and cardiovascular disease. Milk and milk products improve insulin resistance, postprandial hyperglycemia, lower blood pressure, increase nitric oxide, improve endothelial function, and decrease inflammation and oxidative stress (81, 83). All of these effects may reduce the risk of CHD (81–83). Milk proteins, both caseins and whey proteins are rich sources of angiotensin converting enzyme (ACE) inhibitory peptides that significantly reduce blood pressure (81–83). Although milk products appear innocuous, they are pro-inflammatory for some individuals; caution is advised, especially for intolerant subjects.

TABLE 12.2
Types of Fatty Acids in Coconut Oil

Common name	Fatty acid	Percentage
Caproic acid	6:0	0.2–0.5
Caprylic acid	8:0	5.4–9.5
Capric acid	10:0	4.5–9.7
Lauric acid	12:0	44.1–51
Myristic acid	14:0	13.1–18.5
Palmitic acid	16:0	7.5–10.5
Stearic acid	18:0	1.0–3.2
Arachidic acid	20:0	0.2–1.5
Oleic acid	18:1*n*-9	5.0–8.2
Linoleic acid	18:2*n*-6	1.0–2.6

Whey Protein

Several studies show that chronic intake of several grams (typically 20 grams) of whey protein significantly reduces blood pressure (84–87), decreases TG and cholesterol levels (88), increases intracellular glutathione levels, and lowers inflammation in patients with cardiovascular disease (85, 89). These benefits may come from chronic consumption rather than a single dose (90). The type of whey protein may impact results. Clinical trial data indicate that whey protein must be hydrolyzed to ACE inhibitor peptides for it to have antihypertensive properties (84–86, 91, 92). In addition, certain whey protein preparations may result in a relatively higher insulin response relative to other protein sources (93, 94), which may or may not be beneficial in some patient populations.

Eggs

The effect of eggs on serum cholesterol and CHD risk has been a contentious argument over the past few decades, but recent studies have provided scientific guidance. A retrospective review of 17 studies with 556 subjects found that for each 100 mg of dietary cholesterol per day in eggs, the total cholesterol (TC) increased 2.2 mg/dL, low-density cholesterol (LDL-C) increased 1.9 mg/dL, high-density cholesterol (HDL-C) increased 0.3 mg/dL, and the TC/HDL ratio increased 0.2 units (95). A 50-gram egg contains about 200 mg of cholesterol, 6 grams of protein, and 5 grams of fat (36% SFA, 48% MUFA, and 16% PUFA) (95).

Subjects with metabolic syndrome or type 2 diabetes mellitus (T2DM) consuming three whole eggs per day on a carbohydrate-restricted diet of less than 30% energy, compared to an egg substitute, had reductions in tumor necrosis alpha (TNF alpha) and triglycerides (TG), increases in HDL-C—with no change in TC, LDL, or other inflammatory markers—and a lower risk of T2DM or its progression (96, 97). In the HPFS and NHS studies, with almost 18,000 subjects followed for 8–14 years, there was no evidence of any significant association between egg consumption and risk for CHD with the possible exception of T2DM (98). However, in another study, egg consumption was not associated with any cardiovascular outcome in individuals with T2DM either (99). In a prospective study over 13 years of 37,766 men and 32,805 women who were free of CVD, egg consumption was assessed at baseline with a food-frequency questionnaire (99. There was no statistically significant association between egg consumption and risk of MI in either men or women. In the Kuopio Ischemic Heart Disease Risk Factor Study of 1032 men, egg or cholesterol intakes were not associated with increased CHD risk, even in ApoE4 carriers (100). A meta-analysis, including 90,735 participants and a follow-up time from 5.8 to 20.0 years, evaluated the role of egg consumption on CHD risk (101). Comparison of the highest category (≥1 egg/d) of egg consumption with the lowest (<1 egg/week or none) showed no change in the risk of CHD and MI based on the number of eggs eaten per day or week. This meta-analysis suggests that egg consumption is not associated with the risk of cardiovascular disease and cardiac mortality in the general population. The results from many nutrition trials suggest that consumption of 6 to 12 eggs per week, in the context of a diet that is consistent with guidelines on cardiovascular health promotion, has no adverse effect on major CHD risk factors in individuals at risk for developing diabetes or in those with T2DM.

REFINED CARBOHYDRATES, SUGARS, AND SUGAR SUBSTITUTES

Refined carbohydrates are associated with an increased risk of CHD in all studies (31, 32, 51, 102). Sugars, refined carbohydrates, high fructose corn syrup (HFCS) and starches confer significant risk for dyslipidemia, nonalcoholic fatty liver disease (NAFLD) and CHD compared to omega-3 FA, MUFA, fermented foods, fiber, fruits and vegetables, dairy, and the TMD and DASH 2 diets (103). A prospective study of 117,366 subjects without history of diabetes, CHD, or stroke examined intakes of carbohydrates and grains in relation to CHD using food-frequency questionnaires over 7.6 years (102). Carbohydrate intake accounted for about 68% of the total energy intake. Carbohydrate intake and CHD were highly associated with an increased risk of CHD by 80%. In the NHS and HPFS, carbohydrates from refined starches and added sugars increased the risk of CHD by ten percent (51).

In a study of 39,786 subjects, daily diet soft drink consumption increased the risk of total CVA by 21% and the risk of all vascular events by 43%, which includes ischemic CVA, CHD, MI, and vascular death (104). The Japan Public Health Center study showed both sugar-sweetened and low-calorie sodas significantly increased the risk of stroke by 16% per one serving daily and CHD by 20% per one serving daily (105). Sugar substitutes increase the risk for obesity, weight gain, metabolic syndrome, T2DM, and CHD. Sugar substitutes interfere with learned responses that normally regulate consumption of food, have an adverse effect on the microbiome, and alter leptin levels, which increases obesity and decreases satiety (106, 107).

ADVANCED GLYCATION END PRODUCTS

Food preparation needs to be discussed in relationship to nutrition and cardiovascular health. Advanced glycation end products (AGEs) are a group of oxidant and inflammatory compounds known to play a role in the pathogenesis of chronic diseases, including cardiovascular disease and CHD. They are formed when some sugar, proteins, and lipids come together in the presence of heat. Several modern cooking methods, including industrial heat processing, grilling, broiling, roasting, searing, and frying, significantly increase dietary AGE formation and exposure (108). A low-AGE diet may decrease circulating blood AGE levels, improve endothelial function, lower inflammatory mediators, and reduce atherosclerosis development (109–111). Dietary intake of AGEs can be reduced by avoiding foods known to be high in AGEs such as full-fat cheeses, meats, and highly processed foods, while increasing the consumption of fish, grains, low-fat milk products, fruits, and vegetables. Boiling, poaching, and stewing as well as steaming and slower cooking at a lower heat can reduce dietary AGE exposure (108).

PROTEIN

VEGETARIAN DIETS AND PLANT-BASED NUTRITION

Vegetarian diets significantly reduce CVD, CHD, and coronary artery calcium (CAC) score that is proportional to the dietary intake of vegetables (32, 103–115). In the European Prospective Investigation into Cancer and Nutrition (EIPC) study of 44,561 subjects in England and Scotland followed for 11.6 years, the body mass index (BMI), lipids, and blood pressure were all reduced in the vegetarian group and there was a 32% lower incidence of CHD (112). A study of 96,469 Seventh–Day Adventist men and women from 2002 to 2007 demonstrated a 12% decrease in total mortality, 15% in vegans, 9% in lacto-ovo vegetarians, 19% in pesco-vegetarian and 8% in semi-vegetarians, which was primarily related to decreases in CVD (113). The coronary artery calcium (CAC) score is also reduced with chronic dietary intake of fruits and vegetables (114).

A meta-analysis of nine studies of 222,081 men and women found the overall reduction in CHD risk was 4% for each additional portion of fruit and vegetable intake per day and 7% for each additional serving of fruit (115). Dark green leafy vegetables had the most dramatic reduction in CHD risk. In a meta-analysis of 95 studies, for fruits and vegetables combined per 200 grams/day, there was an 8% decrease in CHD. Similar associations were observed for fruits and vegetables separately. Reductions in risk were observed for up to 800 grams/day for CHD with the intake of apples and pears, citrus fruits, green leafy vegetables, cruciferous vegetables, and salads (116). Some vegetarian diets may be deficient in many nutrients which require supplemental B12, vitamin D, omega-3 fatty acids, iron, calcium, carnitine, zinc, and some high-quality amino acids and protein (117). Other studies suggest several other problems, such as decreased sulfur amino acid intake with a low elemental sulfur, increased homocysteine, and oxidative stress. In addition, lean muscle mass was 10% lower and there may be an increased risk of subclinical malnutrition and CVD (117).

ANIMAL PROTEIN DIETS

Recent studies show either no correlation or an inverse correlation of grass-fed beef, wild game, organically fed animals, and other sources of protein with

CHD (118–123). The Paleolithic diet has also shown reductions in total mortality of 23% and CV mortality of 22% in a study of 21,423 subjects (118). All meat (including red meat, fish, seafood, poultry) had an inverse relationship to CVD mortality in men in Asian countries (120). Another meta-analysis showed no association between red meat consumption and CHD but found that processed red meat increased risk of hypertension, total mortality, CHD, and T2DM risk (118–123). A recently published study of over a half a million subjects answering food questionnaires that were followed for 16 years did identify a significant association between all forms of red meat consumption, all-cause and cardiovascular mortality. In the Beef in an Optimal Lean Diet (BOLD) study trial, a low dietary SFA intake heart healthy diet containing lean beef elicits a favorable effect on CHD and blood lipids that are comparable to the DASH diet (119).

SOY PROTEIN

There has been debate about the inclusion of soy protein in the diet and CHD (124–126). Most likely, the variability in results may also be due to the different types of available soy products and their degree of processing, resulting in a variety of byproducts formed, such as fermentation complexes. Thus, not all soy is equal, and the more processed it is, the more it should be avoided. Organic tofu, tempeh, miso, and edamame are important sources of soy protein that one may consider.

FISH

Studies largely support fish consumption for cardiovascular health (127–140). Eating fish one to two times weekly, especially higher omega-3 fatty acid-containing fish, reduces risk of CHD by 36% and total mortality by 17%. You should eat a variety of seafood with limited intake of high mercury-containing fish but with greater fish consumption ($\geq$ five servings/week). Fish intake reduces the risk of congestive heart failure by 6% for each 20 grams of daily fish. Active ingredients in bonito and other coldwater fish may contribute to its cardioprotective qualities, such as the presence of ACE inhibitory peptides (133–135). Intake of sardine muscle protein by mildly hypertensive volunteers led to 9.7/5.3 mm Hg reduction in blood pressure in one week (136).

It is important to consider the type of fish and their relative methylmercury levels as well as the degree to which individuals can transport mercury based on their genetics (137). Methylmercury has detrimental effects that increase the risk of CHD, MI, and hypertension (138–140). Thus, larger fish, such as tilefish, shark, swordfish, large grouper, and tuna may contain higher mercury levels; caution and even omission are advised. However, the benefits of smaller fish consumption likely outweigh the risks from the potential toxins it contains (128).

DIETARY ACID LOAD AND PROTEIN

Diet-induced 'low-grade' metabolic acidosis is thought to play an important role in the development of cardiovascular disease, hypertension, dyslipidemia, and obesity (141, 142). Vegetables, fruits, and some beverages (red wine and coffee) are considered alkaline, while fats and oils are neutral. Meats, especially red meat. has a high acid load, and dairy products and cereal grains are acid-producing (141, 142). Dietary acid load can be improved by increasing intake of fruits and vegetables and decreasing excessively high dietary animal protein intake (143). A ten-day intervention with an alkaline Paleolithic-style diet led to a marked increase in potassium levels and improvements in vascular reactivity, blood pressure, glucose levels, insulin sensitivity, and lipid levels (143). While the definitive effects of dietary acid load on cardiovascular health are not yet clear, it is apparent that such dietary changes are in line with DASH and the TMD.

SPECIFIC DIETARY AND NUTRITIONAL COMPONENTS AND CALORIC RESTRICTION

Several dietary and nutritional components have been shown to decrease inflammation by interrupting the inflammatory vascular receptors (8). These include the following:

- Curcumin (turmeric).
- Cinnamaldehyde (cinnamon).
- Sulforaphane (broccoli).
- Resveratrol (nutritional supplement, red wine, grapes).
- Epigallocatechin gallate (EGCG) (green tea).
- Luteolin (celery, green pepper, rosemary, carrots, oregano, oranges, olives).
- Quercetin (tea, apples, onion, tomatoes, capers).

These interactions between food groups or supplements with the vascular membrane receptors may initiate improved vascular responses, decreased vascular inflammation, oxidative stress, and vascular immune responses that reduce CHD risk.

A prospective study of 42 subjects over two years showed a significant reduction in the progression of CHD as assessed by coronary artery calcium (CAC) compared to historical controls using a phytonutrient concentrate containing a high content of fruit and vegetable extracts. The change in the CAC score was significantly less in the treated patients vs the control patients (19.6% vs 34.7% increase respectively, a 15.1% difference) (144).

CAFFEINE

The cytochrome P-450 CYP1A2 genotype modifies the association between caffeinated coffee intake and the risk of hypertension, CVD, CHD, and MI in a linear relationship (145–152). Caffeine is exclusively metabolized by CYP1A2 (145). Caffeine also blocks vasodilating adenosine receptors (152). The rapid metabolizers of caffeinated coffee IA/IA allele have average BP reduction of 10/7 mm Hg and reduced risk of MI by 17–52% (150). This SNP represents about 40–45% of the population (145–150). The slow metabolizers of caffeine IF/IF or IA/IF allele have higher BP of 8.1/5.7 mm Hg lasting > three hours after consumption, tachycardia, increased risk of MI, increased aortic stiffness, higher pulse wave velocity, vascular inflammation, and increased catecholamines (145–150). Based on age and consumption, the risk of MI will vary. At age 59 there was a 36% increase in MI with two to three cups/day and a 64% increase with four cups/day or more. Under the age of 59, MI increased by 24% (one cup/day), 67% (two cups/day) and 233% (four or more cups/day) (150, 151). This SNP represents about 55–60% of the population. If you are a slow metabolizer of caffeine, then you should completely eliminate it from your diet from all sources.

CALORIC RESTRICTION

Caloric restriction refers to a reduction of energy intake at the individualized level that is sufficient to maintain a slightly low to normal body weight (i.e., body mass index of about 21 kg/m^2) without causing malnutrition (153). Findings from long-term calorie restriction of 40% in animal models have revealed improvements in metabolic health, offsetting chronic disease, and, consequently, extending life span (154). Animal studies, including the rhesus monkey, on caloric restriction, have identified cardiovascular benefits including reductions in oxidative stress and inflammation in the heart and vasculature, beneficial effects on endothelial function and arterial stiffness, and protection against atherosclerosis and less detrimental age-related changes in the heart (155). Limited evidence from human data suggests some of these effects translate to human caloric restriction (156). Alternate-day (ADF) or intermittent fasting is another similar approach with cardiovascular benefit. Typically, ADF involves consuming 25% of energy needs on the fast day and ad libitum food intake on the following day (157). Results indicate weight loss and improvements in cardiometabolic health, such as reductions in aortic vascular stiffness, C reactive protein, adiponectin, leptin, total cholesterol, LDL cholesterol, triglyceride concentration, and systolic blood pressure (158). Caloric restriction could be implemented by constructing a personalized diet based on nutrient-dense, low-energy foods such as vegetables, fruits, whole grains, nuts, fish, low-fat dairy products, and lean meats (159). Another type of fast would be to avoid eating for 12–14 hours overnight. You can also reduce the daily intake of calories by 20% and exercise daily to consume another 20% of calories.

ALCOHOL

The connection between alcohol consumption and CHD is based on a U-shaped curve, such that overconsumption or under-consumption is not as likely to reduce CHD as the base of the U-shaped curve which is associated with the lowest risk of CHD (160–162). An alcoholic drink in most research studies is 14 grams of ethanol or 0.6 fluid ounces of pure alcohol. This equates to a 12-ounce beer, a 5-ounce glass of wine, or 1.5 ounces of hard liquor (160). "Light to moderate" drinking (defined as one drink a day for women, two drinks a day for men) is associated with lower rates of total mortality, CHD morbidity, diabetes mellitus, heart failure, and strokes, especially in people over 50 years of age (160–162). This was confirmed in an analysis of studies combining data on over 1 million people and overall death rates where the U-shaped curve was best at one to two drinks per day for women and two to four drinks per day for men (161). Red wine in particular is rich in polyphenols, with antioxidant, anti-inflammatory, and antiplatelet actions (160, 161).

In a recent review and meta-analysis of alcohol consumption and cardiovascular disease, light to moderate alcohol consumption (162), reduced the risk for CHD by 29%, and all-cause mortality was reduced by 13%. Pinot noir is generally credited with having the highest concentration of the potent polyphenol resveratrol in the grape (160–162).

GLUTEN

About 1% of the public has celiac disease and perhaps another 6–7% have verified gluten sensitivity with dramatic changes in the appearance of their gastrointestinal tract when it is consumed on a regular basis (163). A key consequence of the damage to the intestinal wall lining is that the normally tight junctions that bind cells lining the gastrointestinal (GI) tract become loose. When these junctions are loose, the contents of the GI tract can enter the wall of the bowels and then enter into the bloodstream. Many studies have shown that, after a fatty meal, a wave of inflammation and endotoxins enter the bloodstream and may remain present for hours (6, 9, 10). When gliadin, a component of gluten-containing foods like bread, is present in the intestines of those with celiac or gluten sensitivity, a newly discovered protein called zonulin is released into the gut (163). Zonulin is now thought to have a potential role not only in celiac disease but also Type 1 diabetes, obesity, and other immune illnesses (163). It has been shown to be the "crowbar" that opens tight junctions and leads to autoimmune responses, such as a leaky GI tract (163). The ability to measure blood levels of zonulin may revolutionize our understanding of GI, autoimmune, and other systemic diseases.

There are little data linking gluten and CHD. In an analysis of patients who had suffered an MI in Sweden, those with celiac disease had similar outcomes to those without celiac disease (164). There are case reports of cardiomyopathy being associated with gluten sensitivities that respond to the withdrawal of foods (165). In another case report, a review of gluten antibodies and proven celiac disease was reported in nine additional cases of cardiomyopathy (166).

Generally, 100% whole grains, as opposed to processed white flour-based foods, are to be encouraged for patients with CHD. A meta-analysis that examined whole grain consumption and the risk of developing CHD and MI in more than 400,000 participants found that the highest consumption of whole grains reduced the risk by about 25% (167). The authors indicated that whole grain foods contain fiber, vitamins, minerals, phytoestrogens, phenolic compounds, and have a favorable effect on measures of cholesterol, blood glucose, inflammation, and arterial function. In a 26-year follow-up of 64,714 women in the NHS and 45,303 men in the HPFS, dietary gluten intake was not associated with risk of CHD (fatal or nonfatal MI) (168).

NUTS

Nuts are high in MUFA and PUFA but may also contain some omega-6 FA. The beneficial effects of nut consumption on cardiovascular disease, CV deaths, CHD, and

MI was well documented in the PREDIMED trial with a reduction in total CV death of 28% with nut consumption (15–18). In the Adventist Health Study, which examined obesity and metabolic syndrome in more than 800 people, there was a strong inverse relationship between tree nut consumption and developing both medical conditions (169). Other studies suggest that eating tree nuts does not lead to weight gain, and the high concentration of fiber and nutrients offsets the calories consumed. Nuts may reduce CHD deaths and all-cause mortality as well (15–18, 170). In a larger analysis of the Adventist Health Study examining death in residents over age 84 and consuming nuts >5 times a week, there was a 20% reduction in total mortality and a 40% reduction in CHD mortality (170). The impact of including nuts in the diet has been analyzed in a recent large meta-analysis (171). The habit of eating 28 grams of nuts/day reduced the risk for CHD by 29%. It was estimated that 4 million deaths a year could be avoided worldwide by eating a handful of nuts daily (172). In a study of 40 subjects comparing a walnut-enriched diet to a control diet over eight weeks, the walnut diet reduced total cholesterol and apolipoprotein B, which is the carrier in the blood for LDL cholesterol (173). Walnuts also significantly improve endothelial function (174).

Dietary Sodium, Potassium, and Magnesium

Increased dietary sodium is associated with an increased risk of hypertension, CHD, MI, CHF, CVA, renal insufficiency, and proteinuria (175–179). Approximately 75 million people in the United States and up to 1 billion worldwide have been diagnosed with hypertension (175–179). Up to 50% of cardiovascular related deaths result from hypertension.

The sodium–potassium ratio may be more important that the actual dietary sodium and potassium intake and the risk of CHD (175). A number of population studies demonstrating that higher dietary potassium, as rated by urinary excretion or dietary recall, was generally associated with lower blood pressure and CHD regardless of the level of sodium intake (175–180). According to a report of the Institute of Medicine, adult recommendations are to consume at least 4.7 grams of potassium daily to control blood pressure and reduce dietary sodium intake to about 1.5 to 2 grams per day (2, 4, 175–179). The potassium/sodium ratio should be greater than 2.5 to 3.0 (2, 4, 175–179). Foods that are high in potassium include bran, mushrooms, macadamia nuts and almonds, dark leafy greens, avocados, apricots, fruits, and acorn squash.

The role of dietary magnesium in cardiovascular health is important and supported by many studies. It is estimated that nearly half the US population consumes less than the recommended amount of magnesium in their diets, and magnesium deficiency is a commonly overlooked risk factor for cardiovascular disease (177). The lower the dietary intake of magnesium, the greater the risk of succumbing to cardiovascular disease. Magnesium supplementation can be therapeutic for a range of cardiovascular issues including arrhythmias, hypertension, atherosclerosis, and endothelial dysfunction. Magnesium is critical for tissues that have electrical or mechanical activity, such as nerves, muscles (including the heart), and blood vessels (177). In a six-month study of patients with known CHD, magnesium supplementation led to an impressive *decrease* in angina attacks and a *decrease* in the use of antianginal drugs, such as nitroglycerin, by improving endothelial function (178).

SUMMARY AND KEY TAKE AWAY POINTS

1. The role of nutrition in the prevention and treatment of CHD has been clearly demonstrated in published clinical trials.
2. The top five cardiovascular risk factors, as presently defined, are not an adequate explanation for the current limitations to prevent and to reduce CHD.
3. Proper definition and analysis of the top five CV risk factors, evaluation of the three finite responses, and sound nutritional advice and evaluation based on scientific studies will be required to affect an improvement in risk for CHD.
4. Early detection of CHD coupled with aggressive prevention and treatment of all cardiovascular risk factors will diminish the progression of functional and structural cardiovascular abnormalities and clinical CHD.
5. Utilization of targeted personalized and precision treatments that apply genetics with optimal nutrition coupled with exercise, ideal weight and body composition, and discontinuation of all tobacco use can prevent approximately 80% of CHD.
6. Nutritional studies provide evidence that CHD can be reduced with a weighted plant-based diet with ten servings of fruits and vegetables per day, MUFA, PUFA, nuts, whole grains, coldwater fish, the DASH diets, PREDIMED-TMD diet; and reduction of refined carbohydrates and sugars, sucrose, sugar substitutes, high fructose corn syrup, long-chain SFAs, processed foods, and elimination of all TFA (Table 12.1).
7. Eggs and dairy products are not associated with CHD.
8. Coconut oil is neutral for CHD risk.
9. Organic grass-fed beef and wild game may reduce CHD.
10. High intakes of potassium and magnesium are recommended in conjunction with sodium restriction.
11. Caffeine intake should be adjusted depending on the genetic ability to metabolize it via the CYP 1A2 system.
12. Alcohol is associated with a U-shaped curve and CHD.
13. The roles of gluten, soy, and caloric restriction regarding CHD in humans require further study.

REFERENCES

1. Yusuf S, Hawken S, Ounpuu S, Dans T, Avezum A, Lanas F, McQueen M, Budaj A, Pais P, Varigos J, Lisheng L, INTERHEART Study Investigators. Effect of potentially modifiable risk factors associated with myocardial infarction in 52 countries (the INTERHEART study): case-control study. *Lancet.* 2004; 364(9438):937–52.
2. Houston Mark C. *What Your Doctor May Not Tell You About Heart Disease. The Revolutionary Book that Reveals the Truth Behind Coronary Illnesses and How You Can Fight Them. Grand Central Life and Style.* New York: Hachette Book Group; 2012.

3. O'Donnell CJ, Nabel EG. Genomics of cardiovascular disease. *N Engl J Med.* 2011; 365(22):2098–109.
4. Houston MC. Nutrition and nutraceutical supplements in the treatment of hypertension. *Expert Rev Cardiovasc Ther.* 2010;8:821–33.
5. ACCORD Study Group, Gerstein HC, Miller ME, Genuth S, Ismail-Beigi F, Buse JB, Goff DC Jr, Probstfield JL, Cushman WC, Ginsberg HN, Bigger JT, Grimm RH Jr, Byington RP, Rosenberg YD, Friedewald WT. Long-term effects of intensive glucose lowering on cardiovascular outcomes. *N Engl J Med.* 2011;364(9):818–28.
6. Youssef-Elabd EM, McGee KC, Tripathi G, et al. Acute and chronic saturated fatty acid treatment as a key instigator of the TLR-mediated inflammatory response in human adipose tissue, in vitro. *J Nutr Biochem.* 2012;23: 39–50.
7. El Khatib N, Génieys S, Kazmierczak B, Volpert V. Mathematical modelling of atherosclerosis as an inflammatory disease. *Philos Transact A Math Phys Eng Sci.* 2009;367(1908):4877–86.
8. Zhao L, Lee JY, Hwang DH Inhibition of pattern recognition receptor-mediated inflammation by bioactive phytochemicals. *Nutr Rev.* 2011;69(6):310–20.
9. Mah E, Bruno RS Postprandial hyperglycemia on vascular endothelial function: mechanisms and consequences. *Nutr Res.* 2012;32(10):727–40.
10. Houston M The role of nutraceutical supplements in the treatment of dyslipidemia. *J Clin Hypertens (Greenwich).* 2012;14(2):121–32.
11. Houston Mark C. *Handbook of Hypertension.* Oxford UK: Wiley –Blackwell; 2009.
12. Della Rocca DG, Pepine CJ. Endothelium as a predictor of adverse outcomes. *Clin Cardiol.* 2010; 33(12):730–2.
13. Houston MC The role of cellular micronutrient analysis, nutraceuticals, vitamins, antioxidants and minerals in the prevention and treatment of hypertension and cardiovascular disease. *Ther Adv Cardiovasc Dis.* 2010;4(3):165–83.
14. Freeman AM, Morris PB, Barnard N, Esselstyn CB, Ros E, AgatstonA, Devries S, O'Keefe J, Miller M, Ornish D, KimWilliams PK-E. Trending cardiovascular nutrition controversies. *Journal of the American College of Cardiology.* 2017 Mar;69(9): 1172–1187 DOI: 10.1016/j.jacc.2016.10.086
15. Sofi F, Abbate R, Gensini GF, Casini A. Accruing evidence on benefits of adherence to the Mediterranean diet on health: an updated systematic review and meta-analysis. *Am J Clin Nutr.* 2010;92(5):1189–96.
16. Estruch R, Ros E, Salas-Salvadó J, Covas MI, Corella D, Arós F, Gómez-Gracia E, Ruiz-Gutiérrez V, Fiol M, Lapetra J, Lamuela-Raventos RM, Serra-Majem L, Pintó X, Basora J, Muñoz MA, Sorlí JV, Martínez JA, Martínez-González MA, PREDIMED Study Investigators. Primary prevention of cardiovascular disease with a Mediterranean diet. *N Engl J Med.* 2013;368(14):1279–90.
17. Nadtochiy SM, Redman EK Mediterranean diet and cardioprotection: the role of nitrite, polyunsaturated fatty acids, and polyphenols. *Nutrition.* 2011;27(7–8):733–44.
18. Salas-Salvadó J, Bulló M, Estruch R, Ros E, Covas MI, Ibarrola-Jurado N, Corella D, Arós F, Gómez-Gracia E, Ruiz-Gutiérrez V, Romaguera D, Lapetra J, Lamuela-Raventós RM, Serra-Majem L, Pintó X, Basora J, Muñoz MA, Sorlí JV, Martínez-González MA. Prevention of diabetes with Mediterranean diets: a subgroup analysis of a randomized trial. *Ann Intern Med.* 2014;160(1):1–10.
19. de Lorgeril M, Salen P, Martin JL, Monjaud I, Delaye J, Mamelle N. Mediterranean diet, traditional risk factors, and the rate of cardiovascular complications after myocardial infarction: final report of the Lyon Diet Heart Study. *Circulation.* 1999;99(6):779–85.
20. Buckland G, Mayén AL, Agudo A, Travier N, Navarro C, Huerta JM, Chirlaque MD, Barricarte A, Ardanaz E, Moreno-Iribas C, Marin P, Quirós JR, Redondo ML, Amiano P, Dorronsoro M, Arriola L, Molina E, Sanchez MJ, Gonzalez CA Olive oil intake and mortality within the Spanish population (EPIC-Spain). *Am J Clin Nutr.* 2012;96(1):142–9.

21. Castañer O, Corella D, Covas MI, Sorlí JV, Subirana I, Flores-Mateo G, Nonell L, Bulló M, de la Torre R, Portolés O, Fitó M, PREDIMED study investigators. In vivo transcriptomic profile after a Mediterranean diet in high-cardiovascular risk patients: a randomized controlled trial. *Am J Clin Nutr.* 2013;98(3):845-5.
22. Konstantinidou V, Covas MI, Sola R, Fitó M. Up-to date knowledge on the in vivo transcriptomic effect of the Mediterranean diet in humans. *Mol Nutr Food Res.* 2013;57(5):772-83.
23. Corella D, Ordovás JM. How does the Mediterranean diet promote cardiovascular health? Current progress toward molecular mechanisms: gene-diet interactions at the genomic, transcriptomic, and epigenomic levels provide novel insights into new mechanisms. *Bioessays.* 2014;36(5):526-37.
24. Appel LJ, Moore TJ, Obarzanek E, Vollmer WM, Svetkey LP, Sacks FM, Bray GA, Vogt TM, Cutler JA, Windhauser MM, Lin PH, Karanja N. A clinical trial of the effects of dietary patterns on blood pressure. DASH Collaborative Research Group. *N Engl J Med.* 1997;336(16):1117-24.
25. Sacks FM, Svetkey LP, Vollmer WM, Appel LJ, Bray GA, Harsha D, Obarzanek E, Conlin PR, Miller ER 3rd, Simons-Morton DG, Karanja N, Lin PH, DASH-Sodium Collaborative Research Group. Effects on blood pressure of reduced dietary sodium and the Dietary Approaches to Stop Hypertension (DASH) diet. DASH-Sodium Collaborative Research Group. *N Engl J Med.* 2001;344(1):3-10.
26. Fung TT Chiuve SE, McCullough ML, Rexrode KM, Logroscino G, Hu FB. Adherence to a DASH-style diet and risk of coronary heart disease and stroke in women. *Arch Intern Med.* 2008;168(7):713-20.
27. Chowdhury R, Warnakula S, Kunutsor S, Crowe F, Ward HA, Johnson L, Franco OH, Butterworth AS, Forouhi NG, Thompson SG, Khaw KT, Mozaffarian D, Danesh J, Di Angelantonio E Association of dietary, circulating, and supplement fatty acids with coronary risk: a systematic review and meta-analysis. *Ann Intern Med.* 2014;160(6):398-406.
28. Guasch-Ferré M, Babio N, Martínez-González MA, Corella D, Ros E, Martín-Peláez S, Estruch R, Arós F, Gómez-Gracia E, Fiol M, Santos-Lozano JM, Serra-Majem L, Bulló M, Toledo E, Barragán R, Fitó M, Gea A, Salas-Salvadó J, PREDIMED Study Investigators. Dietary fat intake and risk of cardiovascular disease and all-cause mortality in a population at high risk of cardiovascular disease. *Am J Clin Nutr.* 2015;102(6):1563-73.
29. Ravnskov U, DiNicolantonio JJ, Harcombe Z Kummerow FA, Okuyama H, Worm N. The questionable benefits of exchanging saturated fat with polyunsaturated fat. *Mayo Clin Proc.* 2014;89(4):451-3.
30. Alexander DD, Miller PE, Van Elswyk ME, Kuratko CN, Bylsma LC. A Meta-Analysis of Randomized Controlled Trials and Prospective Cohort Studies of Eicosapentaenoic and Docosahexaenoic Long-Chain Omega-3 Fatty Acids and Coronary Heart Disease Risk. *Mayo Clin Proc.* 2017: 92(1):15-29.
31. DiNicolantonio JJ, Lucan SC, O'Keefe JH The Evidence for Saturated Fat and for Sugar Related to Coronary Heart Disease. *Prog Cardiovasc Dis.* 2016;58(5):464-72.
32. Siri-Tarino PW, Krauss RM. Diet, lipids, and cardiovascular disease. *Curr Opin Lipidol.* 2016;27(4):323-8.
33. Adamson S, Leitinger N. Phenotypic modulation of macrophages in response to plaque lipids. *Curr Opin Lipidol.* 2011;22(5):335-42.
34. Dow CA, Stauffer BL, Greiner JJ, DeSouza CA. Influence of dietary saturated fat intake on endothelial fibrinolytic capacity in adults. *Am J Cardiol.* 2014;114(5):783-8.
35. Santos S, Oliveira A, Lopes C. Systematic review of saturated fatty acids on inflammation and circulating levels of adipokines. *Nutr Res.* 2013;33(9):687-95.

36. Ruiz-Núñez B, Kuipers RS, Luxwolda MF, De Graaf DJ, Breeuwsma BB, Dijck-Brouwer DA, Muskiet FA. Saturated fatty acid (SFA) status and SFA intake exhibit different relations with serum total cholesterol and lipoprotein cholesterol: a mechanistic explanation centered around lifestyle-induced low-grade inflammation. *J Nutr Biochem.* 2014 25(3):304–12.
37. Forsythe CE, Phinney SD, Fernandez ML, Quann EE, Wood RJ, Bibus DM, Kraemer WJ, Feinman RD, Volek JS. Comparison of low fat and low carbohydrate diets on circulating fatty acid composition and markers of inflammation. *Lipids.* 2008;43(1):65–77.
38. Volek JS, Fernandez ML, Feinman RD, Phinney SD. Dietary carbohydrate restriction induces a unique metabolic state positively affecting atherogenic dyslipidemia, fatty acid partitioning, and metabolic syndrome. *Prog Lipid Res.* 2008;47(5):307–18.
39. Peña-Orihuela P, Camargo A, Rangel-Zuñiga OA, Perez-Martinez P, Cruz-Teno C, Delgado-Lista J, Yubero-Serrano EM, Paniagua JA, Tinahones FJ, Malagon MM, Roche HM, Perez-Jimenez F, Lopez-Miranda J. Antioxidant system response is modified by dietary fat in adipose tissue of metabolic syndrome patients. *J Nutr Biochem.* 2013;24(10):1717–23.
40. Devkota S, Wang Y, Musch MW, Leone V, Fehlner-Peach H, Nadimpalli A, Antonopoulos DA, Jabri B, Chang EB Dietary-fat-induced taurocholic acid promotes pathobiont expansion and colitis in Il10−/− mice. *Nature.* 2012;487(7405):104–8.
41. Ma W, Wu JH, Wang Q, Lemaitre RN, Mukamal KJ, Djoussé L King IB, Song X, Biggs ML, Delaney JA, Kizer JR, Siscovick DS, Mozaffarian D. Prospective association of fatty acids in the de novo lipogenesis pathway with risk of type 2 diabetes: the Cardiovascular Health Study. *Am J Clin Nutr.* 2015;101(1):153–63.
42. Praagman J, Beulens JW, Alssema M, Zock PL, Wanders AJ, Sluijs I, van der Schouw YT The association between dietary saturated fatty acids and ischemic heart disease depends on the type and source of fatty acid in the European Prospective Investigation into Cancer and Nutrition-Netherlands cohort. *Am J Clin Nutr.* 2016; 103(2):356–65.
43. Chen M, Li Y, Sun Q, Pan A, Manson JE, Rexrode KM, Willett WC, Rimm EB, Hu FB. Dairy fat and risk of cardiovascular disease in 3 cohorts of US adults. *Am J Clin Nutr.* 2016;104(5):1209–17.
44. Zong G, Li Y, Wanders AJ, Alssema M, Zock PL, Willett WC, Hu FB, Sun Q. Intake of individual saturated fatty acids and risk of coronary heart disease in US men and women: two prospective longitudinal cohort studies. *BMJ.* 2016 Nov 23;355: i5796.
45. Micha R, Mozaffarian D Saturated fat and cardiometabolic risk factors, coronary heart disease, stroke, and diabetes: a fresh look at the evidence. *Lipids.* 2010;45(10):893–905.
46. de Souza RJ, Mente A, Maroleanu A, Cozma AI, Ha V, Kishibe T, Uleryk E, Budylowski P, Schünemann H, Beyene J, Anand SS. Intake of saturated and trans unsaturated fatty acids and risk of all-cause mortality, cardiovascular disease, and type 2 diabetes: systematic review and meta-analysis of observational studies. *BMJ.* 2015;351:h3978.
47. Ruiz-Núñez B, Dijck-Brouwer DA, Muskiet FA The relation of saturated fatty acids with low-grade inflammation and cardiovascular disease. *J Nutr Biochem.* 2016;36:1–20.
48. Chang LF, Vethakkan SR, Nesaretnam K, Sanders TA, Teng KT Adverse effects on insulin secretion of replacing saturated fat with refined carbohydrate but not with monounsaturated fat: A randomized controlled trial in centrally obese subjects.. *J Clin Lipidol.* 2016;10(6):1431–41.
49. Zock PL, Blom WA, Nettleton JA, Hornstra G Progressing insights into the role of dietary fats in the prevention of cardiovascular disease. *Curr Cardiol Rep.* 2016 Nov;18(11):111.
50. Ros E, López-Miranda J, Picó C, Rubio MÁ, Babio N, Sala-Vila A, Pérez-Jiménez F, Escrich E, Bulló M, Solanas M, Gil Hernández AS-SJ. Consensus on fats and oils in the diet of s ish adults; position paper of the spanish federation of food, nutrition and dietetics societies]. *Nutr* Hosp. 2015 Aug 1;32(2):435–77.

51. Li Y, Hruby A, Bernstein AM, Ley SH, Wang DD, Chiuve SE, Sampson L, Rexrode KM, Rimm EB, Willett WC, Hu FB. Saturated fats compared with unsaturated fats and sources of carbohydrates in relation to risk of coronary heart disease: A prospective cohort study. *J Am Coll Cardiol*. 2015;66(14):1538–48.
52. CFlock MR, Kris-Etherton PM. Diverse physiological effects of long-chain saturated fatty acids: implications for cardiovascular disease. *Curr Opin Clin Nutr Metab Care*. 2013;16(2):133–40.
53. Hooper L, Summerbell CD, Thompson R, Sills D, Roberts FG, Moore HJ, Smith GD. Reduced or modified dietary fat for preventing cardiovascular disease. *Sao Paulo Med J*. 2016;134(2):182–3.
54. Björck L, Rosengren A Winkvist A Capewell S, Adiels M Bandosz P Critchley J, Boman K, Guzman-Castillo M, O'Flaherty M, Johansson I . Changes in dietary fat intake and projections for coronary heart disease mortality in Sweden: A simulation study. *PLoS One*. 2016;11(8)e0160474. doi:10.1371/journal.pone.0160474. eCollection 2016.
55. Williams CM, Salter A Saturated fatty acids and coronary heart disease risk: the debate goes on. *Curr Opin Clin Nutr Metab Care*. 2016;19(2):97–10.
56. Dawczynski C, Kleber ME, März W, Jahreis G, Lorkowski S Saturated fatty acids are not off the hook. *Nutr Metab Cardiovasc Dis*. 2015;25(12):1071–8.
57. Finzi AA, Latini R, Barlera S, Rossi MG, Ruggeri A, Mezzani A, Favero C, Franzosi MG, Serra D, Lucci D, Bianchini F, Bernasconi R, Maggioni AP, Nicolosi G, Porcu M, Tognoni G, Tavazzi L, Marchioli R. Effects of n-3 polyunsaturated fatty acids on malignant ventricular arrhythmias in patients with chronic heart failure and implantable cardioverter-defibrillators: A substudy of the Gruppo Italiano per lo Studio della Sopravvivenza nell'Insufficienza Cardiaca (GISSI-HF) trial. *Am Heart J*. 2011;161(2):338–43.
58. Mozaffarian D, Lemaitre RN, King IB, Song X, Huang H, Sacks FM, Rimm EB, Wang M, Siscovick DSPlasma phospholipid long-chain ω-3 fatty acids and total and cause-specific mortality in older adults: a cohort study. *Ann Intern Med*. 2013;158(7):515–25.
59. Gajos G, Zalewski J, Rostoff P, Nessler J, Piwowarska W, Undas A Reduced thrombin formation and altered fibrin clot properties induced by polyunsaturated omega-3 fatty acids on top of dual antiplatelet therapy in patients undergoing percutaneous coronary intervention (OMEGA-PCI clot). *Arterioscler Thromb Vasc Biol*. 2011;31(7):1696–702.
60. Davis W, Rockway S, Kwasny M. Effect of a combined therapeutic approach of intensive lipid management, omega-3 fatty acid supplementation, and increased serum 25 (OH) vitamin D on coronary calcium scores in asymptomatic adults. *Am J Ther*. 2009;16(4):326–32.
61. Nozue T, Yamamoto S, Tohyama S, Fukui K, Umezawa S, Onishi Y, Kunishima T, Sato A, Nozato T, Miyake S, Takeyama Y, Morino Y, Yamauchi T, Muramatsu T, Hibi K, Terashima M, Michishita I. Effects of serum n-3 to n-6 polyunsaturated fatty acids ratios on coronary atherosclerosis in statin-treated patients with coronary artery disease. Am J Cardiol. 2013;111(1):6–11.
62. Greene SJ, Temporelli PL, Campia U, Vaduganathan M, Degli Esposti L Buda S, Veronesi C, Butler J, Nodari S Effects of Polyunsaturated Fatty Acid Treatment on Postdischarge Outcomes After Acute Myocardial Infarction. *Am J Cardiol*. 2016; 117(3):340–6.
63. Arnesen H. n-3 Fatty acids and revascularization procedures. *Lipids*. 2001;36(Suppl): S103–6.
64. Arnesen H, Seljeflot I Studies on very long chain marine n-3 fatty acids in patients with atherosclerotic heart disease with special focus on mechanisms, dosage and formulas of supplementation. *Cell Mol Biol*. 2010;56(1):18–27.

65. Sekikawa A, Miura K, Lee S, Fujiyoshi A, Edmundowicz D, Kadowaki T, Evans RW, Kadowaki S, Sutton-Tyrrell K, Okamura T, Bertolet M, Masaki KH, Nakamura Y, Barinas-Mitchell EJ, Willcox BJ, Kadota A, Seto TB, Maegawa H, Kuller LH, Ueshima H, ERA JUMP Study Group. Long chain n-3 polyunsaturated fatty acids and incidence rate of coronary artery calcification in Japanese men in Japan and white men in the USA: population based prospective cohort study. *Heart.* 2014;100(7):569–73.
66. Abedin M, Lim J, Tang TB, Park D, Demer LL, Tintut Y. N-3 fatty acids inhibit vascular calcification via the p38-mitogen-activated protein kinase and peroxisome proliferator-activated receptor-gamma pathways. *Circ Res.* 2006;98(6):727–9.
67. Jacobson TA, Glickstein SB, Rowe JD, Soni PN. Effects of eicosapentaenoic acid and docosahexaenoic acid on low-density lipoprotein cholesterol and other lipids: a review. *J Clin Lipidol.* 2012 6(1):5–18.
68. Jans A, Konings E, Goossens GH, Bouwman FG, Moors CC, Boekschoten MV, Afman LA, Müller M, Mariman EC, Blaak EE. PUFAs acutely affect triacylglycerol-derived skeletal muscle fatty acid uptake and increase postprandial insulin sensitivity. *Am J Clin Nutr.* 2012;95(4):825–36.
70. García-López S, Villanueva Arriaga RE, Nájera Medina O, Rodríguez López CP, Figueroa-Valverde L, Cervera EG, Muñozcano Skidmore O, Rosas-Nexticapa M. One month of omega-3 fatty acid supplementation improves lipid profiles, glucose levels and blood pressure in overweight schoolchildren with metabolic syndrome. *J Pediatr Endocrinol Metab.* 2016;29(10):1143–50.73.
71. Sawada T, Tsubata H Hashimoto N, Takabe M, Miyata T, Aoki K Yamashita S, Oishi S, Osue T, Yokoi K, Tsukishiro Y, Onishi T, Shimane A, Taniguchi Y, Yasaka Y Ohara T Kawai H, Yokoyama M. Effects of 6-month eicosapentaenoic acid treatment on postprandial hyperglycemia, hyperlipidemia, insulin secretion ability, and concomitant endothelial dysfunction among newly-diagnosed impaired glucose metabolism patients with coronary artery disease. An open label, single blinded, prospective randomized controlled trial. *Cardiovasc Diabetol.* 2016;15(1):1–13.
72. Houston M. The role of nutrition and nutraceutical supplements in the treatment of hypertension. *World J Cardiol.* 2014;6(2):38–66.
73. Joris PJ, Mensink RP Role of cis-Monounsaturated Fatty Acids in the Prevention of Coronary Heart Disease. *Curr Atheroscler Rep.* 2016;18(7):38.
74. Abdullah MM, Jew S, Jones PJ. Health benefits and evaluation of healthcare cost savings if oils rich in monounsaturated fatty acids were substituted for conventional dietary oils in the United States. Nutr Rev. 2017 Mar;75(3):163–174. doi: 10.1093/nutrit/nuw062. [Epub ahead of print].
75. Mölenberg FJ, de Goede J, Wanders AJ, Zock PL, Kromhout D, Geleijnse JM. Dietary fatty acid intake after myocardial infarction: a theoretical substitution analysis of the Alpha Omega Cohort*Am J Clin Nutrition.* 2017 Sep 1;106(3):895–901pii: ajcn157826. doi: 10.3945/ajcn.117.157826. [Epub ahead of print].
76. Wand DD, Li Y, Chiuve S, et al. Specific dietary fats in relation to total and cause-specific mortality, *JAMA Internal Medicine.* 2016;176(8):1134–45.
77. Trumbo PR, Shimakawa T Tolerable upper intake levels for trans fat, saturated fat, and cholesterol. *Nutr Rev.* 2011;69(5):270–8.
78. Nestel P Trans fatty acids: are its cardiovascular risks fully appreciated? *Clin Ther.* 2014;36(3):315–21.
79. Eyres L, Eyres MF, Chisholm A,Brown RC. Coconut oil consumption and cardiovascular risk factors in humans. *Nutr Rev.* 2016;74(4):267–80.
80. DeLany JP, Windhauser MM, Champagne CM, Bray GA. Differential oxidation of individual dietary fatty acids in humans. *Am J Clin Nutr.* 2000; 72(4):905–11.

81. Feranil AB, Duazo PL, Kuzawa CW, Adair LS. Coconut oil is associated with a beneficial lipid profile in pre-menopausal women in the Philippines. *Asia Pac J Clin Nutr.* 2011;20(2):190–5.
82. Ballard KD, Bruno RS Protective role of dairy and its constituents on vascular function independent of blood pressure-lowering activities. *Nutr Rev.* 2015;73(1):36–50.
83. Khoramdad M, Esmailnasab N, Moradi G, Nouri B, Safiri S, Alimohamadi Y. The effect of dairy consumption on the prevention of cardiovascular diseases: A meta-analysis of prospective studies. *J Cardiovasc Thorac Res.* 2017;9(1):1–11. doi: 10.15171/jcvtr.2017.01. Epub 2017 Mar 18.
84. Chrysant SG, Chrysant GS. An update on the cardiovascular pleiotropic effects of milk and milk products. *J Clin Hypertens.* 2013;15(7):503–10.
85. FitzGerald RJ, Murray BA, Walsh DJ. Hypotensive peptides from milk proteins. *J Nutr.* 2004 Apr;134(4):980S–8S.
86. Pins JJ, Keenan JM. Effects of whey peptides on cardiovascular disease risk factors. *J Clin Hypertens.* 2006 Nov;8(11):775–82. Retraction in: J Clin Hypertens (Greenwich). 2008 Aug;10(8):631.
87. Aihara K, Kajimoto O, Takahashi R, Nakamura Y. Effect of powdered fermented milk with Lactobacillus helveticus on subjects with high-normal blood pressure or mild hypertension. *J. Am. Coll. Nutr.* 2005;24(4), 257–65.
88. Sousa GT, Lira FS, Rosa JC, de Oliveira EP, Oyama LM, Santos RV, Pimentel GD. Dietary whey protein lessens several risk factors for metabolic diseases: a review. *Lipids Health Dis.* 2012 Jul 10;11:67.
89. Berthold HK, Schulte DM, Lapointe JF, Lemieux P, Krone W, Gouni-Berthold I. The whey fermentation product malleable protein matrix decreases triglyceride concentrations in subjects with hypercholesterolemia: a randomized placebo-controlled trial. *J Dairy Sci.* 2011 Feb;94(2):589–601
90. de Aguilar-Nascimento JE, Prado Silveira BR, Dock-Nascimento DB. Early enteral nutrition with whey protein or casein in elderly patients with acute ischemic stroke: a double-blind randomized trial. *Nutrition.* 2011 Apr;27(4):440–4.
91. Pal S, Ellis V. Acute effects of whey protein isolate on blood pressure, vascular function and inflammatory markers in overweight postmenopausal women. *Br J Nutr.* 2011 May;105(10):1512–9.
92. Tavares T, Sevilla MÁ, Montero MJ, Carrón R, Malcata FX. Acute effect of whey peptides upon blood pressure of hypertensive rats, and relationship with their angiotensin-converting enzyme inhibitory activity. *Mol Nutr Food Res.* 2012 Feb;56(2):316-24.
93. Pins JJ, Geleva D, Keenan JM, Frazel C, O'Connor PJ, Cherney LM. Do whole-grain oat cereals reduce the need for antihypertensive medications and improve blood pressure control? *J Fam Pract.* 2002 Apr;51(4):353–9.
94. Mortensen LS, Holmer-Jensen J, Hartvigsen ML, Jensen VK, Astrup A, de Vrese M, Holst JJ, Thomsen C, Hermansen K. Effects of different fractions of whey protein on postprandial lipid and hormone responses in type 2 diabetes. *Eur J Clin Nutr.* 2012 Jul;66(7):799–805.
95. Esteves de Oliveira FC, Pinheiro Volp AC, Alfenas RC. Impact of different protein sources in the glycemic and insulinemic responses. *Nutr Hosp.* 2011 Jul-Aug;26(4):669–76.
96. Weggemans RM, Zock PL, Katan MB. Dietary cholesterol from eggs increases the ratio of total cholesterol to high-density lipoprotein cholesterol in humans: a meta-analysis. *Am J Clin Nutr.* 2001;73(5):885–91.
97. Blesso CN, Andersen CJ, Barona J, Volk B, Volek JS, Fernandez ML. Effects of carbohydrate restriction and dietary cholesterol provided by eggs on clinical risk factors in metabolic syndrome. *J Clin Lipidol.* 2013;7(5):463–71.

98. Virtanen JK, Mursu J, Tuomainen TP, Virtanen HE, Voutilainen S Egg consumption and risk of incident type 2 diabetes in men: the Kuopio Ischaemic Heart Disease Risk Factor Study. *Am J Clin Nutr.* 2015;101(5):1088–96.
99. Hu FB, Stampfer MJ, Rimm EB, Manson JE, Ascherio A, Colditz GA, Rosner BA, Spiegelman D, Speizer FE, Sacks FM, Hennekens CH, Willett WC A prospective study of egg consumption and risk of cardiovascular disease in men and women. *JAMA.* 1999;281(15):1387–94
100. Virtanen JK, Mursu J, Virtanen HE, Fogelholm M, Salonen JT, Koskinen TT, Voutilainen S, Tuomainen TP Associations of egg and cholesterol intakes with carotid intima-media thickness and risk of incident coronary artery disease according to apolipoprotein E phenotype in men: the Kuopio Ischaemic Heart Disease Risk Factor Study *Am J Clin Nutr.* 2016.;103(3):895–901.
101. Shin JY, Xun P, Nakamura Y, He K Egg consumption in relation to risk of cardiovascular disease and diabetes: a systematic review and meta-analysis. *Am J Clin Nutr.* 2013;98(1):146–59.
102. Yu D, Shu XO, Li H, Xiang YB, Yang G, Gao YT, Zheng W, Zhang X. Dietary carbohydrates, refined grains, glycemic load, and risk of coronary heart disease in Chinese adults. *Am J Epidemiol.* n.d.;178(10):1542–9.
103. Keller A, Heitmann BL, Olsen N. Sugar-sweetened beverages, vascular risk factors and events: a systematic literature review. *Public Health Nutr.* 2015;18(7):1145–54.
104. Bernstein AM, de Koning L, Flint AJ, Rexrode KM, Willett WC. Soda consumption and the risk of stroke in men and women. *Am J Clin Nutr.* 2012;95(5):1190–9.
105. Eshak ES, Iso H, Kokubo Y, Saito I, Yamagishi K, Inoue M, Tsugane S. Soft drink intake in relation to incident ischemic heart disease, stroke, and stroke subtypes in Japanese men and women: the Japan Public Health Centre-based study cohort I. *Am J Clin Nutr.* 2012;96(6):1390–7.
106. Swithers SE Artificial sweeteners produce the counterintuitive effect of inducing metabolic derangements. *Trends Endocrinol Metab.* 2013;24(9):431–41.
107. Shankar P, Ahuja S, Sriram K. Non-nutritive sweeteners: review and update. *Nutrition.* 2013;29(11–12):1293–9.
108. Uribarri J, Woodruff S, Goodman S, Cai W, Chen X, Pyzik R, Yong A, Striker GE, Vlassara H., Advanced glycation end products in foods and a practical guide to their reduction in the diet. *J Am Diet Assoc.* 2010 Jun;110(6):911-16.e12.
109. Lin RY, Choudhury RP, Cai W, Lu M, Fallon JT, Fisher EA, Vlassara H. Dietary glycotoxins promote diabetic atherosclerosis in apolipoprotein E-deficient mice. *Atherosclerosis.* 2003 Jun;168(2):213–20.
110. Uribarri J, Stirban A, Sander D, Cai W, Negrean M, Buenting CE, Koschinsky T, Vlassara H. Single oral challenge by advanced glycation end products acutely impairs endothelial function in diabetic and nondiabetic subjects. *Diabetes Care.* 2007 Oct;30(10):2579–82.
111. Luévano-Contreras C, Garay-Sevilla ME, Wrobel K, Malacara JM, Wrobel K. Dietary advanced glycation end products restriction diminishes inflammation markers and oxidative stress in patients with type 2 diabetes mellitus. *J Clin Biochem Nutr.* 2013 Jan;52(1):22–6.
112. Crowe FL Appleby PN, Travis RC, Key TJ. Risk of hospitalization or death from ischemic heart disease among British vegetarians and nonvegetarians: results from the EPIC-Oxford cohort study. *Am J Clin Nutr.* 2013;97(3):597–603.
113. Orlich MJ, Singh PN, Sabaté J, Jaceldo-Siegl K, Fan J, Knutsen S, Beeson WL, Fraser GE Vegetarian dietary patterns and mortality in Adventist Health Study 2. *JAMA Intern Med.* 2013;173(13):1230–8.

114. Miedema MD, Petrone A, Shikany JM, Greenland P, Lewis CE, Pletcher MJ, Gaziano JM, Djousse L. Association of fruit and vegetable consumption during early adulthood with the prevalence of coronary artery calcium after 20 years of follow-up: The coronary artery risk development in young adults (CARDIA) study. *Circulation.* 2015;132(21):1990–8.
115. Dauchet L, Amouyel P, Hercberg S, Dallongeville J Fruit and vegetable consumption and risk of coronary heart disease: a meta-analysis of cohort studies. *J Nutr.* 2006;136(10):2588–93.
116. Aune D, Giovannucci E, Boffetta P Fadnes LT, Keum N, Norat T Greenwood DC, Riboli E, Vatten LJ, Tonstad S. Fruit and vegetable intake and the risk of cardiovascular disease, total cancer and all-cause mortality-a systematic review and dose-response meta-analysis of prospective studies. *Int J Epidemiol.* 2017 Feb 22. doi: 10.1093/ije/dyw319. [Epub ahead of print]
117. Ingenbleek Y, McCully KS Vegetarianism produces subclinical malnutrition, hyperhomocysteinemia and atherogenesis. *Nutrition.* 2012;28(2):148–53.
118. Whalen KA, Judd S, McCullough ML, Flanders WD, Hartman TJ, Bostick RM. Paleolithic and mediterranean diet pattern scores are inversely associated with all-cause and cause-specific mortality in adults. *J Nutr.* 2017 Feb 8;pii:jn241919. doi: 10.3945/jn.116.241919. [Epub ahead of print].
119. Roussell MA, Hill AM, Gaugler TL, West SG, Heuvel JP, Alaupovic P, Gillies PJ, Kris-Etherton PM. Beef in an Optimal Lean Diet study: effects on lipids, lipoproteins, and apolipoproteins. *Am J Clin Nutr.* 2012;95(1):9–16.
120. Lee JE, McLerran DF, Rolland B, Chen Y, Grant EJ, Vedanthan R, Inoue M, Tsugane S, Gao YT, Tsuji I, Kakizaki M, Ahsan H, Ahn YO, Pan WH, Ozasa K, Yoo KY, Sasazuki S, Yang G, Watanabe T, Sugawara Y, Parvez F, Kim DH, Chuang SY, Ohishi W, Park SK, Feng Z, Thornquist M, Boffetta P, Zheng W, Kang D, Potter J, Sinha R Meat intake and cause-specific mortality: a pooled analysis of Asian prospective cohort studies. *Am J Clin Nutr.* 2013;98(4):1032–41.
121. Micha R, Wallace SK, Mozaffarian D. Red and processed meat consumption and risk of incident coronary heart disease, stroke, and diabetes mellitus: a systematic review and meta-analysis. *Circulation.* 2010;121(21):2271–83.
122. Bellavia A, Larsson SC Bottai M, Wolk A, Orsini N. Differences in survival associated with processed and with nonprocessed red meat consumption. *Am J Clin Nutr.* 2014;100(3):924–9.
123. Lajous M, Bijon A, Fagherazzi G, Rossignol E, Boutron-Ruault MC, Clavel-Chapelon F. Processed and unprocessed red meat consumption and hypertension in women. *Am J Clin Nutr.* 2014;100(3):948–52.
124. Anderson JW, Bush HM. Soy protein effects on serum lipoproteins: a quality assessment and meta-analysis of randomized, controlled studies. *J Am Coll Nutr.* 2011 Apr;30(2):79–91.
125. Campbell SC, Khalil DA, Payton ME, Arjmandi BH. One-year soy protein supplementation does not improve lipid profile in postmenopausal women. *Menopause.* 2010 May-Jun;17(3):587–93.
126. Rebholz CM, Reynolds K, Wofford MR, Chen J, Kelly TN, Mei H, Whelton PK, He J. Effect of soybean protein on novel cardiovascular disease risk factors: a randomized controlled trial. *Eur J Clin Nutr.* 2013 Jan;67(1):58–63.
127. Roughead ZK, Hunt JR, Johnson LK, Badger TM, Lykken GI. Controlled substitution of soy protein for meat protein: effects on calcium retention, bone, and cardiovascular health indices in postmenopausal women. *J Clin Endocrinol Metab.* 2005 Jan;90(1):181–9. Epub 2004 Oct 13.

128. Park K, Mozaffarian D. Omega-3 fatty acids, mercury, and selenium in fish and the risk of cardiovascular diseases. *Curr Atheroscler Rep.* 2010 Nov;12(6):414–22.
129. Mozaffarian D, Rimm EB. Fish intake, contaminants, and human health: evaluating the risks and the benefits. *JAMA.* 2006 Oct 18;296(15):1885–99. Review. Erratum in: JAMA. 2007 Feb 14;297(6):590.
130. Li YH, Zhou CH, Pei HJ, Zhou XL, Li LH, Wu YJ, Hui RT. Fish consumption and incidence of heart failure: a meta-analysis of prospective cohort studies. *Chin Med J.* 2013 Mar;126(5):942–8.
131. Watanabe Y, Tatsuno I Omega-3 polyunsaturated fatty acids for cardiovascular diseases: present, past and future. *Expert Rev Clin Pharmacol.* 2017 Aug;10(8):865–73.
132. Chowdhury R, Stevens S, Gorman D, Pan A, Warnakula S, Chowdhury S, Ward H, Johnson L, Crowe F, Hu FB, Franco OH. Association between fish consumption, long chain omega-3 fatty acids, and risk of cerebrovascular disease: systematic review and meta-analysis. *BMJ.* 2012 Oct 30;345:e6698
133. de Goede J, Verschuren WM, Boer JM, Kromhout D, Geleijnse JM. Gender-specific associations of marine n-3 fatty acids and fish consumption with 10-year incidence of stroke. *PLoS One.* 2012;7(4):e33866
134. Curtis JM, Dennis D, Waddell DS, MacGillivray T, Ewart HS. Determination of angiotensin-converting enzyme inhibitory peptide Leu-Lys-Pro-Asn-Met (LKPNM) in bonito muscle hydrolysates by LC-MS/MS. *J Agric Food Chem.* 2002 Jul 3;50(14):3919–25.
135. Qian ZJ, Je JY, Kim SK. Antihypertensive effect of angiotensin i converting enzyme-inhibitory peptide from hydrolysates of Bigeye tuna dark muscle, Thunnus obesus. *J Agric Food Chem.* 2007 Oct 17;55(21):8398–403. Epub 2007 Sep 26.
136. Otani L, Ninomiya T, Murakami M, Osajima K, Kato H, Murakami T. Sardine peptide with angiotensin I-converting enzyme inhibitory activity improves glucose tolerance in stroke-prone spontaneously hypertensive rats. *Biosci Biotechnol Biochem.* 2009 Oct;73(10):2203–9. Epub 2009 Oct 7.
137. Kawasaki T, Seki E, Osajima K, Yoshida M, Asada K, Matsui T, Osajima Y. Antihypertensive effect of valyl-tyrosine, a short chain peptide derived from sardine muscle hydrolyzate, on mild hypertensive subjects. *J Hum Hypertens.* 2000 Aug;14(8):519–23.
138. Schläwicke Engström K, Strömberg U, Lundh T, Johansson I, Vessby B, Hallmans G, Skerfving S, Broberg K. Genetic variation in glutathione-related genes and body burden of methylmercury. *Environ Health Perspect.* 2008 Jun;116(6):734–9
139. Valera B, Dewailly E, Poirier P. Association between methylmercury and cardiovascular risk factors in a native population of Quebec (Canada): a retrospective evaluation. *Environ Res.* 2013 Jan;120:102–8.
140. Choi AL, Weihe P, Budtz-Jørgensen E, Jørgensen PJ, Salonen JT, Tuomainen TP, Murata K, Nielsen HP, Petersen MS, Askham J, Grandjean P. Methylmercury exposure and adverse cardiovascular effects in Faroese whaling men. *Environ Health Perspect.* 2009 Mar;117(3):367–72.
141. Roman HA, Walsh TL, Coull BA, Dewailly É, Guallar E, Hattis D, Mariën K, Schwartz J, Stern AH, Virtanen JK, Rice G. Evaluation of the cardiovascular effects of methylmercury exposures: current evidence supports development of a dose-response function for regulatory benefits analysis. *Environ Health Perspect.* 2011 May;119(5):607–14.
142. Zhang L, Curhan GC, Forman JP. Diet-dependent net acid load and risk of incident hypertension in United States women. *Hypertension.* 2009 Oct;54(4):751–5.
143. Engberink MF, Bakker SJ, Brink EJ, van Baak MA, van Rooij FJ, Hofman A, Witteman JC, Geleijnse JM. Dietary acid load and risk of hypertension: the Rotterdam Study. *Am J Clin Nutr.* 2012 Jun;95(6):1438–44.

144. Pizzorno J, Frassetto LA, Katzinger J. Diet-induced acidosis: is it real and clinically relevant? *Br J Nutr.* 2010 Apr;103(8):1185–94.
145. Houston MC, Cooil B, Olafsson BJ, Raggi P Juice powder concentrate and systemic blood pressure, progression of coronary artery calcium and antioxidant status in hypertensive subjects: a pilot study. *Evid Based Complement Alternat Med.* 2007;4(4): 455–62.
146. Palatini P, Ceolotto G, Ragazzo F, Dorigatti F, Saladini F, Papparella I, Mos L, Zanata G, Santonastaso M. CYP1A2 genotype modifies the association between coffee intake and the risk of hypertension. *Hypertens.* 2009 Aug;27(8):1594–601.
147. Hu G, Jou ilahti P, Nissinen A, Bidel S, Antikainen R, Tuomilehto J. Coffee consumption and the incidence of antihypertensive drug treatment in Finnish men and women. *Am J Clin Nutr.* 2007;86(2):457–64.
148. Vlachopoulos CV, Vyssoulis GG, Alexopoulos NA, Zervoudaki AI, Pietri PG, Aznaouridis KA, Stefanadis CI . Effect of chronic coffee consumption on aortic stiffness and wave reflections in hypertensive patients. *Eur J Clin Nutr.* 2007;61(6):796–802.
149. Mesas AE, Leon-Muñoz LM, Rodriguez-Artalejo F, Lopez-Garcia E. The effect of coffee on blood pressure and cardiovascular disease in hypertensive individuals: a systematic review and meta-analysis. *Am J Clin Nutr.* 2011;94(4):1113–26.
150. Liu J, Sui X, Lavie CJ, Hebert JR, Earnest CP, Zhang J, Blair SN. Association of coffee consumption with all-cause and cardiovascular disease mortality. *Mayo Clin Proc.* 2013;88(10):1066–74.
151. Renda G, Zimarino M, Antonucci I, Tatasciore A, Ruggieri B, Bucciarelli T, Prontera T, Stuppia L. De Caterina R8 determinants of blood pressure responses to caffeine drinking *Am J Clin Nutr.* 2012;95(1):241–8.
152. Omodei D, Fontana L. Calorie restriction and prevention of age-associated chronic disease. *FEBS Lett.* 2011 Jun 6; 585(11):1537–42.
153. Fontana L. Modulating human aging and age-associated diseases. *Biochim Biophys Acta.* 2009 Oct;1790(10):1133–8.
154. Weiss EP, Fontana L. Caloric restriction: powerful protection for the aging heart and vasculature. *Am J Physiol Heart Circ Physiol.* 2011 Oct;301(4):H1205–19.
155. Longo VD, Antebi A, Bartke A, Barzilai N, Brown-Borg HM, Caruso C, Curiel TJ, de Cabo R, Franceschi C, Gems D, Ingram DK, Johnson TE, Kennedy BK, Kenyon C, Klein S, Kopchick JJ, Lepperdinger G, Madeo F, Mirisola MG, Mitchell JR, Passarino G, Rudolph KL, Sedivy JM, Shadel GS, Sinclair DA, Spindler SR, Suh Y, Vijg J, Vinciguerra M, Fontana L. Interventions to slow aging in humans: Are we ready? *Aging Cell.* 2015 Aug;14(4):497–510.
156. Varady KA, Hellerstein MK. Alternate-day fasting and chronic disease prevention: a review of human and animal trials. *Am J Clin Nutr.* 2007 Jul;86(1):7–13.
157. Varady KA, Bhutani S, Church EC, Klempel MC. Short-term modified alternate-day fasting: a novel dietary strategy for weight loss and cardioprotection in obese adults. *Am J Clin Nutr.* 2009 Nov;90(5):1138–43. doi: 10.3945/ajcn.2009.28380. Epub 2009 Sep 30.
158. Jakicic JM, Tate DF, Lang W, Davis KK, Polzien K, Rickman AD, Erickson K, Neiberg RH, Finkelstein EA. Effect of a stepped-care intervention approach on weight loss in adults: a randomized clinical trial. *JAMA.* 2012 Jun 27;307(24):2617–26.
159. O'Keefe JH, Bhatti SK, Bajwa A, et al. Alcohol and cardiovascular health: the dose makes the poison…or the remedy. *Mayo Clin Proc.* 2013;89 (3):382–93.
160. Di Castelnuevo A, Costanzo S, Bagnardi V, et al. Alcohol dosing and total mortality in men and women: an updated meta-analysis of 34 prospective studies. *Arch Intern Med.* 2006;166 (22):2437–45.

161. Ronksley PE, Brien SE, Turner BJ, et al. Association of alcohol consumption with selected cardiovascular disease outcomes: a systematic review and meta-analysis. *BMJ.* 2011 Feb 22;342: d67.
162. Sturgeon C, Fasano A. Zonulin, a regulator of epithelial and endothelial barrier functions, and its involvement in chronic inflammatory diseases. *Tissue Barriers.* 2016 Oct 21;4(4): e1251384.
163. Emilsson L Carlsson R, James S, et al. Follow-up of ischaemic heart disease in patients with coeliac disease. *Eur J Prev Cardiol.* 2015 Jan;22(1):83–90.
164. McGrath S, Thomas A, Gorard DA Cardiomyopathy responsive to gluten withdrawal in a patient with coeliac disease. *BMJ Case Rep.* 2016 Mar 14;2016.
165. Milisavljević N, Cvetković M, Nikolić G, et al. Dilated cardiomyopathy associated with celiac disease: case report and literature review. *Srp Arh Celok Lek.* 2012 Sep-Oct;140(9–10):641–3.
166. Tang G, Wang D, Long J, et al. Meta-analysis of the association between whole grain intake and coronary heart disease risk. *Am J Cardiol.* 2015 Mar 1;115(5):625–9.
167. Lebwohl B, Cao Y, Zong G, Hu FB, Green PHR, Neugut AI, Rimm EB, Sampson L, Dougherty LW, Giovannucci E, Willett WC Sun Q, Chan AT Long term gluten consumption in adults without celiac disease and risk of coronary heart disease: prospective cohort study. *BMJ.* 2017 May 2;357:j1892. doi: 10.1136/bmj.j1892.
168. Jaceldo-Siegel K, Haddad E, Fraser GE, et al. Tree nuts are inversely associated with metabolic syndrome and obesity: The Adventist health study-2. *PLoS One.* 2014 Jan 8;9(1): e85133.
169. Fraser GE, Shavlik DJ. Risk factors for all-cause and coronary heart disease mortality in the oldest-old. The Adventist Health Study. *Arch Intern Med.* 1997 Oct 27;157(19):2249–58.
170. Aune D, Keum N, Giovannucci E, et al. Nut consumption and risk of cardiovascular disease, total cancer, all-cause and cause-specific mortality: a systematic review and dose-response meta-analysis of prospective studies. *BMC Med.* 2016 Dec 5;14(1):207.
171. Micha R, Peñalvo JL, Cudhea F et al. Association Between Dietary Factors and Mortality From Heart Disease, Stroke, and Type 2 Diabetes in the United States. *JAMA.* 2017 Mar 7;317(9):912–924.
172. Wu L, Piotrowski K, Rau T et al. Walnut-enriched diet reduces fasting non-HDL cholesterol and apolipoprotein B in healthy Caucasian subjects: a randomized controlled cross-over clinical trial. *Metabolism.* 2014 Mar;63(3):382–91.
173. Xiao Y, Huang W, Peng C, Zhang J, Wong C, Kim JH, Yeoh EK, Su X Effect of nut consumption on vascular endothelial function: A systematic review and meta-analysis of randomized controlled trials. 4*Clin Nutr.* 2017 Apr 20;pii: S0261-5614(17)30150-4. doi: 10.1016/j.clnu.2017.04.011. [Epub ahead of print]
174. McDonough AA, Veiras LC, Guevara CA, et al. Cardiovascular benefits associated with higher dietary K+ vs. lower dietary Na+: evidence from population and mechanistic studies. *Am J Physiol Endocrinol Metab.* 2017 Apr 1;312(4):E348–56.
175. Park J, Kwock CK, Yang YJ. The effect of the sodium to potassium ratio on hypertension prevalence: A propensity score matching approach. *Nutrients.* 2016 Aug 6;8(8). pii: E482.
176. Cunha AR, Umbelino B, Correia ML, et al. Magnesium and vascular changes in hypertension. *Int J Hypertens.* 2012:75425.
177. Pokan R, Hofmann P, von Duvillard SP, et al. Oral magnesium therapy, exercise heart rate, exercise tolerance, and myocardial function in coronary artery disease patients. *Br J Sports Med.* 2006 Sep;40(9):773–8.

178. Turgut F, Kanbay M, Metin MR, et al. Magnesium supplementation helps to improve carotid intima media thickness in patients on hemodialysis. *Int Urol Nephrol.* 2008;40(4):1075–82.
179. Cunha AR, D'El-Rei J, Medeiros F, et al. Oral magnesium supplementation improves endothelial function and attenuates subclinical atherosclerosis in thiazide-treated hypertensive women. *J Hypertens.* 2017 Jan;35(1):89–97.
180. Zhang W, Iso H, Ohira T, et al. Associations of dietary magnesium intake with mortality from cardiovascular disease: the JACC study. *Atherosclerosis.* 2012 Apr;221(2):587–95. doi: 10.1016/j.atherosclerosis.2012.01.034. Epub 2012 Jan 28.

13 Nutrition Part II
The Practice of Nutrition in Your Daily Life to Prevent and Treat CHD

Now that we have looked at an overview of all the clinical nutrition studies that demonstrate how to prevent and treat CHD, let's look at how these nutrition programs work in real terms of your daily life so that you can practice them. By addressing everything, from rituals to recipes, I will provide you with a truly comprehensive way to reframe and rethink your relationship to food, health, wellness, and CHD prevention and treatment. I want to thank Lee Bell NC, BCHN, my nutritionist in the Hypertension Institute, for her assistance with this chapter.

Types of Food that You Should Avoid

1. The "big four" foods to avoid:
   a. Unfamiliar.
   b. Unpronounceable.
   c. More than five ingredients.
   d. Includes high fructose corn syrup.
2. Don't eat anything your great-grandmother wouldn't recognize as food.
3. Avoid food products that make health claims.
4. Don't eat anything incapable of rotting.
5. Shop the outer aisles of the supermarket, and stay out of the middle aisles.
6. Get out of the supermarket and into the farmer's market whenever possible.
7. Eat mostly plants, especially leaves.
8. You are what you eat.
9. Eat well-grown food from healthy soils.
10. Eat wild foods when you can.
11. Regard nontraditional foods with skepticism.
12. Don't look for a "magic bullet" in the traditional diet.
13. Eat meals—not snacks.
14. Pay more, eat less.
15. Do all your eating at a table.
16. Don't get your fuel from the same place your car does.
17. Try not to eat alone, and eat slowly.
18. Cook your own meals and, if you can, plant a garden.

CREATE YOUR PLATE

50-60% non-starchy vegetables; the rest is negotiable.

25–30% clean-sourced fish, turkey, chicken, and proteins (organic meat; grass-fed beef). Avoid any animal product with pesticides, toxins, or hormones. Avoid farm-raised fish. Wild coldwater fish is best, like cod, mackerel, salmon, halibut, and tuna.

15–20% health-promoting fats, like monounsaturated and omega-3 fats.

Add some starch root vegetables, such as carrots, parsnips, and turnips.

NON-STARCHY VEGETABLES

Broccoli.
Brussels sprouts.
Bean sprouts.
Alfalfa sprouts.
Cauliflower.
Bok choy.
Cabbage (green, red, Chinese, Napa).
Swiss chard.
Kohlrabi.
Mushrooms.
Swiss chard.
Arugula.
Watercress.
Microgreens.
Salad greens (chicory, endive, escarole, lettuce, romaine, spinach, arugula, radicchio, watercress).
Sprouts.
Squash (summer, crookneck, spaghetti, zucchini).
Greens (collard, kale, mustard, turnip).
Kale
Asparagus.
Celery.
Radishes.
Hearts of palm.
Jerusalem artichokes, sun chokes.
Onions.
Leeks.
Scallions.
Shallots.

CLEAN-SOURCED PROTEINS

Alaskan halibut.
Alaskan salmon (fresh, canned).

Canned tuna.
Freshwater bass.
Hawaiian fish.
Sardines.
Whitefish.
Chicken.
Turkey.
Duck.
Beef.
Bison.
Lamb.
Pork.
Wild game (elk, venison, boar).

A note about plant-based "meats": These are processed and *not* healthy.

HEALTH-PROMOTING FATS

Algae oil.
Avocado.
Avocado oil.
Olives.
Olive oil.
Coconut, shredded, flakes, unsweetened.

Note on coconut oil: Coconut oil has mixed effects on HDL and LDL cholesterol and neutral effects on CHD. It has more short-chain fatty acids but some long-chain fatty acids. Thus, it's more of a neutral fat. See Chapter 12, Part I on nutrition regarding coconut oil.

Ghee.
Grass-fed, cultured butter.
Macadamia oil.
MCT oil.
Perilla oil.
Walnut oil.
Red palm oil.
Sesame oil.
Cod liver oil.
Almonds.
Chia seed.
Macadamia nuts.
Walnuts.
Pistachios.
Pecans.
Hazelnuts.
Flaxseeds.
Hemp seeds.

Pine nuts.
Brazil nuts.

STARCHY AND ROOT VEGETABLES TO CONSIDER

Cassava.
Carrots.
Sweet potatoes or yams.
Winter squash (acorn, butternut, hubbard, kabocha, delicata).
Parsnips.
Celery root.
Green plantains.
Green bananas.
"Siete" brand tortillas.
Jicama.
Turnips.
Tiger nuts.
Green mango.
Green papaya.

SOMETHING SWEET

Allulose.
Erythritol (Swerve).
Monk Fruit (also called Luo han guo).
Stevia.
Xylitol.

FRUITS

Blueberries.
Blackberries.
Raspberries.
Strawberries.
Pomegranate.
Pineapple.
Mango.
Kiwi.
Guava.
Apples.
Cranberries.
Canteloupe.
Oranges.
Plums.
Cherries.
Grapefruit.

PROTEINS AND FOODS TO SAY "YES" TO

Wild caught fish.
Alaskan halibut.
Alaskan salmon (fresh, canned).
Canned tuna.
Freshwater bass.
Hawaiian fish.
Sardines.
Whitefish.

PASTURED POULTRY

Chicken.
Turkey.
Duck.

SWEETENERS

Allulose.
Erythritol (Swerve).
Monk Fruit, Luo han guo.
Stevia.
Xylitol.

FLOURS

Almond.
Sweet potato.
Cassava.
Tiger nut.
Arrowroot.

HERBS

Basil.
Mint.
Parsley.
Cilantro.
Rosemary.
Thyme.
Sage.
Chives.
Dill.
Oregano.

STARCHES

Cassava.
Carrots.
Sweet potatoes or yams.
Winter squash (acorn, butternut, hubbard, kabocha, delicata).
Parsnips.
Celery root.
Green plantains.
Green bananas.
"Siete" brand tortillas.
Jicama.
Turnips.
Tiger nuts.
Green mango.
Green papaya.

FERMENTED FOODS

Raw sauerkraut.
Raw fermented vegetables.
Kombuchat.
Almond milk yogurt.
Sheep and goat milk yogurt.
Vinegar: any without sugar.

NUTS AND SEEDS

Almonds, sprouted or raw.
Macadamia nuts.
Walnuts.
Pistachios.
Pecans.
Hazelnuts.
Flaxseed.
Hemp seed.
Pine nuts.
Brazil nuts.

FOODS TO AVOID OR EAT IN MODERATION

Refined, starchy foods.
Rice.
Pasta.
White potatoes.

Potato chips.
Bread.
Pastry.
Cookies.
Crackers.
Pretzels.
Cereal.
Products made from grains and pseudo-grains (amaranth, buckwheat, quinoa, etc.).

SWEETENERS TO AVOID

Agave.
Splenda (sucralose).
Sweet One (acesulfame K).
Sugar.
NutraSweet (aspartame).
Splenda (sucralose).
Sweet'N Low (saccharin).
Diet drinks.
Crystal Light.
Maltodextrin.

LEGUMES TO AVOID

Alfalfa.
Beans (pinto, black, white, navy).
Lentils.
Peas.
Chickpeas.
Carob.
Soybeans.
Peanuts.

OILS TO AVOID

Soy.
Grapeseed.
Corn.
Peanut.
Cottonseed.
Safflower.
Sunflower.
Vegetable.
Canola.
Partially hydrogenated.
Margarine.

A Word about Sodium

- When dining out, ask for information on calories, fat, and sodium. Many chains offer nutritional profiles of their menu options online.
- When possible, ask for meals to be prepared without salt, then add a pinch or two.
- At the market, read nutrition labels and don't buy foods that include a lot of sodium. Processed foods with high sodium tend to be breads, cold cuts, cured meats, frozen meals, and packaged soups. Better yet, remember real food doesn't have a label. So, opt for fresh food, which is typically found on the perimeter of the store. Focus less on foods that come in a box, bag, or can.
- Cook at home more and base those meals on fresh and whole foods, not highly processed, heat-and-serve products that tend to be high in sodium.
- Keep portions in check. To attract customers, most restaurants serve portions that are 2–3 times larger than sensible dietary guidelines recommend.
- Instead of prepackaged snack foods, opt for these low sodium snacks:
- Organic, air-popped popcorn.
  - Low glycemic fruit such as berries.
  - Sprouted or raw, unsalted nuts.
  - Steamed edamame.
  - Homemade kale chips.
  - Homemade roasted chickpeas.
  - Homemade sweet potato chips.
  - Guacamole.
  - Protein smoothie.
  - Vegetables: baby carrots, celery sticks, cherry tomatoes, cucumber slices, red and green peppers, steamed broccoli, cauliflower florets.
  - Coconut milk yogurt with berries (freeze yogurt and blend with berries).

Food prepared with little or no salt should still be delicious and flavorful. Try to load up on fresh herbs and spices such as cumin, paprika, oregano, lemon peel, garlic, onion powder, and rubbed sage to season food. Vinegars, citrus juices, and zests are good, too. I am not a fan of salt substitutes or 'lite salt', as most of them contain potassium chloride. While these have no sodium, they may cause the body to retain potassium.

And remember, most of the foods we eat regularly today are not the normal foods of our species. Instead, they're foods that have been created to elicit an unnatural, elevated taste response; as a result, these foods are high in processed sugar, fat, and salt. Our taste buds and the pleasure centers in our brain find these unnatural foods very appealing, making whole natural foods less palatable by comparison. But don't despair: "neuroadaptation" is a normalizing process, whereby the taste buds will change once you adjust your diet to whole foods.

Based on the patient's need, I recommend somewhere between 1500 and 2000 mg of sodium a day for those with cardiovascular disease or high blood pressure. This is the equivalent to about 3/4 teaspoon of table salt. (Note that the average American takes in over 5000 mg of sodium per day.)

RECIPES

1. **Crispy Cauliflower with Gremolata**
   1 head cauliflower, quartered, cored, and cut into bite-size florets
   3 to 4 tablespoons extra-virgin olive oil, plus extra for drizzling
   Salt and freshly cracked pepper
   1 lemon
   1 large handful of fresh parsley (about 1/2 cup/25 g), roughly chopped
   Sea salt, for serving
   Preheat the oven to 425°F (220°C). Spread the cauliflower on a baking sheet in a single layer. Drizzle with the oil, season generously with salt and pepper, and toss to coat. Roast the cauliflower, tossing the florets halfway through, until they are deep golden and crispy, 30 to 35 minutes total.
   While the cauliflower is roasting, prepare gremolata. Transfer the roasted cauliflower to a serving bowl and mix with gremolata to taste. For some it's a couple tablespoons, for others they like more sauce!!!

   Gremolata (Italian herb sauce)
   1 cup packed Italian parsley (small stems are okay)
   1–2 garlic cloves
   Zest of one small lemon, plus 1–2 teaspoons lemon juice (Meyer lemon is especially nice)
   ½ cup olive oil
   ⅛ teaspoon kosher salt and pepper, to taste
   Pinch chili flakes (optional)
   On a cutting board or mat, chop everything very finely and place it into a bowl. Stir in olive oil, salt, and pepper.
   Add chili flakes for a touch of heat if you like. Store in a jar in the fridge for up to one week. An excellent sauce for eggs, fish, or chicken!

2. **Avocado Egg "Toast"**
   1/4 medium avocado
   1/2 clove garlic, mashed
   1/4 teaspoon ground pepper
   1 slice sweet potato toast
   1 large egg, poached
   1 teaspoon store-bought hot sauce (optional)
   1 tablespoon sliced red onion
   Combine avocado, pepper, garlic, and hot sauce.
   Top sweet potato toast with mashed avocado and poached egg. Garnish with red onions.

3. **Salmon with Fresh Herbs**
   3 tablespoons olive oil
   3/4 teaspoon kosher salt

1/2 teaspoon freshly ground black pepper
4 (6-ounce) skin-on salmon fillets
1/4 teaspoon smoked paprika
2 tablespoons chopped fresh tarragon
2 tablespoons chopped fresh dill
2 tablespoons chopped fresh sage

Preheat the oven to 450 degrees. Rinse fish, pat dry. Mix herbs together. Brush each salmon fillet with olive oil sprinkle with salt and pepper. Place prepared salmon skin side down on baking sheet. Top with herbs. Bake until salmon is firm but still pink in the center 12–14 minutes.

4. **Lemon Asparagus**

1 lb fresh asparagus, trimmed
2 Tbsp olive oil
2 cloves garlic, minced
1 lemon, thinly sliced
2 Tbsp freshly squeezed lemon juice (approx. 1 lemon)
1/2 tsp sea salt
1/4 tsp ground black pepper

Preheat your oven to 400 degrees F and line a rimmed baking sheet with parchment paper.

Add the asparagus, lemon slices, olive oil, freshly squeezed lemon juice, sea salt, ground black pepper, minced garlic, to the baking sheet. Toss to evenly coat. Place in the oven and roast for eight to ten minutes or until the asparagus is crisp on the outside and tender in the center.

5. **Homemade Sweet Potato Chips**

2 medium sweet potatoes
4 tablespoons olive oil
1 teaspoon sea salt
1/2 teaspoon black pepper

Preheat the oven to 375 degrees. Use a mandoline to thinly slice potatoes. In a large bowl, toss sweet potatoes with oil, salt, and pepper. Coat a wire rack with cooking spray. Place rack in a shallow baking pan and arrange half of the slices on the rack. Bake 30 minutes until crispy at edges. Repeat with remaining potatoes.

6. **Mexican Cod**

4 (4 oz.) frozen, skinless cod filets
2 tablespoons lime juice
2 teaspoons chili powder
1 teaspoon cumin
1/2 teaspoon sea salt
3 tablespoons avocado oil
Lime wedges

Rinse fish, pat dry. In a small bowl, combine lime juice, chili powder, cumin, and salt. Brush both sides of fish with lime mixture. In a skillet, heat 3 teaspoons avocado oil over medium-high heat. Add fish, cook four to six minutes per 1/2 thickness or until fish flakes easily, turning once. Serve with lime wedges.

7. **Cauliflower Tabbouleh**

    3 tablespoons extra virgin olive oil
    1 1/2-lb. (5 cups) head cauliflower, riced, finely chopped, or grated
    1 1/2 teaspoon sea salt
    1 lemon, juiced
    1/2 cup red onion, chopped
    1/2 cup chopped parsley
    1/2 cup chopped dill or mint
    1 cup cherry tomatoes, halved
    1 cup seeded and chopped cucumber
    Lemon wedges
    Olive oil for drizzle

    Heat the olive oil in an extra-large skillet over medium-high heat. Add cauliflower and 1 teaspoon salt to the hot skillet. Cook, stirring occasionally, about five minutes or until crisp-tender. Spread cauliflower out on a large baking sheet to cool.

    In a large bowl, stir together the remaining salt and lemon. Add cooled cauliflower, red onion, herbs, tomatoes, and cucumber. Cover and let stand at room temperature for one hour, stirring occasionally. Drizzle with olive oil, salt, and pepper, if desired.

8. **Thai Cucumber Salad**

    2 medium cucumbers, peeled, cut in half lengthwise, seeded, cut into ¼-inch slices
    3 tablespoons seasoned rice vinegar
    1 teaspoon red Thai chili pepper
    1/2 teaspoon salt
    1/2 teaspoon grated lime zest
    1/2 teaspoon freshly grated ginger
    3 tablespoons fresh basil, sliced thinly

    In a medium bowl, combine vinegar, chili pepper, salt, lime, and ginger. Add cucumbers. Allow to sit for 15 minutes. Add basil. Refrigerate for up to three days.

9. **Arugula and Fennel Salad**

    1 medium fennel bulb
    3 cups baby arugula
    3 tablespoons extra virgin olive oil
    3 tablespoons fresh lemon juice
    1/2 teaspoon sea salt
    1/4 teaspoon freshly ground black pepper

    Trim and discard the outer layers and fronds from the fennel. Slice it paper-thin using a mandolin and place in a large bowl. In a small bowl, whisk olive oil, lemon juice, salt, and pepper. Pour over the salad and toss to coat evenly.

10. **Avocado Egg Cups**

    2 avocados

    4 eggs

    Salt and pepper to taste

    1 tablespoon green onions, chopped

    With a knife, cut avocados lengthwise into halves and remove pit. If needed, slightly hollow out the avocados to make room for the eggs.

    Arrange avocados in a single layer on a baking dish. Break an egg into each avocado half and season with salt and pepper to taste. Bake in a 425°F oven for about 10–15 minute or until eggs are cooked to your liking. Remove from the oven and garnish with green onions. Serve hot.

11. **Green Beans with Walnuts**

    3 cups fresh green beans, trimmed

    1 cup yellow or red onion, halved, sliced thin

    1/2 cup walnuts or pecans, chopped

    1/2 teaspoon salt

    1/4 teaspoon black pepper

    2 tablespoons extra virgin olive oil

    In a skillet, add olive oil and, over medium heat, sauté green beans, onions, salt, and pepper.

    Heat 15 minutes or until green beans are crisp-tender, stirring occasionally. Add walnuts.

12. **Black Cod with Miso**

    1 lb. black cod, cut into four filets

    1/2 white cup of miso paste

    1/8 cup mirin

    1/8 cup sake or other sweet wine

    1 tablespoon allulose or erythritol

    Combine last four ingredients in a blender until smooth.

    Pat the black cod fillets thoroughly dry with paper towels and discard. Slather the fish with the miso marinade and place in a non-reactive dish or bowl and cover tightly with plastic wrap. Leave to marinate in the refrigerator for two to three days.

    Preheat the oven to 400°F. Heat an oven-proof skillet over high heat on the stovetop. Lightly wipe off any excess miso clinging to the fillets, but don't rinse. Swirl the pan with a little olive oil, then place the fish skin-side up in the pan and cook until the bottom of the fish browns and blackens in spots, about three minutes. Flip and continue cooking until the other side is browned, two to three minutes. Transfer to the oven and bake for five to ten minutes, until fish is opaque and flakes easily.

13. **White Bean Dip**

    1 cup cooked cannellini beans

    1/4 cup extra virgin olive oil

    3 cloves garlic, peeled

    6 fresh basil leaves, divided

2 teaspoons grated lemon zest
1/4 teaspoon ground black pepper
1/4 teaspoon sea salt

Place beans, olive oil, garlic, 5 basil leaves, lemon zest, pepper, and salt in a high-speed food processor. Blend for about 45 seconds, until creamy. Transfer into covered glass dish and refrigerate for at least 30 minutes before serving. Serve with basil leaf and drizzled with additional olive oil. Serve with freshly cut peppers, steamed asparagus, cauliflower, or strips of grilled chicken.

14. **Zucchini Noodles with Garlic and Basil**

    4 zucchini, medium
    4 cloves fresh garlic, minced
    4 Tbsp. Olive or avocado oil
    Lemon zest, optional
    Celtic salt and freshly ground pepper

    Spiral cut the zucchini. Add oil to pan on low/medium heat and sauté the garlic until soft, about 30 seconds.

    Add zucchini to pan and cook just until heated through, not mushy. Add salt, pepper, and lemon zest to taste.

    *This is a very versatile dish which works well with the addition of tomatoes, spices, or chicken, fish, or beef. Add your favorite spices and ingredients for a great side dish or entree.

15. **Shaved Brussels Sprouts with Shiitake**

    12 ounces organic Brussels sprouts
    2 tablespoons fat (coconut oil or olive oil)
    1 leek, thinly sliced, green and white
    2 cloves garlic, minced
    ½ cup shiitake mushrooms
    Sea salt and black pepper to taste

    Wash Brussels sprouts and remove stems. Shave sprouts with a food processor or slice thinly. Heat oil in a large pan over medium heat. Add onion, and let soften for one to two minutes.

    Add garlic and cook until onions are translucent. Add Brussels sprouts and cook for about two minutes. Add mushrooms and sea salt. Cook until sprouts and mushrooms are soft.

16. **Curried Coconut Carrot Soup**

    1 Tbsp. olive oil
    1 Medium onion, chopped
    1 Large shallot, sliced
    2 Lbs. carrots, peeled and sliced
    1 Tbsp. garlic, chopped
    1 Heaping Tbsp. ginger, minced
    1 Tsp. minced serrano or jalapeno chili
    1 Tsp. ground coriander
    3/4 Tsp. Madras style hot curry powder

1 Quart organic vegetable or chicken broth
1/2 Tbsp. lime juice
1 Tbsp. basil chiffonade
Heat oil in soup pot over medium heat. Add onion, shallot, and carrots. Add garlic, ginger, chilies, coriander, and curry powder and continue to sauté until fragrant, another minute. Add broth, cover partially, and bring to a boil over high heat. Reduce heat and gently simmer 20 minutes. Remove from heat. Puree soup in blender, working in batches or blend smooth with an immersion blender. Stir in coconut milk and lime juice, season to taste with salt and pepper. If soup is too thick, thin soup with water. To serve, ladle into cups and chilies garnish with basil and pepitas (pumpkin seeds). Can be served warm or chilled.

17. **Egg Drop Soup (serves 2)**
   4 cups bone broth, preferably homemade
   4 eggs, pasture-raised, whole or whisked
   1/2 cup red onion or scallions, sliced thinly
   1 cup spinach, chard, collards or kale, sliced
   2 teaspoons tamari, gluten-free
   1 tablespoon fresh basil, chiffonade
   1/2 teaspoon dark sesame oil
   Salt and pepper, to taste
   - Place bone broth and tamari in a medium saucepan on medium heat until close to boil.
   - Add onions and chard and cook for two minutes.
   - Drop eggs into soup until cooked, about two minutes, with a lid in place.
   - Pour into a bowl and add sesame oil and top with fresh basil.

18. **Pumpkin Pie Spiced Seeds**
   Yield: 2 cups
   2 cups pumpkin seeds, organic preferred
   2 teaspoons pumpkin pie spice (Trader Joe's)
   1 tablespoon real maple syrup
   Preheat the oven to 375 degrees and line baking sheet with parchment paper. Mix all three ingredients and spread onto the prepared baking pan. Roast for 10 to 15 minutes, checking every five minutes to make sure they don't burn. Let cool.

19. **Middle Eastern Spiced Seeds**
   2 cups pumpkin seeds, organic preferred
   1 teaspoon dried thyme
   1 teaspoon dried marjoram
   2 teaspoon za'atar or sumac
   1 1/2 teaspoons Celtic or Himalayan salt
   Zest of 1 lemon, grated
   1 tablespoon olive oil
   Preheat the oven to 375 degrees and line baking sheet with parchment paper. Mix all ingredients and spread onto prepared baking pan. Roast for 10–15 minutes. Let cool.

20. **Instant Pot Garlic-Lime Shredded Chicken**
    1/2 cup low-sodium vegetable or chicken broth
    1/4 cup olive oil
    1/4 cup lime juice
    4 cloves garlic, peeled and smashed
    1 jalapeno, seeded and chopped (see notes)
    1–2 tsp sea or kosher salt (see notes)
    2 tsp cumin
    1 tsp paprika
    1 tsp cracked black pepper
    2 lb boneless, skinless chicken breasts
    1/2 cup chopped cilantro
    Place the chicken broth, olive oil, lime juice, garlic, jalapeno, salt, cumin, paprika, and pepper into a food processor or blender. Blend until smooth. Place chicken in Instant Pot and set on slow cook mode. Pour the sauce over the chicken. Cover and cook on low for eight hours for best results (or high for four hours).
    Using two forks, shred the chicken. Add the cilantro to the chicken, and toss again.
21. **Oven Roasted Vegetables with Chimichurri**
    Asparagus
    Zucchini
    Pear or Roma tomatoes, cut in half, lengthwise
    Carrots
    Fennel
    Red onion
    Green, red, and yellow peppers
    1 Tbsp. fresh thyme
    1/2 cup extra virgin olive oil
    Sea salt
    Freshly ground pepper
    Fresh basil, cut in chiffonade, to taste
    Preheat the oven to 350 degrees. Wash the vegetables and cut into the preferred sizes and shapes. Place in a bowl and toss with the olive oil, thyme, salt, and pepper. Place all vegetables on a baking sheet, or two, depending on quantity making sure not to overcrowd. Baking time will depend on the size of the vegetables but expect 15–25 minutes.
    Remove from the oven and toss with fresh basil and a drizzle of more olive oil, if needed.
    Feel free to experiment with a variety of seasonal produce.
    Serve with or without chimichurri.

Chimichurri
   1 cup extra virgin olive oil
   1 cup flat leaf, or Italian parsley, trimmed and packed
   1 cup cilantro, trimmed and packed
   8 cloves fresh garlic

¼ small red onion
¼ cup lime juice, freshly squeezed
1 tablespoon oregano, dried
1 ½ teaspoon Celtic or Himalayan salt, or to taste
¾ teaspoon black pepper
1 tablespoon red pepper flakes, optional

Place ingredients into a blender or food processor and pulse until all ingredients are finely chopped and combined. Serve immediately or store in airtight container in the refrigerator for up to a week. Bring to room temperature before serving.

22. **Balsamic Skirt Steak with Cherry Tomatoes and Arugula**

    Marinade:

    1/2 cup balsamic vinegar
    1/4 cup olive oil
    3 garlic cloves, minced
    2 tablespoons fresh herbs such as dill, rosemary, basil, sage, or a combination
    3 lbs. grass-fed skirt or hanger steak
    1 basket small cherry tomatoes
    1 red chili pepper, optional
    1 tablespoon balsamic vinegar
    2 cloves garlic, minced
    2 tablespoons extra-virgin olive oil
    1 cup fresh basil, roughly chopped
    3 cups wild arugula

    Add all marinade ingredients to a blender until smooth. Season steaks with salt and pepper, and pour contents of blender over steak in a glass dish and allow to marinate over night. When ready to grill, season the cherry tomatoes with dusting of salt and chilies, if including. Stir in the vinegar, garlic and olive oil. Grill steak over high heat for a few minutes per side. Let rest. Toss the basil and arugula with the tomatoes and put on platter. Add sliced steak to top of salad.

23. **Clean Salad Dressings—Four Ways**
    - **Avocado Dressing**
        - 1/2 cup filtered water
        - 1 large avocado, peeled and pitted
        - 2 tablespoons cilantro, chopped
        - 2 tablespoons, basil, chopped
        - 2 cloves garlic, fresh
        - Squeeze of lime
        - Pinch of salt

    *Add 1/2 small jalapeño if you like it spicy

    Add ingredients to blender.
    Blend until smooth.
    Refrigerate in an airtight glass container, but use within three days.

- **Tahini**
  - 1 cup tahini sesame seed paste (I prefer the paste made from light colored seeds)
  - 3/4 cup lukewarm water, or more for consistency
  - 3 cloves garlic, minced
  - 1/4 cup fresh lemon juice, or more to taste
  - 1/4 tsp salt, or more to taste
  - 2 tsp fresh parsley, minced

  Place all ingredients except for parsley into a food processor and blend until smooth and light colored. Scrape down the sides every 30 seconds or so. Add water depending on your desired consistency. Refrigerate in an airtight glass container, but use within three days. Eat the parsley separately.

- **Mustard Vinaigrette**
  - 1 rounded tablespoon Dijon mustard
  - 2-1/2 tablespoons fresh lemon juice
  - Salt
  - Freshly ground pepper
  - 1 cup extra virgin olive oil
  - 1 garlic clove, lightly crushed but intact
  - 1/2 teaspoon fresh thyme (dill works nicely, too)

  In a small bowl or measuring cup, combine the mustard, lemon juice, salt, and pepper. Whisk in the oil.

  Once blended, add the thyme. Place the garlic clove in the dressing and allow to marinate for at least 30 minutes. Remove from dressing before serving. Refrigerate in an airtight glass container, but use within three days.

- **Raw Ranch**
  - 1/2 cup raw cashews
  - 1/4 cup filtered water
  - 1/4 cup cashew milk, unsweetened
  - 1 garlic clove, minced
  - 2 tablespoon fresh dill, chopped finely
  - 1 teaspoon fresh lemon juice, or to taste
  - 1/2 teaspoon salt
  - Freshly ground pepper

  Blend cashews in a food processor to flour consistency. Blend in water and cashew milk. Add herbs, salt, and pepper, and blend. Refrigerate in an airtight glass container, but use within three days.

24. **Four Ingredient Banana Bread**

    1-1/2 lbs. ripe bananas (approx. four to five medium bananas)
    2 cups sprouted rolled oats
    1 cup almond, cashew, or peanut butter, unsweetened (nuts only)
    1 cup mini chocolate chips, sweetened with Erythritol

*Optional: Add 1/2 cup chopped crystallized ginger when you add the chocolate chips

Preheat the oven to 350 degrees. Lightly grease 9X5 inch loaf pan with olive oil or line with parchment paper; set aside. Add oats to blender and blend until fine powder, and then transfer to a medium mixing bowl. Add four of the five bananas to the blender with nut butter and blend until smooth. Add to bowl with oat powder and stir just until blended; do not over mix. Add chocolate chips. Batter will be thick. Place in loaf pan and smooth to fill edges of pan. Garnish with remaining banana and sprinkle chocolate chips. Bake for 30 min or until toothpick inserted in the center comes out clean. Let cool completely in loaf pan.

25. **Protein Truffles**

1 cup sprouted rolled oats
1/2 cup unsweetened nut butter
1/3 cup raw honey
1/2 cup dark chocolate
2 tablespoons flax seeds
2 tablespoons chia seeds
1 tablespoon vanilla collagen or whey powder

Stir oats, nut butter, honey, chocolate, flax, chia, and protein powder in a bowl until blended. Cover bowl with lid or plastic wrap and refrigerate for 30 min. Scoop chilled mixture into balls. Keep cold until ready to eat.

BREAKFAST IDEAS

Golden Apple Oatmeal

Preparation Time: 15 minutes
Number of Servings: 1
1 golden delicious apple, diced
2/3 cup water
Dash of cinnamon
Dash of nutmeg
1/3 cup quick-cook rolled oats, uncooked

Combine apples, water, and seasonings in a pot and bring to a boil. Stir in rolled oats; cook for three to five minutes. Cover and let stand several minutes before serving.

Swiss Muesli

Preparation Time: 20 minutes
Number of Servings: 3
3/4 cups rolled oats
3/4 cups water
1 cup shredded, unpeeled apples
1 dried fig
1 Tbsp lemon juice

1/4 tsp cinnamon
1/4 cup chopped almonds
1 Tbsp ground flaxseed

Combine oats, water, shredded apples, lemon juice, cinnamon, and flaxseeds. Cover and refrigerate overnight. In the morning, spoon some of the muesli into a cereal bowl. Top with your choice of fresh fruits and nuts. Serve with a dollop of plain yogurt or almond milk, if desired. Muesli can be stored in covered container in refrigerator for several days.

Black and Blue Berry Smoothie

Smoothies are a quick and easy way to ensure a healthy breakfast. This recipe and the one that follows are loaded with beta-carotene, protein, flavonoids, pectin, and vitamin C. Try frozen strawberries, blueberries, mixed berries, mango, or peaches. If juice is added, you might try pineapple juice, orange-tangerine juice, and other 100 percent juice blends.

Preparation Time: 10 minutes
Number of Servings: 2
1 scoop whey protein powder (look for one with about 20 grams protein per serving)
1/2 cup blackberries
1/2 cup blueberries
1/2 cup fat free plain yogurt
1/2 cup organic whole milk
1/2 tsp vanilla extract
1 cup ice

Place all ingredients into blender and blend until smooth. Serve immediately.

Raspberry Mango Smoothie

Preparation Time: 5 minutes
Number of Servings: 1
1 cup unsweetened, frozen raspberries
1 scoop vanilla whey protein powder
1 cup almond milk
5 ice cubes
1 tsp cinnamon

Blend well in blender and drink!

Salad as a Side Dish

Jicama and Asian Pear Salad

Preparation Time: 15 minutes
Number of Servings: 6
2 cups shredded romaine lettuce
2 cups julienned jicama
2 cored and chopped Asian pears
1 Tbsp golden raisins
1/4 cup white wine vinaigrette salad dressing
1/4 cup apple cider vinegar
1/4 tsp Chinese five-spice powder or ground allspice

In a bowl, toss the shredded lettuce, jicama, Asian pears, and golden raisins until combined. For the dressing, whisk together the salad dressing, apple cider or juice, and five-spice powder or allspice until well mixed. Drizzle over salad and toss well. Serve immediately.

Pineapple Slaw

Preparation Time: 10 minutes
Number of Servings: 6
2-1/2 cups shredded cabbage
1 cup shredded carrots
1 cup pineapple chunks
1 Tbsp raisins
2-1/2 Tbsp pineapple juice
1 Tbsp olive or grapeseed oil

Combine all ingredients in a large bowl. Toss and serve or put in the refrigerator covered until serving time.

Apple-Hazelnut Salad in a Cup

Preparation Time: 10 minutes
Number of Servings: 1
2 Tbsp non-fat bottled raspberry vinaigrette
1 apple, diced
2 Tbsp chopped hazelnuts
1 cup pre-cut mixed greens, rinsed and drained

Layer ingredients, in order with dressing on the bottom, in a large, travel-proof, lidded, insulated cup. When ready to eat, shake the cup well, grab a fork and enjoy!

Entrees

Lentil Soup—Indian Style

Preparation Time: 2 hours
Number of Servings: 8
1 lb dry lentils
10 cups water
2 onions, chopped
1 green pepper, chopped
2 cloves garlic, finely minced
2 tsp salt
½ tsp black pepper
1 15-oz. can low-sodium tomato sauce
½ tsp cinnamon
2 cardamom pods
1 tsp. turmeric
1 tsp. coriander
3/8 tsp. crushed red pepper flakes
2 tsp. curry powder

Combine lentils in water with onion, green pepper, garlic, salt, and pepper. Bring to a boil and simmer for 30 minutes.

Add other ingredients; simmer for one hour. Strain out cardamom pods. Blend about ¾ of the soup in the blender, ¼ unblended for texture. Return blended soup to pot.

Salmon Tacos

Preparation Time: 30 minutes
Number of Servings: 6
1/2 cup nonfat sour cream
1/4 cup fat-free mayonnaise
1/2 cup chopped fresh cilantro
1/2 package low-sodium taco seasoning, divided
1 lb salmon fillets, cut into 1-inch pieces
1 Tbsp olive oil
2 Tbsp lemon juice
2 cups shredded red and green cabbage
2 cups diced tomato
12 6-inch, warmed Trader Joe's low-carb wheat tortillas
Lime wedges for serving

In a small bowl, combine sour cream, mayonnaise, cilantro, and 2 Tbsp seasoning mix. In a medium bowl, combine salmon, vegetable oil, lemon juice, and remaining seasoning mix; pour into large skillet. Cook, stirring constantly, over medium-high heat for four to five minutes or until salmon flakes easily when tested with a fork. Fill warm tortillas with fish mixture. Top with cabbage, tomato, sour cream mixture, lime wedges, and taco sauce.

Tuna Bean Main Dish Salad

Preparation Time: 4 hours, 15 minutes
Number of Servings: 6

Dressing

1/2 tsp grated lemon peel
1/3 cup lemon juice
1/4 cup olive oil
2 Tbsp fresh chopped parsley
1 tsp rosemary
1 Tbsp Dijon mustard

Mix all dressing ingredients thoroughly and store in a tightly covered container in the refrigerator until ready to be used.

Salad

3 medium green bell peppers
3 medium red bell peppers
2 (15-oz.) cans white beans, rinsed and drained
2 (6-oz.) cans low-sodium water-packed tuna, drained
1/2 cup sliced ripe olives
1 head lettuce
2 medium tomatoes, cut into wedges

Set the oven to broil. Place bell peppers on a broiler pan. Broil with tops 4 to 5 inches from heat for about minutes on each side or until skin blisters and browns. Remove from the oven. Wrap in a towel; let it stand for five minutes. Remove the skin, stems, seeds, and membranes of the peppers. Cut peppers into 1/4-inch slices. Toss peppers, beans, tuna, olives, and dressing in a bowl. Cover and chill for four hours, stirring occasionally. Spoon salad onto lettuce leaves and garnish with tomato wedges.

Lasagna

Preparation Time: 1 hour, 30 minutes
Number of Servings: 9
1 lb ground, range-fed turkey
1 lb low-fat cottage cheese
1/2 lb low-fat or part-skim ricotta cheese
2 egg whites
2 Tbsp grated low-fat Parmesan cheese
1 Tbsp fresh minced chives
1 Tbsp fresh minced parsley
1/4 tsp freshly ground black pepper
8 oz. whole wheat lasagna noodles, uncooked
1 large onion, minced
1/4 cup dry red wine
1/2 lb. sliced mushrooms
1 cup chopped zucchini
4 cups low-sodium red sauce of your choice

In a nonstick frying pan, cook ground turkey until it is no longer pink; drain juices and set aside.

Puree cottage cheese, ricotta, egg whites, and Parmesan cheese. Blend in chives, parsley, and pepper by hand.

In a large pot of lightly salted boiling water, cook lasagna noodles until just tender but not mushy, about ten minutes. Remove noodles with a slotted spoon, dip into cold water, and lay out flat on clean kitchen towels (not a paper towel, or they will stick).

In a covered skillet, simmer onions in wine for about five minutes until very soft. Stir frequently, but keep the pot covered in between stirrings. Add mushrooms and zucchini and cook about five minutes until vegetables are soft and half their original volume. Drain the vegetables.

Preheat the oven to 375°F. Combine the cheese mixture and all but 1/4 cup of the vegetable mixture. Spread two cups of red sauce over the bottom of a 9" × 14" baking pan. Alternate layers of noodles with cheese mixture and ground turkey ending with a final layer of noodles. Cover noodles with remaining sauce and distribute reserved vegetables over the top.

Cover and bake for one hour. Remove cover and bake for an additional five minutes. Remove from the oven and let the lasagna sit for ten minutes before cutting.

SUMMARY, CONCLUSIONS, AND TAKE-AWAY POINTS

Eating a heart-healthy diet is easy, tastes good, and can prevent CHD.

- Vegetables: eat 8 to 12 servings per day of non-starchy vegetables such as broccoli, cauliflower, leafy greens, asparagus.
- Fruit: limit to four servings. ½ cup of pieces = one serving. One small whole fruit = one serving.
- Fats: eat good fats and avoid some saturated fats and all trans fats. Omega-3 fatty acids (1–2 grams per day) Extra virgin olive oil (4 tablespoons per day), and olive products.
- Protein: 25–30% clean-sourced fish, turkey, chicken, and proteins. Organic meat, grass-fed beef. Avoid any animal product with pesticides, toxins, and hormones. Avoid farm-raised fish. Wild coldwater fish like cod, mackerel, salmon, halibut, and tuna are best.
- Water: 50% of body weight (pounds) in ounces of filtered or distilled water per day. If you weigh 150 pounds that is 75 ounces of water, herbal teas, broths daily. Do not drink from plastic containers or bottles.
- Tea: 16 ounces of decaffeinated green tea per day.
- Pomegranate: ¼ cup of pomegranate seeds per day.
- Detox + elimination diet, followed by an anti-inflammatory, blood-sugar regulating protocol. Emphasize high-fiber foods and add servings of cruciferous vegetables, such as broccoli, cauliflower, Brussels sprouts, cabbage, etc.
- Caloric restriction (CR): 12–14 hours overnight fast with 12.5% caloric restriction and 12.5% increase in energy expenditure with exercise. PREDIMED (Mediterranean)-type diet.
- FMD (Fasting Mimicking Diet (PROLON) three months on, then three months off and repeat).

BIBLIOGRAPHY

1. https://www.apa.org/ptsd-guideline/patients-and-families/cognitive-behavioral
2. https://www.ncbi.nlm.nih.gov/pmc/articles/PMC5764193/
3. https://www.researchgate.net/publication/14137648_Dietary_under-reporting_What_people_say_about_recording_their_food_intake
4. Macdiarmid JI, Blundell JE. *BioPsychology Group.* Leeds: Department of Psychology, University of Leeds.
5. https://newsroom.ucla.edu/releases/Dieting-Does-Not-Work-UCLA-Researchers-7832#:~:text=People%20on%20diets%20typically%20lose,be%20significantly%20higher%2C%20they%20said
6. https://www.hsph.harvard.edu/obesity-prevention-source/obesity-causes/food-environment-and-obesity/
7. https://www.ecoandbeyond.co/articles/big-food/

14 The Blood Vessel, Brain, and Immune System Connections

All of us have experienced a fast heart rate, skipping of the heart with palpitations—perhaps some chest pain or even shortness of breath—sweating, dizziness, headache, high blood pressure, or other symptoms when we have anxiety, stress, a panic attack, or get angry. These reactions are very rapid, can last for a while, and be very frightening. You imagine that something is wrong with your heart. You may even go to the emergency department to be seen by a physician. What you experience are the very fast connections that occur through your brain and nervous system to your heart, arteries, gut, and immune system. This is the "brain-heart-immune system connection". Chronic stress, anxiety, and depression increase the risk of coronary heart disease (CHD) and myocardial infarction (MI) through a complex messaging system that increases the "sympathetic nervous system activity". This will now be discussed in detail.

THE AUTONOMIC NERVOUS SYSTEM

You have heard about the "flight or fight reaction" or the "rest and relax reaction". These opposing reactions are related to the three distinct and major parts of the autonomic nervous system (ANS), called the "sympathetic nervous system" (SNS), the "parasympathetic nervous system" (PNS), and the "enteric nervous system" (ENS) (related to our gut) (Figure 14.1). The SNS and PNS systems oppose each other to give us a balance that is important to regulate the brain and entire nervous system with the arteries, heart, endocrine system, gut, and immune system. This internal communication is very important. If any of the SNS, PNS, or ENS predominates, it can lead to many cardiovascular problems and diseases. The autonomic nervous system is a component of the peripheral nervous system that regulates involuntary processes of our physiology and daily functions, including heart rate, blood pressure, respiration, digestion, and sexual arousal (Table 14.1).

The SNS and the PNS contain nerve fibers that go in two opposing directions and are called "afferent and efferent fibers". They provide our ability to feel sensation and determine motor function and output, respectively, to the brain. The ENS is an extensive, web-like structure that contains over 100 million neurons and is chiefly responsible for the regulation of digestive processes.

The SNS supplies nerve energy to nearly every living tissue in the body. The SNS increases heart rate, blood pressure, sweating, constricts the arteries, causes

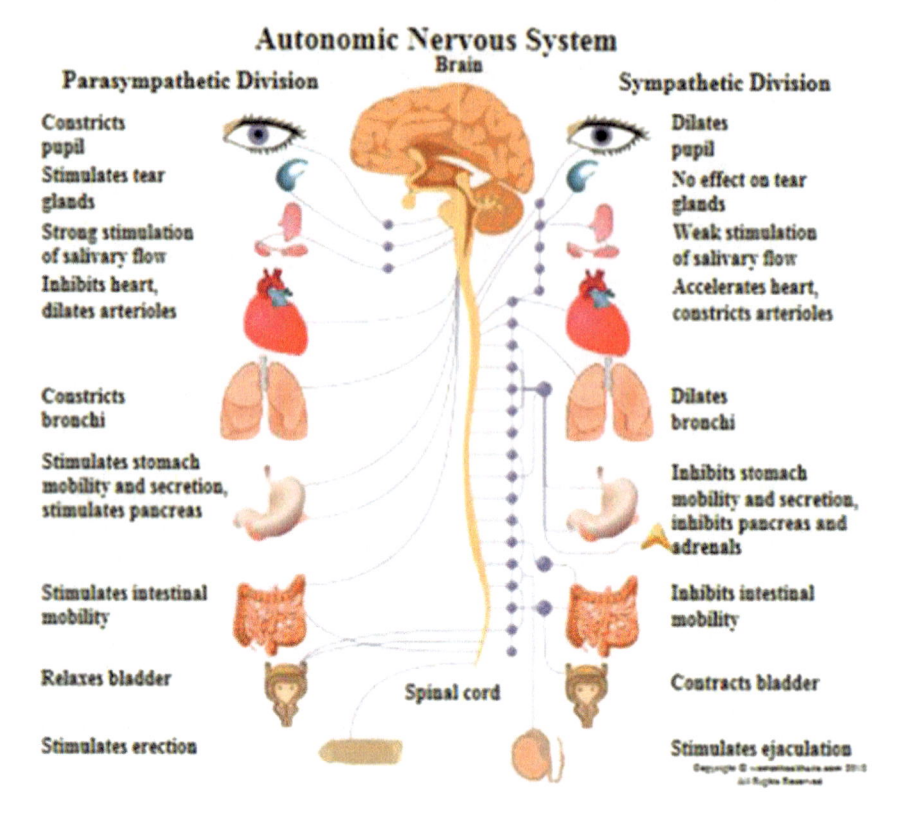

FIGURE 14.1 Autonomic nervous system.

TABLE 14.1
Effects of the Two Branches of the ANS

Organ	Sympathetic Effect	Parasympathetic Effect
Pupil	Dilation	Constriction
Lens	Far focus (lower curvature)	Near focus (increased curvature)
Salivary gland secretion	High in viscosity	Serous
Heart	Increased rate and pressure	Lower rate and pressure
Lungs	Dilation of respiratory passages	Constriction of respiratory passages
Gastrointestinal	Decreased motility	Increased motility
Kidneys	Decreased filtration rate	Increased filtration rate
Male genitalia	Ejaculation	Erection
Vascular smooth muscle	Variable depending on the neurotransmitter	Relaxation
Sweat glands	Increased activity	No innervation
Arteries to skeletal muscle	Dilation	No innervation
Veins	Variable depending on the neurotransmitter	No innervation

palpitations, and much more (Table 14.1 and Figure 14.1). The PNS promotes the "rest and digest" processes, with a lower heart rate and blood pressure, improved digestive functions, lower blood sugar, and regulates the levels of hormones, such as adrenaline and cortisol, in the blood (Table 14.1 and Figure 14.1). The PNS supplies nerve energy to the arteries, head, gut, immune system, and external sexual organs, but is notably vacant in much of the musculoskeletal system and skin, making it significantly smaller than the SNS. The ENS is composed of reflex pathways that control digestive functions such as muscle contraction/relaxation, secretion/absorption, and blood flow (Table 14.1 and Figure 14.1).

The SNS and PNS utilize chemicals called norepinephrine (NE) and acetylcholine (Ach), respectively, to transmit all the nerve information. The sympathetic neurons generally produce NE as their chemical transmitter to act upon target tissues, while the parasympathetic neurons use Ach throughout. Enteric neurons use several major nerved substances to transmit information such as Ach, nitric oxide, and serotonin.

The SNS controls our responses from the brain to the heart and arteries by nerves that cause instantaneous changes in heart rate and blood pressure, the "flight or fight reaction". This results in a fast heart rate, palpitations, constriction of the arteries, high blood pressure, headache, sweating, dizziness, increased blood sugar, cortisol, and NE, and slows down our gut function resulting in such symptoms as indigestion and constipation. This allows us to respond immediately to danger and try to escape. This is the correct and needed reaction for our survival, in some circumstances, but should be short-lived and resolve slowly until complete recovery. However, if the SNS is chronically activated, then many cardiovascular problems occur, such as abnormal heart rhythms, palpitations, high blood pressure, shortness of breath, chest pain, angina, CHD, MI, and sudden death. Changes also occur with the endocrine system and include increased blood sugar, adrenaline, and cortisol levels.

Various immune mechanisms are involved in heart and arterial damage, as I discussed in Chapters 5 and 6. The ANS is critical in increasing or reducing immune dysfunction and inflammation. Nerve connections send signals to the brain through the important vagal nerve in the PNS. Efferent nerve connections are mediated by Ach and are anti-inflammatory and anti-immune pathways that include the spleen and immune cells called T cells, monocytes, and macrophages (as discussed in Chapters 5 and 6). This means that the immune system is innervated (has direct nerve connections from the brain!)

There are receptors in the brain that will cause constriction in the arteries with high blood pressure, resulting in T-cell activation and development of vascular and heart inflammation.

Local brain inflammation or reductions in blood or oxygen supply may mediate vascular inflammation and vice versa. For example, oxidative stress in the brain increases nerve firing. Inflammation in the body will send nerve signals to the brain by the afferent nerves to attempt to control this inflammation. All of these messages are very fast and instantaneous. On the other hand, the nerve transmissions of Ach through a nerve that goes from the brain, called the vagal nerve, will block inflammation and decrease many inflammation cells called proinflammatory cytokines (Figure 14.2). There are inflammatory substances in the body that stimulate the cell to make the proinflammatory cytokines. These are named in Figure 14.2. As

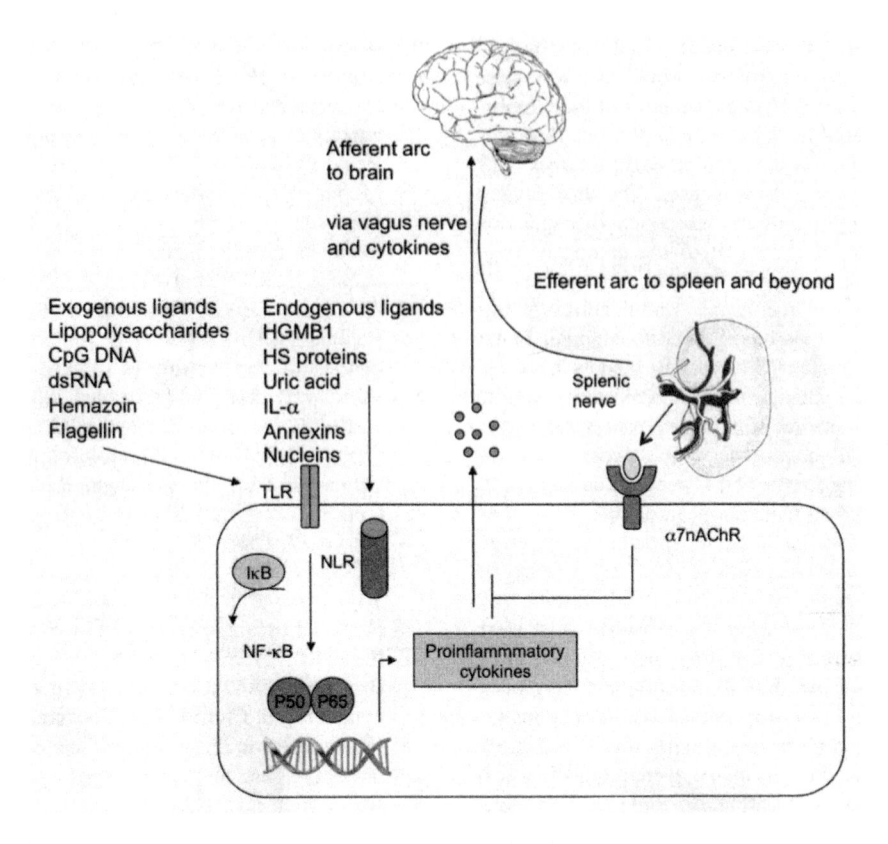

FIGURE 14.2 The direct nerve pathways from the brain and the body that regulate arterial inflammation, oxidative stress, and immune vascular function.

was discussed earlier in Chapters 5 and 6, these substances stimulate inflammation receptors on the arteries that allow the cell to make these cytokines.

TREATMENT FOR AN OVERACTIVE SYMPATHETIC NERVOUS SYSTEM

The importance of the ANS is now clear. Abnormal and long-term activation of the SNS will increase the risk of CHD, MI, and other cardiovascular complications. The SNS activation is a CHD risk factor as much as high blood pressure, high cholesterol, or diabetes mellitus. We have to control our stress, anxiety, and anger to have a reduction in CHD. This reduction in SNS activity can be achieved in a number of ways including taking walks in nature, regular mild to moderate exercises, and disconnecting from technology; practicing guided relaxation and mindfulness meditation techniques; massage, prayer and spirituality, restorative or slow-paced yoga, tai chi, being with animals, deep breathing exercises, and guided imagery; getting plenty of rest; sleep, and optimal nutrition. Some natural compounds, herbs, and teas may also help such as ashwagandha, chamomile, valerian, lavender, passionflower, gamma-aminobutyric acid (GABA), Rhodiola, magnesium, glycine, melatonin, B vitamins, L theanine, zinc, inositol, Bacopa, omega-3 fatty acids, lemon balm, ginkgo, curcumin, and holy basil. You should avoid alcohol, caffeine, and sugar.

SUMMARY AND KEY TAKE AWAY POINTS

1. Chronic stress, anxiety, and depression increase the risk of CHD and MI through an imbalance of the ANS with increased SNS activity.
2. The autonomic nervous system is a balance of the SNS, PNS, and ENS.
3. Increased and chronic activation of the SNS ("flight or fight") is a CHD risk factor and needs to be reduced and controlled.
4. Increased SNS activity causes heart and arterial inflammation, oxidative stress, and vascular immune dysfunction.
5. Increased SNS via NE activity causes elevated blood pressure and heart rate, other cardiovascular problems, and many endocrine problems such as elevated blood sugar, adrenaline, and cortisol.
6. The PNS ("rest and relax and digest") via Ach reduces inflammation, oxidative stress, and vascular immune dysfunction.
7. The SNS and PNS must be balanced to reduce the risk of CHD.
8. There are many treatments to balance the autonomic nervous system including lifestyle, nutrition, exercise, herbs, and supplements.

BIBLIOGRAPHY

1. Sternini C. Organization of the peripheral nervous system: Autonomic and sensory ganglia. *Journal of Investigative Dermatology Symposium Proceedings*. 1997 Aug;2(1):1–7.
2. Karemaker JM. An introduction into autonomic nervous function. *Physiological Measurement*. 2017 May;38(5):R89–R118.
3. Leuckeroth RO. Enteric nervous system development: Migration, differentiation, and disease. *American Journal of Physiology: Gastrointestinal and Liver Physiology*. 2013 Jul 01;305(1): G1–24.
4. Siéssere S, Vitti M, Sousa LG, Semprini M, Iyomasa MM, Regalo SC. Anatomic variation of cranial parasympathetic ganglia. *Brazilian Oral Research*. 2008 Apr-Jun; 22(2):101–5.
5. Koopman FA, Stoof SP, Straub RH, Van Maanen MA, Vervoordeldonk MJ, Tak PP. Restoring the balance of the autonomic nervous system as an innovative approach to the treatment of rheumatoid arthritis. *Molecular Medicine*. 2011 Sep-Oct;17(9–10):937–48.
6. Kenney MJ, Ganta CK. Autonomic nervous system and immune system interactions. *Comprehensive Physiology*. 2014 Jul;4(3):1177–200.

15 What Is Plugging Your Heart Arteries? Plaque Formation, Types of Plaque, and Plaque Rupture

WHY DO YOU HAVE A MYOCARDIAL INFARCTION?

How many times have you heard of a friend or a relative who appeared to be healthy, was previously told by their doctor that they had normal examinations, and never complained of chest pain or shortness of breath or any other cardiovascular symptoms—but suddenly had an MI. Perhaps someone you know had been jogging for years but one day was found dead in the park near their home. The autopsy showed a massive heart attack. How is this possible? Why and how did it occur? Was it preventable? In this chapter, I will explain why and how MI happens and how to prevent it. Most patients who have an MI have not had any previous chest pain, shortness of breath, or other cardiovascular signs or symptoms. This is one of the reasons regular examinations that perform the proper and most up-to-date testing are mandatory to prevent an MI. Now let us discuss the concepts that cause MI: atherosclerosis, plaque, obstruction of the coronary arteries, and plaque rupture (Figure 15.1).

ATHEROSCLEROSIS

Atherosclerosis is a hardening and narrowing of your arteries due to obstruction with plaque, which is made up of fatty material, oxidized cholesterol and fats, inflammatory cells, white blood cells, immune cells, T lymphocytes, activated smooth muscle cells, cell debris, and other substances. As the plaque grows, it blocks the artery, thereby threatening blood flow. The severity of the blockage will determine the distal blood flow from the area of plaque obstruction. This blockage can be up to 99% and patients may have minimal symptoms of chest pain or shortness of breath (unless they are engaged in heavy exercise). This is referred to as atherosclerosis or atherosclerotic cardiovascular disease (Figure 15.1). Atherosclerosis is a slow process that starts very early in life (Figure 15.2). The older you become, the more likely you are to have CHD with plaque formation. This can be evaluated with many different tests, such as a coronary computerized tomographic angiogram (CTA), cardiac

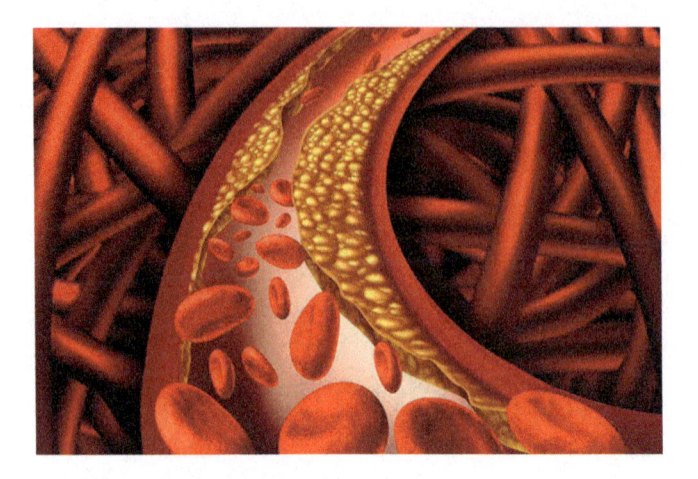

FIGURE 15.1 The inside of the coronary artery (the lumen) is shown here with the red blood cells (in red) and white blood cells (in white) and arterial plaque (in yellow).

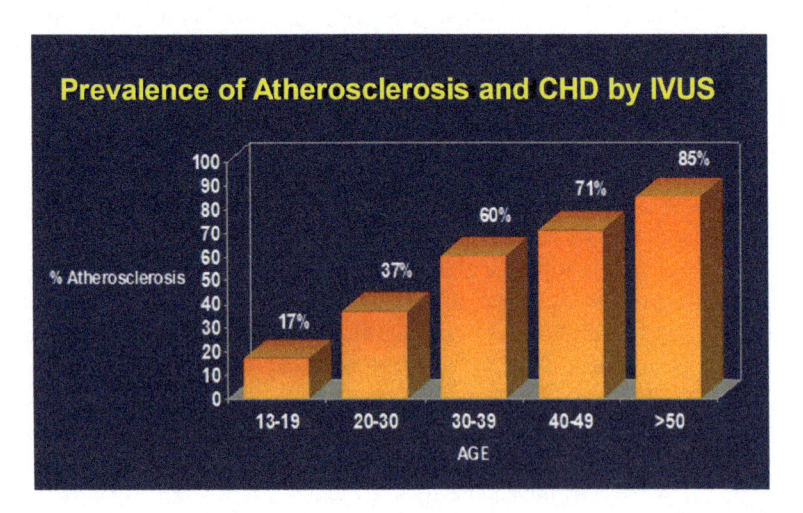

FIGURE 15.2 Atherosclerosis with plaque formation starts early in life. This can be evaluated with ultrasound of the coronary arteries, called intravascular ultrasound (IVUS) as shown below. The age of the patient is on the horizontal axis and the percent of the population that have atherosclerosis or CHD is shown on the vertical axis. At ages 13–19 years about 17% of the population in the US have CHD. By the age of 50, about 85% of the population has CHD.

magnetic resonance imaging or angiogram (MRI or MRA), or coronary arteriogram. Another type of test is an ultrasound of the coronary arteries, called intravascular ultrasound (IVUS) (Figures 15.2 and 15.3). The IVUS has shown us a lot of new information about the natural history of CHD. It can determine the age of onset, presence, location, and severity of CHD both in the lumen and in the subendothelial

What Is Plugging Your Heart Arteries? 157

layer of the coronary arteries (Figure 15.3). By the age of 50, over 85% of patients will have CHD to some extent.

Eventually, over the decades of your life, this results in an accumulation of fatty material, oxidized cholesterol and fats, inflammatory cells, white blood cells, immune cells, T lymphocytes, activated smooth muscle cells, cell debris, and other substances inside the artery that block the blood flow, oxygen, and nutrients to the heart muscle (Figure 15.4). Another event that may occur is that the plaque may literally "explode", releasing all of these very nasty substances into the blood and forming a blood clot (thrombosis) in the lumen of the coronary artery.

Atherosclerosis begins as a disease in the subendothelium of the artery as shown by IVUS (Figure 15.3). CHD is *not* initially a disease of the lumen or the inside of the artery (Figures 15.3, 15.4, and 15.5). The first abnormality, as we discussed in chapter 5, is "glycocalyx and endothelial dysfunction", then the accumulation of the subendothelial plaque (Figures 15.3, 15.4, and 15.5). You may recall that the glycocalyx and the endothelium are very thin layers of single cells that separate the blood from subendothelium and the wall of the artery. If they are damaged by high blood pressure, high cholesterol, diabetes, smoking, obesity, or other CHD risk factors, then normal function becomes abnormal and is called "glycocalyx and endothelial dysfunction". There is a loss of nitric oxide with increased vascular inflammation, vascular oxidative stress, and vascular immune function that damages the blood cells, the subendothelial area, and the artery wall. The artery wall, the glycocalyx, and the endothelium are now dysfunctional, and all of them begin to get thicker. This is called the intimal medial thickness (IMT) and can be measured by different types of machines, like an ultrasound. The IMT is simply a thickening of the glycocalyx and endothelium (intima) and the coronary artery muscle (media).

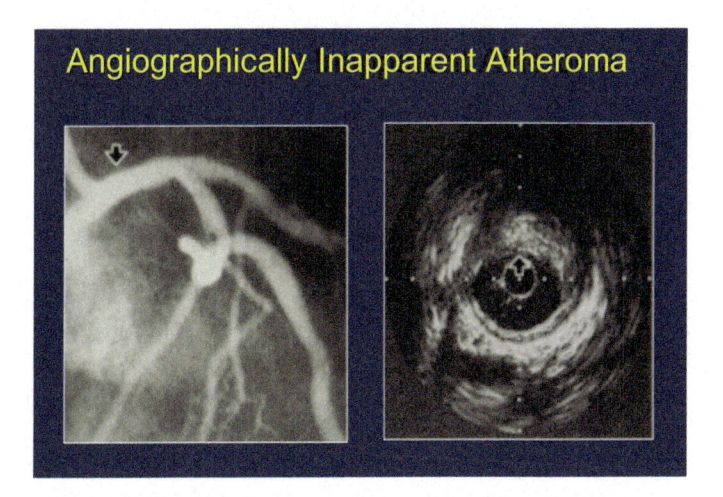

FIGURE 15.3 The left panel shows the coronary artery during a coronary arteriogram. Note the site of the arrow. This shows a normal lumen in the coronary artery without any blockage or plaque. However, in the right panel, the IVUS shows white material in the subendothelial layer that, in early plaque, is outside the lumen or extraluminal. This can progress over time to obstructive coronary artery plaque.

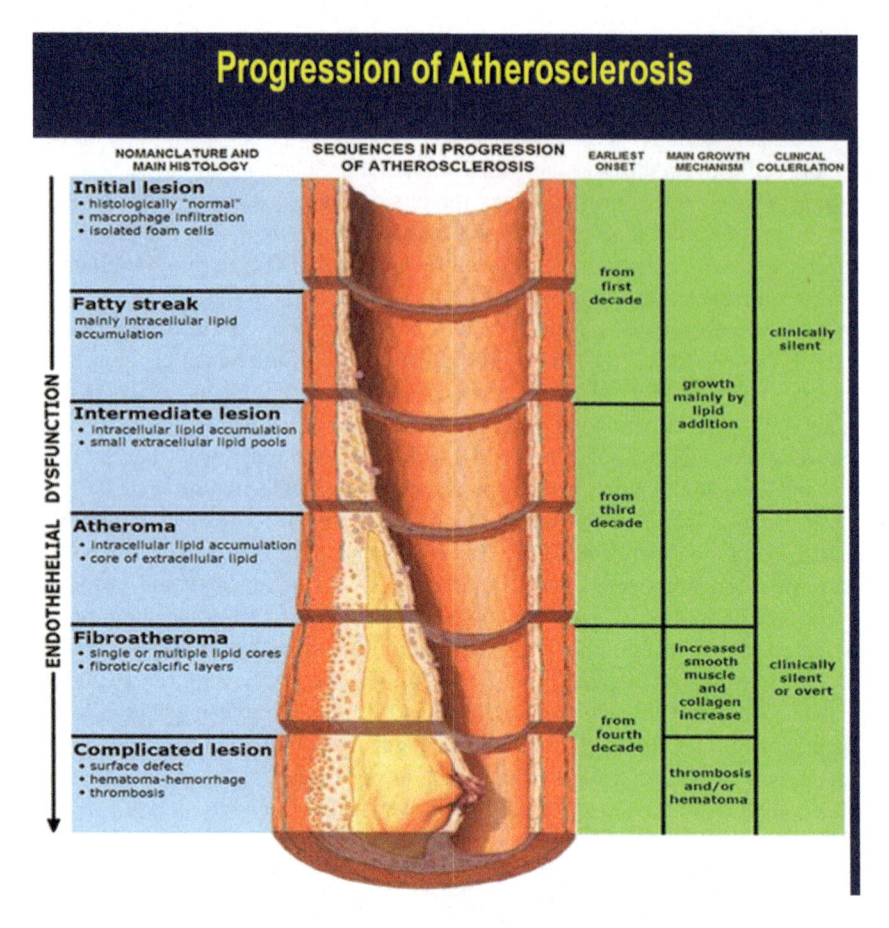

FIGURE 15.4 Atherosclerosis progression from the initial lesion to the end as a complicated lesion (see text for a description of this figure).

This initial lesion in the subendothelium progresses over years with infiltration by cells called monocytes and macrophages to form cells that look like "foam". These are naturally called "foam cells" (Figure 15.4). Later, these clump together to make a streak of fat, called the "fatty streak" (Figure 15.4). This lesion enlarges to an intermediate lesion with more foam cells and fatty streaks and then becomes an "atheroma" or "fibroatheroma" which is a larger collection of fat, inflammatory cells, cell debris, and fibrous tissue. This is the medical term for plaque (Figure 15.4). Once this plaque is formed, it can have different compositions, sizes, shapes, and locations with variable outcomes related to CHD and MI. The plaque may be hard and stable with calcium and obstruct blood flow in the coronary artery, or it may be soft and unstable with minimal obstruction, but a rupture with a clot or thrombosis in the coronary artery could cause complete or partial obstruction.

Now let us discuss more about plaques in the coronary arteries.

What Is Plugging Your Heart Arteries?	159

CHD: Extraluminal Disease: Glagov Principal

Minimal to mild CHD
Lumen normal
Mild extraluminal atheroma

Moderate CHD
Lumen normal size
Mild extraluminal atheroma

Severe CHD
Lumen stenosis
Severe extraluminal and intraluminal atheroma

← 95 - 99% →

← 1 - 5% →

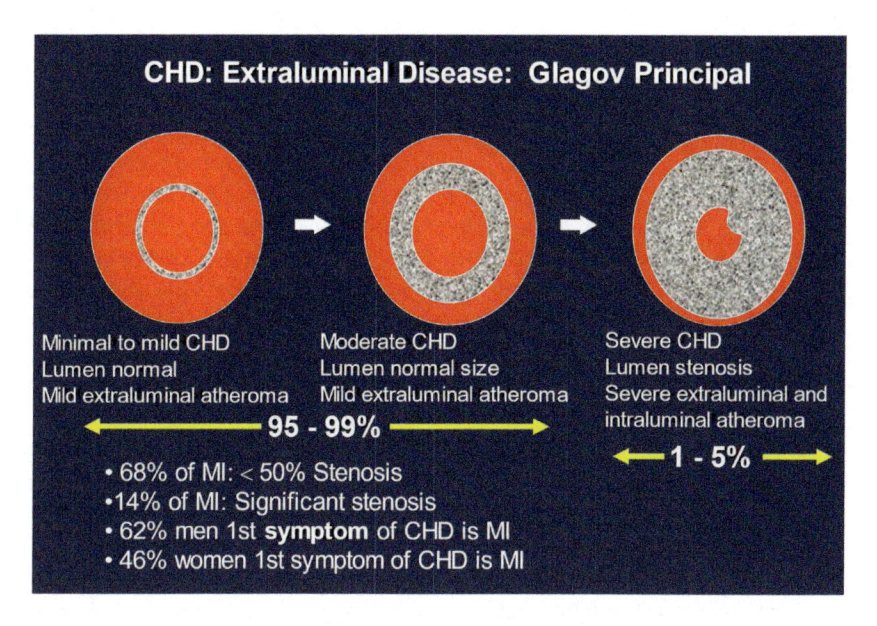

- 68% of MI: < 50% Stenosis
- 14% of MI: Significant stenosis
- 62% men 1st **symptom** of CHD is MI
- 46% women 1st symptom of CHD is MI

FIGURE 15.5 CHD begins as a disease that is outside the lumen (the innermost part) of the coronary artery. This is called "extraluminal disease" and was first described Dr. Glagov. The gray represents the subendothelial material and the red is the artery lumen and artery wall. Notice that as the gray material enlarges, it eventually starts to block the artery lumen. However, in the early stages the artery lumen may be normal despite a large amount of subendothelial plaque formation. Also note the percentages of men and women in whose first symptom of CHD is an MI.

TYPES OF PLAQUES IN THE CORONARY ARTERIES

As I discussed earlier in this chapter, atherosclerosis and CHD begin at an early age as an "extraluminal disease", meaning that it starts outside of the artery lumen (the innermost part of the artery) (Figures 15.3, 15.4, and 15.5). This means that detection of CHD in the early phases requires special testing to determine the extent and location of what is outside the coronary artery lumen in the subendothelium. These tests include coronary artery calcium (CAC) scans, computerized tomographic angiogram (CTA), magnetic resonance arteriograms, (MRA) of the heart, IVUS, and other tests that I will discuss later in the book. In other words, during this early phase of atherosclerosis, a special imaging with a material called contrast dye of the coronary arteries, called a coronary angiogram, arteriogram or cauterization could show a normal artery lumen without obstruction or plaque, but the patient clearly has atherosclerosis with plaque-like material that still remains in the subendothelial layer of the coronary artery. It is estimated that about 68% of MIs occur with less than 50% blockage of a coronary artery. In 62% of men and in 46% of women, the first symptom of CHD is an MI (Figure 15.4).

WHAT DOES A CORONARY ARTERY PLAQUE LOOK LIKE? THE ANATOMY OF A PLAQUE

Let us take a look at how a plaque in the coronary artery looks, its composition, and how it causes an MI (Figure 15.6). When the plaque gets larger, it has very definitive components and different materials that are characteristic (Figure 15.6). The first layer separates the blood from the rest of the artery and is called the intima and includes the endothelium and vascular smooth muscle (media) shown below in Figure 15.6. Under the endothelium is the beginning of the plaque with a fibrous cap, then the lipid core which contains fatty material, oxidized cholesterol, and fats, inflammatory cells, white blood cells, immune cells, T lymphocytes, activated smooth muscle cells, cell debris, etc. The size and structure of the fibrous cap and the underlying lipid core help to determine whether this is a hard plaque or a soft plaque (will be discussed later in this chapter). As the plaque grows larger, it will slowly start to block the coronary artery, leading to reduction of blood flow, oxygen, and nutrients to the heart muscle. This will cause chest pain (angina), shortness of breath, fatigue, or other symptoms of CHD. The other potential result could be a rupture of the plaque into the artery lumen. All of the toxic lipid core and other materials contained within it spill into the coronary artery lumen and cause a clot (thrombosis) that may completely block blood flow. This is referred to as a coronary thrombosis,

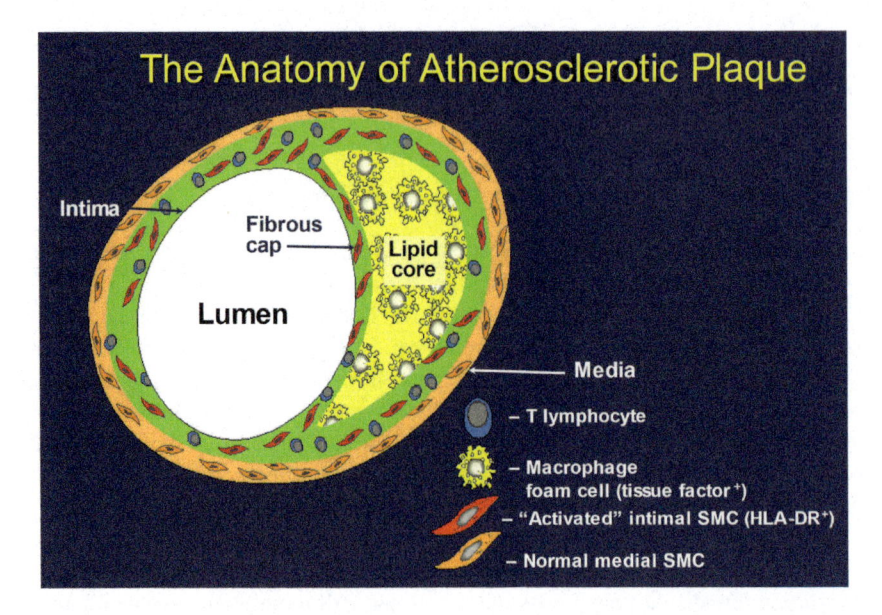

FIGURE 15.6 The anatomy of a coronary artery plaque: what is it made of? Note the lumen in white, the fibrous cap in green, and the lipid core in yellow. In this lipid core are cholesterol, other lipids, T lymphocytes, monocytes, macrophages, activated smooth muscle, and other cell debris.

What Is Plugging Your Heart Arteries? 161

which is another medical name for an acute MI. Depending on the degree of blockage and which artery is blocked, the patient will have variable degrees of damage to the heart muscle that range from mild to severe. The most severe event would be a massive MI with immediate death.

TYPES OF PLAQUES: HARD AND SOFT PLAQUE

Plaques are of two major types: hard plaque and soft plaque. At the far left of Figure 15.7 is the early non-obstructive plaque with only the subendothelial lipid core plaque. This type of lesion progresses over time into the hard or soft plaques in the upper and lower portion of the figure. The hard plaque or stable plaque is shown at the top of the figure and consists of a thick fibrous cap and a small lipid core that is not as active or dangerous. This plaque is less likely to rupture into the lumen of the artery. It may cause obstruction to blood flow but does not usually rupture. It often has calcium deposits in the plaque that can be seen on a coronary artery calcium (CAC) scan. The thin cap or rupture-prone plaque is shown in the lower part of the figure with a very active and dangerous lipid core. To the right of this lesion is the coronary artery after it ruptures. The lipid core is released into the lumen of the coronary artery where a clot or thrombus forms (noted in red).

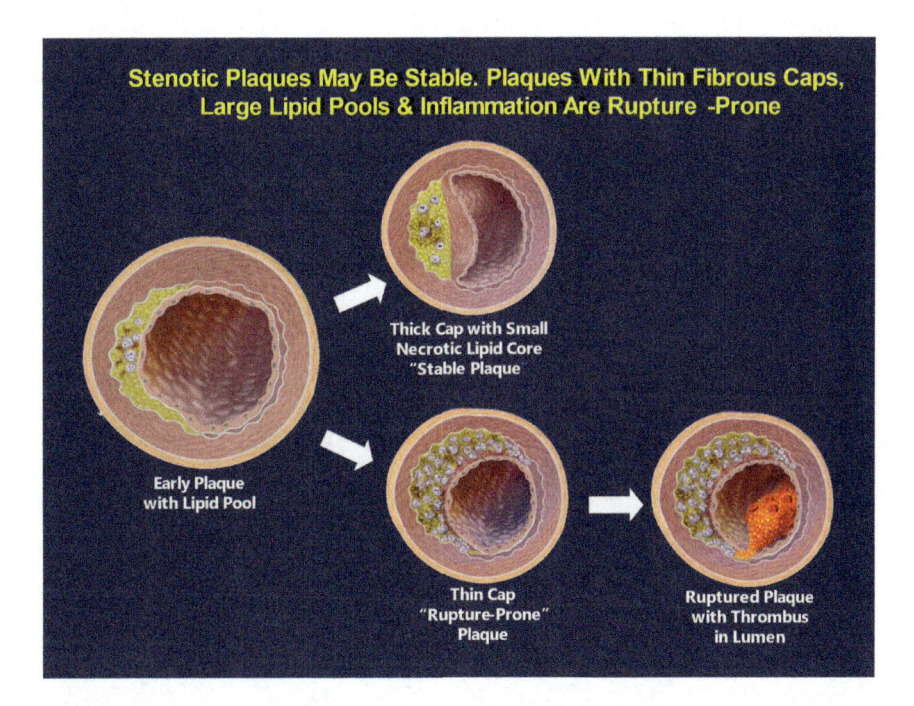

FIGURE 15.7 The anatomy and composition of hard and soft plaque.

WHAT MAKES A PLAQUE PRONE TO RUPTURE? THESE CHARACTERISTICS ARE LISTED BELOW (FIGURES 15.8, 15.9A, AND 15.9B)

1. Has a thin fibrous cap with decreased collagen production and increased collagen breakdown.
2. Contains a lipid core that is large and necrotic with lots of fats and oxidized cholesterol.
3. Has many activated vascular smooth muscle cells.
4. Contains a large number of inflammatory cells and T lymphocytes.
5. Has decreased normal vascular smooth muscle cells.
6. Contains less calcium.
7. The macrophage population of cells is both inflammatory and immunologic.

Most of the soft plaques do not limit coronary artery blood, and they often do not contain calcium; thus, they can be missed on a coronary artery calcium scan. There is often more coronary artery vasospasm with angina in patients with soft plaque. Also, during the early phase of atherosclerosis, a special imaging with a material called contrast dye of the coronary arteries, a coronary angiogram, arteriogram, or cauterization could show a normal lumen without obstruction or plaque but the patient clearly has atherosclerosis with the early plaque that still remains in the subendothelial layer of the coronary artery (Figure 15.3). An arteriogram of the coronary arteries is shown on the left side of Figure 15.3. Note that here is a normal lumen of the artery without any evidence of plaque or obstruction. However, in the

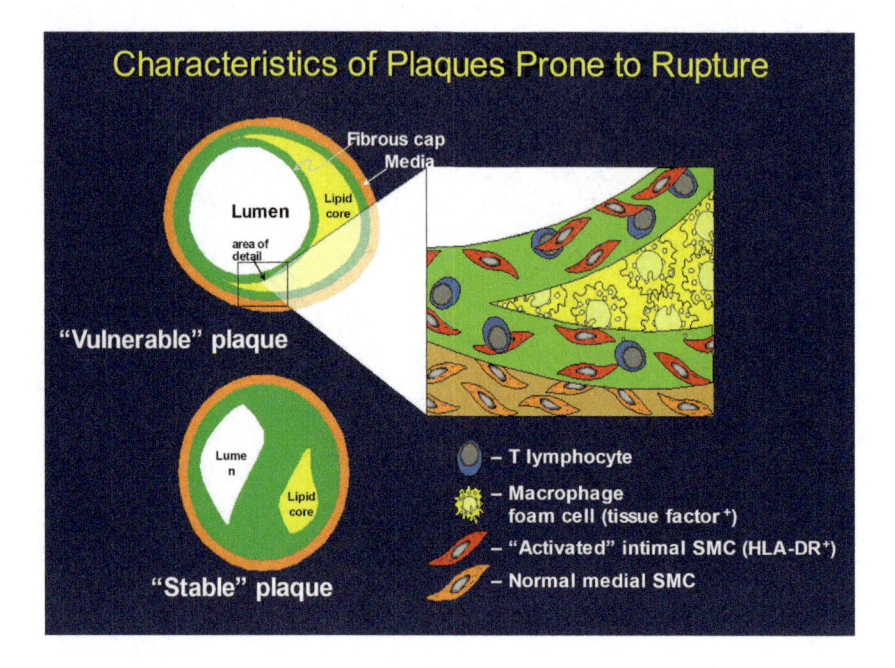

FIGURE 15.8 Characteristics of plaques prone to rupture.

What Is Plugging Your Heart Arteries? 163

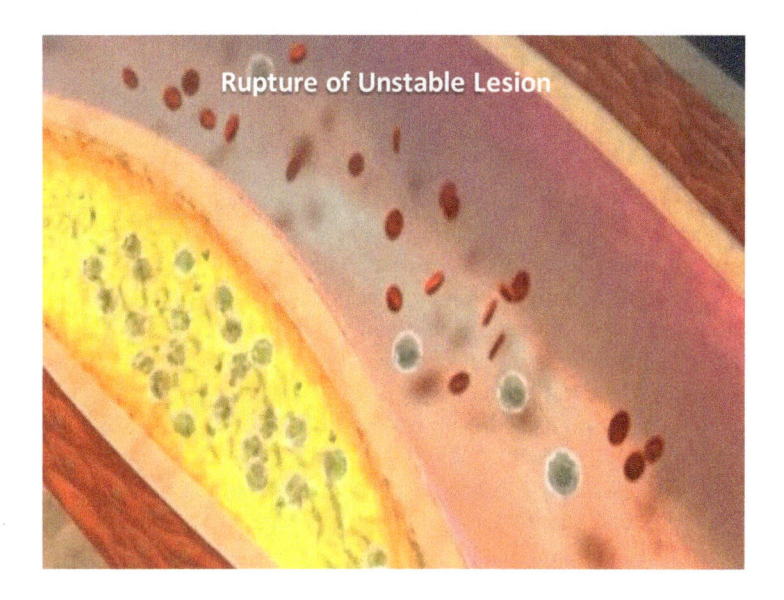

FIGURE 15.9A Rupture of the thin cap unstable plaque (yellow) with platelets, red cells, and clot (red).

75% of Heart Attacks are Caused by Unstable lesion Rupture and a Blood Clot in a Coronary Artery without previous angina symptoms

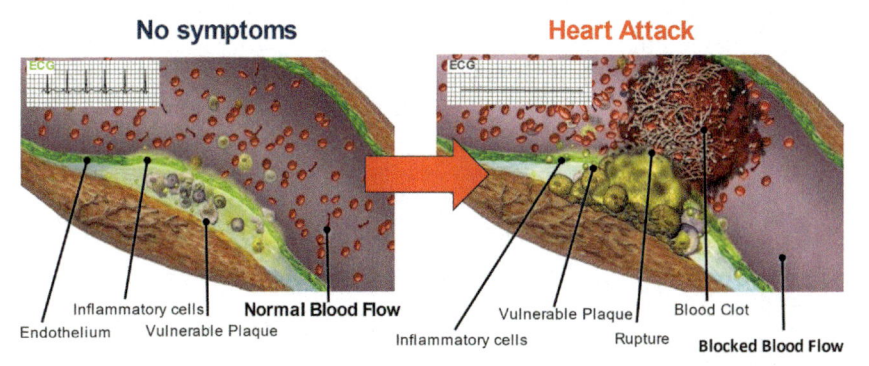

Detecting an unstable lesion before it ruptures identifies individuals with a heart attack risk and allows time to take action

FIGURE 15.9B Heart attack with the unstable thin cap rupture-prone plaque.

right panel of Figure 15.3, using an intravascular ultrasound (IVUS), the subendothelial material can be seen in white. This material will grow over time and cause an obstructive plaque.

Proximally located plaques are larger and have a greater necrotic core and fibrofatty plaque components, whereas distally located lesions were smaller and had a greater calcified plaque component. The proximally located lesions are a higher risk for rapid plaque progression.

PREVENTION AND TREATMENT OF PLAQUE

Plaque regression and change from vulnerable to non-vulnerable plaque are possible with various treatments such as aspirin, statins, blood pressure medications (such as angiotensin-converting enzyme inhibitors (ACEI) and angiotensin receptor blockers (ARB)), and high levels of the good high-density lipoprotein (HDL) that functions well. Various supplements, such as vitamin K2 MK 7, omega-3 fatty acids, niacin, luteolin (celery, green pepper, rosemary, carrots, oregano, oranges, olives), curcumin, quercetin, magnesium, grape seed extract, N acetyl cysteine (NAC), and aged garlic. In addition, some foods, such as pomegranate, broccoli, green tea, or green tea extract (EGCG) are effective. Two other natural proprietary compounds called NEO 40 and Arterosil will improve nitric oxide levels, dilate the coronary arteries, and stabilize and reduce plaque (Arterosil). The combination of many of these will stabilize plaque, reduce plaque volume, reduce the lipid core, and decrease CAC.

Most myocardial infarctions (75%) are caused by the unstable thin cap rupture-prone plaque. These patients often have no previous symptoms such as chest pain, angina, shortness of breath, or other cardiovascular symptoms (Figure 15.9B).

SUMMARY AND KEY TAKE AWAY POINTS

1. Myocardial infarction is due to plaque in the coronary arteries. The final stage of MI is a thrombosis in the coronary artery.
2. The primary types of plaque are hard plaque (or stable plaque), which is more obstructive, and soft rupture-prone plaque, which is not as obstructive. They have different compositions and clinical presentations.
3. About 75% of MIs are caused by soft rupture-prone plaque.
4. The formation of plaque starts as glycocalyx and endothelial dysfunction and progresses over decades through many stages to become complicated plaque.
5. Most MIs occur without any previous symptoms of chest pain, shortness of breath, or other cardiovascular symptoms.
6. Early subendothelial plaque is extraluminal and may not be detectable with a routine coronary arteriogram but can be seen with a CAC scan, MRA of the heart, IVUS, and CTA of the heart.
7. There are many supplements and medications that will reduce plaque size and stabilize the plaque by stabilizing and reducing the lipid core.

BIBLIOGRAPHY

1. Mori H, Torii S, Kutyna M, Sakamoto A, Finn AV, Virmani R. Coronary artery calcification and its progression: What does it really mean? *JACC Cardiovascular Imaging.* 2018 Jan;11(1):127–42.

2. Sharma B, Chang A, Red-Horse K. Coronary artery development: Progenitor cells and differentiation pathways. *Annual Review of Physiology.* 2017 Feb 10;79:1–19.
3. Arzani A. Coronary artery plaque growth: A two-way coupled shear stress-driven model. *International Journal for Numerical Methods in Biomedical Engineering.* 2020 Jan;36(1):e3293.
4. Nakahara T, Dweck MR, Narula N, Pisapia D, Narula J, Strauss HW. Coronary artery calcification: From mechanism to molecular imaging. *JACC Cardiovasc Imaging.* 2017 May;10(5):582–93.
5. Panh L, Lairez O, Ruidavets JB, Galinier M, Carrié D, Ferrières J. Coronary artery calcification: From crystal to plaque rupture. *Archives of Cardiovascular Diseases.* 2017 Oct;110(10):550–61.
6. Rognoni A, Cavallino C, Veia A, Bacchini S, Rosso R, Facchini M, Secco GG, Lupi A, Nardi F, Rametta F, Bongo AS. Pathophysiology of atherosclerotic plaque development. *Cardiovascular & Hematological Agents in Medicinal Chemistry (Formerly Current Medicinal Chemistry-Cardiovascular & Hematological Agents).* 2015 Apr 1;13(1):10–3.
7. Kochergin NA, Kochergina AM, Ganyukov VI, Barbarash OL. [Predictors of coronary plaque vulnerability in patients with stable coronary artery disease]. *Kardiologiia.* 2020 Nov 12;60(10):20–6.
8. Libby P, Pasterkamp G, Crea F, Jang IK. Andrews J, Psaltis PJ, Bartolo BAD, Nicholls SJ, Puri R. Coronary arterial calcification: A review of mechanisms, promoters and imaging. *Trends in Cardiovascular Medicine.* 2018;28(8):491–501.
9. Pérez Sorí Y, Herrera Moya VA, Puig Reyes I, Moreno-Martínez FL, Bermúdez Alemán R, Rodríguez Millares T, Fleites Medina A. Histology of atherosclerotic plaque from coronary arteries of deceased patients after coronary artery bypass graft surgery. *Clínica e Investigación en Arteriosclerosis.* 2019 Mar–Apr;31(2):63–72.

16 Nonobstructive Coronary Heart Disease and Coronary Artery Vasospasm

INTRODUCTION

Nonobstructive coronary heart disease (NO-CHD) is an atherosclerotic plaque in the coronary arteries that would not be expected to obstruct blood flow to the heart muscle or result in anginal symptoms but can cause chest pain, tightness, pressure, shortness of breath, or myocardial infarction (MI). Coronary artery vasospasm (CA-VS) is anginal chest pain, tightness, pressure, shortness of breath, or MI with completely normal coronary arteries and no visible plaque by coronary angiogram.

In NO-CHD, although such lesions are relatively common, occurring in 10–25% of patients undergoing coronary angiography, their presence has been characterized as "insignificant" or "not significant CHD" in the medical literature. However, this perception of NO-CHD may be incorrect, because prior studies have noted that the majority of plaque ruptures and resultant MIs arise from nonobstructive plaques as I explained in Chapter 15. In addition, coronary artery spasm with a temporary tightening (constriction) of the muscles in the wall of one of the coronary arteries can decrease or completely block blood flow to the heart muscle. This is likely due to the inflammation, oxidative stress, and vascular immune dysfunction around the nonobstructive plaque and low levels of nitric oxide in the artery (this will be discussed in more detail later in this chapter).

In patients with CA-VS, the lumen of the coronary artery has no visible or documented plaque by coronary angiography, but the artery may have intense constriction or vasospasm that obstructs blood flow and causes anginal chest pain, tightness, pressure, shortness of breath, and, sometimes, an MI. A coronary artery spasm is a temporary tightening (constriction) of the muscles in the wall of one of the coronary arteries which can decrease or completely block blood flow to sections of the heart (Figure 16.1).

If a spasm lasts long enough, you can have anginal chest pain and tightness and even an MI. Most chest pain or tightness lasts about 15 minutes. Unlike typical angina, which usually occurs with physical activity, coronary artery spasms often occur at rest, typically between midnight and early morning. However, chest pain with exertion also occurs. The other names for coronary artery spasm are Prinzmetal's angina, vasospastic angina, or variant angina.

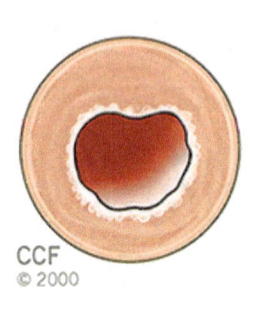

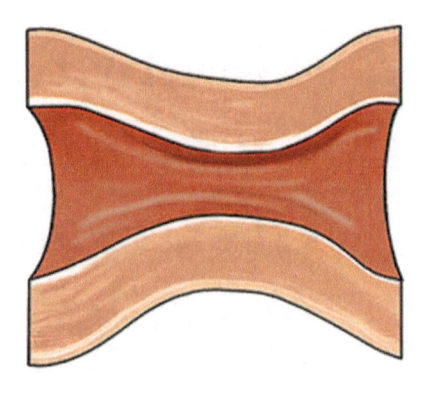

FIGURE 16.1 Coronary artery spasm showing coronary artery in cross section on the left and longitudinal section on the right. The wall of the artery (light red) is constricting into the lumen of the artery on the right (dark red). This indicates partial obstruction of the artery.

Many people who have coronary artery spasms do not always have the usual or common risk factors for CHD, such as high cholesterol and high blood pressure. But they are often smokers and may have diabetes mellitus. If coronary artery spasm results in a dangerously fast heartbeat (ventricular arrhythmia), it can result in sudden death.

WHAT CAUSES NO-CHD AND CA-VS?

In patients with angina and **NO-CHD or CA-VS,** the coronary artery spasm is due to a reduction in nitric oxide levels in the artery which increases the resistance in the artery and reduces dilation. Other factors such as vascular inflammation, oxidative stress, and immune dysfunction may also contribute. Depending on the severity and duration of the vasospasm, the patient may have anginal chest pain, tightness, pressure, shortness of breath, or even an MI. This is associated with a worse prognosis. The vasospasm is often triggered by extreme emotional stress, anxiety, anger, high blood pressure, fatigue, high levels of adrenaline or cortisol, tobacco use, exposure to cold, use of illegal stimulant drugs (such as amphetamines and cocaine), hyperventilation, or administration of provocative medications, such as acetylcholine, ergonovine, histamine, or serotonin.

In patients undergoing elective coronary angiography, NO-CHD compared with patients that had no apparent CHD, there is a significantly greater one-year risk of recurrent angina and MI. This was also true of patients with CA-VS in which one of the provocative tests was done with acetyl choline or ergonovine during coronary angiography.

In patients with CA-VS, the vascular spasm is most frequently observed in the left anterior descending artery (LAD) (90%) and is most frequently diffuse (62 %).

DIAGNOSIS OF CORONARY SPASM

To diagnose CA-VS, you may need to wear an ambulatory monitor for up to 48 hours. The monitor records your heart's electrical impulses, even during sleep. If you have chest pain, this usually will show up on the electrocardiogram (EKG) that indicates coronary spasm. However, not all patients show EKG changes during every episode.

To diagnose coronary spasm, doctors may prescribe a special type of test called an acetyl choline or ergonovine stress test. These are drugs that are injected into the coronary artery during a coronary angiogram. This can trigger coronary spasm, usually within minutes, at which point that is easily visualized. Then another medication is injected into the coronary artery to relieve the spasm. Your EKG is recorded before, during, and after the test.

TREATMENT

- **Nitrates** are used to prevent spasms, improve blood flow, and quickly relieve chest pain and tightness. These can be given in various forms, such as oral capsules, under the tongue, sprays, and transdermal or intravenously. Some are short acting and others are long acting.
- **Calcium channel blockers (CCBs)** relax the arteries, decrease the spasm, and increase coronary blood flow. These are used for high blood pressure and CHD treatment. The best CCBs are nifedipine, amlodipine, and diltiazem.
- **Statins** are medications that lower cholesterol but also may prevent spasms, increase nitric oxide, and reduce vascular inflammation.
- **Alpha-blockers**, especially phentolamine.
- **Natural compounds** that increase nitric oxide in the arteries, such as NEO 40 and Arterosil.

SUMMARY AND KEY TAKE AWAY POINTS

1. In addition to severe obstructive CHD, patients may have NO-CHD or CA-VS, both of which cause anginal chest pain, tightness, pressure, shortness of breath, and MI.
2. The underlying cause for both NO-CHD and CA-VS is a reduction in arterial nitric oxide and possible vascular inflammation, oxidative stress, and immune dysfunction.
3. The diagnosis is made with cardiac monitors and by injection of certain drugs into the coronary artery during a coronary angiogram.
4. The treatment is with nitrates, calcium channel blockers, statins, alpha-blockers, and some natural compounds that increase arterial nitric oxide.

BIBLIOGRAPHY

1. Montone RA, Niccoli G, Russo M, Giaccari M, Del Buono MG, Meucci MC, Gurguglione F, Vergallo R, D'Amario D, Buffon A, Leone AM, Burzotta F, Aurigemma C, Trani C, Liuzzo G, Lanza GA, Crea F. Clinical, angiographic and echocardiographic correlates of epicardial and microvascular spasm in patients with myocardial ischaemia and non-obstructive coronary arteries. *Clinical Research in Cardiology.* 2020 Apr;109(4):435–43.

2. De Vita A, Manfredonia L, Lamendola P, Villano A, Ravenna SE, Bisignani A, Niccoli G, Lanza GA, Crea F. Coronary microvascular dysfunction in patients with acute coronary syndrome and no obstructive coronary artery disease. *Clinical Research in Cardiology*. 2019 Dec;108(12):1364–70.
3. Konst RE, Damman P, Pellegrini D, van Royen N, Maas AHEM, Elias-Smale SE. Diagnostic approach in patients with angina and no obstructive coronary artery disease: emphasising the role of the coronary function test. *Netherlands Heart Journal*. 2021 Mar;29(3):121–8.
4. Najib K, Boateng S, Sangodkar S, Mahmood S, Whitney H, Wang CE, Racsa P, Sanborn TA. Catheter incidence and characteristics of patients presenting with acute myocardial infarction and non-obstructive coronary artery disease. *Cardiovascular Interventions*. 2015 Oct;86(Suppl 1):S23–7.
5. Hjort M, Lindahl B, Baron T, Jernberg T, Tornvall P, Eggers KM. Prognosis in relation to high-sensitivity cardiac troponin T levels in patients with myocardial infarction and non-obstructive coronary arteries. *American Heart Journal*. 2018 Jun;200:60–6.
6. Seitz A, Morár N, Pirozzolo G, Athanasiadis A, Bekeredjian R, Sechtem U, Ong P. Prognostic implications of coronary artery stenosis and coronary spasm in patients with stable angina: 5-year follow-up of the Abnormal Coronary Vasomotion in patients with stable angina and unobstructed coronary arteries (ACOVA) study. *Coronary Artery Disease*. 2020 Sep;31(6):530–7.

17 Women and Coronary Heart Disease (CHD)

INTRODUCTION

Coronary heart disease remains the leading cause of death in men and women in the US. Coronary heart disease (CHD) kills a higher percentage of women (55%) than men (43%). Over 300,000 women die each year from CHD or about one in every five female deaths. CHD is the leading cause of death for both African American and White women in the US. Among American Indian and Alaska Native women, CHD and cancer cause roughly the same number of deaths each year. For Hispanic, Asian, or Pacific Islander women, CHD is second only to cancer as a cause of death. About 1 in 16 women age 20 and older (6.2%) has CHD. About 1 in 16 White women (6.1%), Black women (6.5%), and Hispanic women (6%) and 1 in 30 Asian women (3.2%) have CHD.

Yet CHD is still considered a "disease of men". Many women are unaware that CHD is their main killer. In addition, there is a lack of awareness of CHD in women among doctors. At the time of presentation with CHD, women are ten years older than men, and at the time of their first MI, they are 20 years older. CHD increases in women after menopause.

Women and men with heart disease tend to differ in their presenting symptoms, diagnosis, and treatment as well as their overall prognosis. Women may have more atypical symptoms than men, such as burning in the chest, back pain, abdominal discomfort, nausea, or fatigue. Women seek medical help less often, which means they present later in the process of their CHD or MI. They are also less likely to have procedures and testing, such as coronary angiography which will delay effective treatment. There are also gender differences in those that undergo coronary artery bypass grafting (CABG). At the time of presentation of CHD, women are more likely to have comorbid factors, such as hypertension, abnormal lipids, diabetes mellitus, peripheral vascular disease; systolic and diastolic heart failure; and often need more urgent intervention due to a delayed presentation or diagnosis. The coronary arteries in women are smaller than those of men which makes surgical intervention with CABG, angioplasty PCTA, and stenting more difficult and is associated with a higher mortality rate.

PRESENTING SYMPTOMS OF CHD IN WOMEN

These symptoms of CHD in women may be vague and not as noticeable as the usual chest pain. Women tend to have blockages not only in their main coronary arteries but also in the smaller coronary arteries that supply the heart muscle, a condition

DOI: 10.1201/b22808-18

called "small vessel heart disease" or "coronary microvascular disease". Women have symptoms more often when resting or asleep. Emotional stress can play a role in triggering CHD symptoms or MI in women. Below are the usual symptoms:

- Discomfort, burning, pressure, or pain or tingling in the chest, arm, throat, and back that may occur at night, when sleeping, or with exertion.
- Shortness of breath with exertion.
- Dizziness or light-headedness with exercise.
- Sudden onset of fatigue.
- Nausea and vomiting.
- Indigestion, gas, and belching.
- Feelings of dread and anxiety.
- Sweating.
- Abdominal pain in the upper abdomen.

CHD RISK FACTORS FOR CHD IN WOMEN

CHD risk factors for women differ from men. Women with diabetes mellitus have 2.6 times the death rate from CHD compared to women without diabetes mellitus and a 1.8-fold risk compared to men with diabetes. Also, hypertension is associated with a two to three times increased risk of CHD in women. Reduced levels of high-density lipoprotein (HDL) cholesterol and triglycerides may be a better predictor of CHD in women compared to high levels of low-density lipoprotein (LDL) cholesterol.

Heart disease risk factors for women that are more common than in men include the following:

- Diabetes mellitus. Women with diabetes are more likely to have CHD than men with diabetes mellitus.
- Mental stress, anxiety, and depression.
- Smoking.
- Menopause. Low levels of estrogen after menopause pose a significant risk of developing CHD in the small coronary arteries.
- Pregnancy complications. High blood pressure or diabetes during pregnancy can increase the mother's long-term risk of both hypertension and CHD.
- Family history of early heart disease. This may be a greater risk factor in women than in men.
- Inflammatory diseases. Rheumatoid arthritis, systemic lupus, and other autoimmune diseases will increase CHD.

MENOPAUSE AND CHD IN WOMEN

After menopause in women who have undergone early natural menopause (before age 50) or a surgically induced menopause, a woman's risk of CHD increases. Estrogen reduces the risk of CHD. In other words, prior to menopause, the natural production of estrogen and progesterone is protective against CHD. After menopause, estrogen and progesterone replacement with bioidentical hormones reduces CHD. After menopause, CHD becomes more of a risk for women due to lower estrogen levels.

Lower levels of estrogen cause the following:

- Changes in the coronary arteries that may cause plaque and blood clots to form.
- Changes in the level of lipids (fats) in the blood will increase CHD risk. After menopause, levels of LDL (the "bad" cholesterol) increase, and levels of HDL (the "good" cholesterol) decrease. This leads to more inflammation and plaque formation.
- An increase in fibrinogen (a substance in the blood that helps the blood to clot). If these levels are elevated then coronary artery clots may form.

SUMMARY AND KEY TAKE AWAY POINTS

1. CHD and MI are the number-one causes of death in women in the US.
2. The symptoms of CHD and MI in women are often different than they are in men.
3. The CHD risk factors may also be different in women than in men.
4. Women are often diagnosed later than men and may have a poorer prognosis.
5. Women tend to have worse results from PTCA, stents, and CABG compared to men.
6. After menopause, the risk for CHD and MI increase due to a reduction in estrogen, increased inflammation, changes in blood lipids, and other factors.

BIBLIOGRAPHY

1. Greenwood BN, Carnahan S, Huang L. Patient-physician gender concordance and increased mortality among female heart attack patients. *Proceedings of the National Academy of Sciences.* 2018 Aug;115(34):8569–74.
2. Meyer MR, Bernheim AM, Kurz DJ. Gender differences in patient and system delay for primary percutaneous coronary intervention: current trends in a Swiss ST-segment elevation myocardial infarction population. *Eur Heart J Acute Cardiovasc Care.* 2019 Apr;8(3):283–290.
3. Wenger NK, Hayes SN, Pepine CJ, et al. Cardiovascular care for women: The 10-Q report and beyond. *American Journal of Cardiology.* 2013 Aug 15;112(4): S2.
4. Bai M-F, Wang X. Risk factors associated with coronary heart disease in women: a systematic review. *Herz.* 2020 Dec;45(Suppl 1):52–57.
5. Michael C. Coronary heart disease in women. *Journal of Family Practice.* 2014 Feb;63(2 Suppl): S9–14.

18 Coronary Heart Disease Risk Factors
The Traditional Top Five Risk Factors and the Other 400

INTRODUCTION

There are many CHD risk factors that I will review and discuss in this chapter. You hear mostly about the top five CHD risk factors that include hypertension (high blood pressure), dyslipidemia (abnormal fats and cholesterol), high blood glucose, diabetes mellitus, obesity, and smoking. However, there are other CHD risk factors that you need to know about and have these measured by your doctor. I will divide these into three groups:

1. Top five CHD risk factors.
2. Top 25 modifiable CHD risk factors.
3. The other 400 CHD risk factors.

TOP FIVE CHD RISK FACTORS

The top five CHD risk factors need to be better defined as I will explain below. The older definitions and criteria are obsolete and do not evaluate your CHD risk accurately and therefore do not tell you the best prevention and treatment options for CHD.

HYPERTENSION

A 24-hour ambulatory blood pressure monitor (ABM) is the gold standard in blood pressure (BP) measurement. Measurements of nocturnal BP and dipping status (normal is a 10% reduction from the average daytime BP during the night), BP surges, BP load (normal is <140/90 mm Hg in 15% of the total BP measurements), and BP variability are superior to random (medical) office BP readings or home BP readings as a predictor of CHD risk. Excessive dipping is associated with an increased risk of ischemic stroke and reverse dipping is associated with an increased risk of intracerebral (brain) hemorrhage. Nocturnal BP is clinically more important than daytime BP because it drives CHD and cardiovascular damage (27/15 mm Hg difference is optimal). Morning BP surges (both the level and rate of BP rise) increase the risk of ischemic stroke, MI, and CHD. This often means that arterial nitric oxide has decreased. The items that should always be considered when evaluating BP include the following:

1. Normal BP is 120/80 mm Hg, but there is a continuum of risk for CHD starting at 110/70 mm Hg.
2. Each increase of BP of 20/10 mm Hg doubles CHD.
3. Before 50 years of age, diastolic BP is a better predictor of CHD risk.
4. After 50 years of age, systolic BP is a better predictor of CHD risk.
5. Twenty-four-hour ABM is more accurate than office BP measurements and should be the standard of care for defining BP, CHD risk, and treatment options, as well as the type of BP medication, the dose, and the time of administration of antihypertensive drugs.
6. Mercury sphygmomanometers are preferred, as they are more accurate and not subject to the wear and tear of needle BP monitors. Electronic arm units are accurate if done correctly and validated. Wrist or finger monitors are not as accurate and should not be used as the basis for a hypertension diagnosis or management until better technologies become available.

DYSLIPIDEMIA

Dyslipidemia should be evaluated using advanced lipid profiles to determine treatment and predict individual CHD risk more accurately. An advanced lipid profile can be obtained from almost any lab. This will include the following:

- Total cholesterol.
- LDL-C.
- LDL particle number (LDL-P), which drives CHD risk.
- LDL sizes—the small dense type B LDL is more atherogenic compared to the large type A LDL.
- Modified LDL (oxidized, glycated, glyco-oxidized, and acetylated).
- Apolipoprotein (APO) B and APO A.
- Lipoprotein (a) (Lp(a)).
- Total high-density lipoprotein (HDL-C).
- HDL particle number (HDL-P).
- HDL size and HDL mapping with the five types of HDL (pre-beta, alpha 1–4).
- Dysfunctional HDL (HDL that does not perform all of its normal functions such as reverse cholesterol transport (RCT), etc.).
- Myeloperoxidase (MPO).
- APO C-III.
- Very low-density lipoprotein cholesterol (VLDL-C) and triglycerides.
- Large VLDL.
- VLDL-P (VLDL particle number).
- Remnant particles.

The primary CHD risk related to LDL-C is LDL-P, APO B, and dense small LDL. HDL-P is more protective against CHD and more predictive than total HDL for reducing CHD risk. The larger HDL type 2b is also an important protective mechanism. Although the smaller pre-beta HDL are more efficient at reverse cholesterol transport. The larger HDL may be more protective against CHD, as it will decrease inflammation, oxidative stress, and immune dysfunction. However, HDL size is now more controversial as to its relative protection for CHD. Patients who have an HDL of 85 mg/dL often have dysfunctional HDL, which does not perform its normal functions (such as reverse cholesterol transport), does not have the same antioxidant, anti-inflammatory, and anti-immune properties, and may not be protective. HDL function is the most important measurement for HDL in predicting CHD. The very low-density lipoprotein (VLDL), especially large VLDL, TGs, and remnant particles, is very atherogenic and thrombogenic.

DYSGLYCEMIA, HYPERGLYCEMIA, INSULIN RESISTANCE, METABOLIC SYNDROME, AND DIABETES MELLITUS

A fasting blood sugar (FBS) of >75 mg/dL increases CHD by 1% for each 1 mg/dL increase in FBS and induces endothelial dysfunction. The current upper limit of blood sugar by most labs is 100 mg/dL which is too high. A two-hour glucose tolerance test (GTT) >110 mg/dL increases CHD by 2% for each 1 mg/dL >110 mg/dL. The current definition of an abnormal two-hour GTT is >140 mg/dL, which is too high and not accurate. High levels of blood insulin (hyperinsulinemia) are an independent risk factor for CHD. Calculating a Homeostatic Model Assessment of Insulin Resistance (HOMA-IR) score will provide additional insight into the clinical presence of insulin resistance and metabolic syndrome. Multiplying the FBS by the fasting insulin level and dividing by 405 will give an excellent estimate of insulin resistance by the HOMA-IR. A normal HOMA-IR is <1.0, mild insulin resistance is 1.0–2.0, and diabetes mellitus is over 2.0. Diabetes mellitus is defined as three FBS in three separate blood samples over 125 mg/dL and a hemoglobin A1C (HbA1C) over 7%. The HbA1C is composed of the post-meal blood glucose (2/3) and fasting glucose (1/3). In men and women without DM between 50 and 75 years of age, HbA1C is an independent risk for CHD as a continuous variable starting with HbA1C as low as 5.1%. For each increase of HbA1C of 1% the risk for CHD increases by 40% in men and 240% in women.

OBESITY AND BODY FAT

Body weight and body mass index (BMI) correlate with CHD and MI. However, the total body fat and especially belly fat (visceral fat) correlates with CVD, CHD, and MI much better. Body fat and regional fat can be measured with a machine called body impedance analysis (BIA) (Figure 18.1). Normal body fat is 22% or less for women and 16% or less for men. At a total body fat (TBF) of 29%, most men will have metabolic syndrome. At a TBF of 37%, most women will have metabolic syndrome.

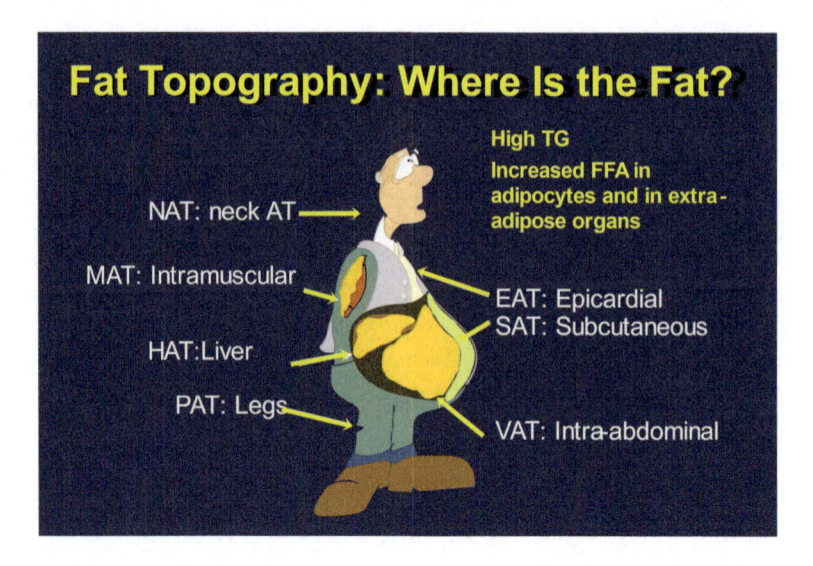

FIGURE 18.1 The types and location of the various fats. AT is adipose or fat tissue.

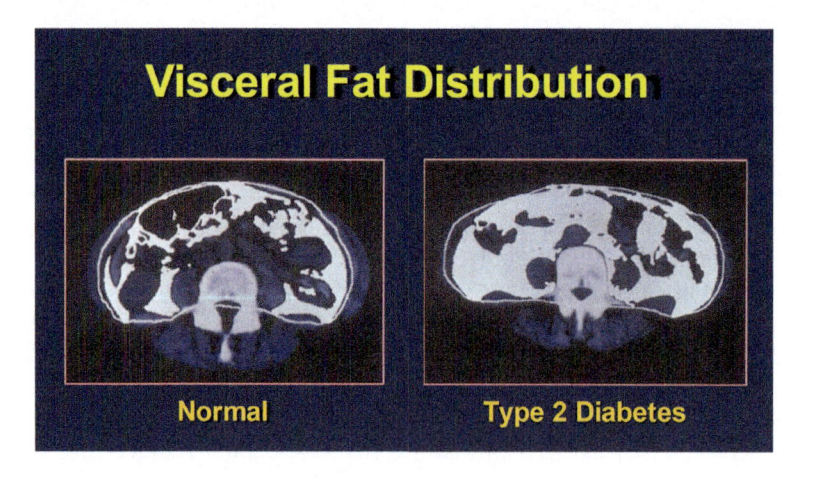

FIGURE 18.2 Distribution of visceral fat shown in white on an abdominal CT scan in a normal patient and one with Type 2 diabetes.

Visceral or abdominal fat has the highest correlation with CHD and MI, as it produces over 45 substances called adipokines (Figure 18.2). These adipokines cause vascular inflammation, oxidative stress, immune dysfunction, thrombosis, high blood pressure, high LDL cholesterol (with low HDL and high triglycerides), high blood sugar, diabetes, hyperinsulinemia, as well as other CHD risk factors, and increases the risk for CHD and MI. Reduction in body and visceral fat while preserving the lean muscle mass will correct many of the CHD risk factors and the CHD risk.

Coronary Heart Disease Risk Factors

SMOKING

Tobacco smoking is the leading preventable cause of death in the United States and an important cause of CHD and MI. Smoking raises your risk of getting CHD and dying early from CHD. Apo E 4 carriers have the highest risk of CHD with smoking. Carbon monoxide, nicotine, and other substances in tobacco smoke can promote atherosclerosis and trigger symptoms of CHD.

Smoking causes the following problems:

- Inflammation, oxidative stress, and increased **high-sensitivity C-reactive protein** (HS-CRP).
- The platelets in your blood clump together more easily and are more likely to form clots. Clumping platelets can then block your coronary arteries and cause an MI.
- Precipitate spasms occur in your coronary arteries, which reduces the blood flow to your heart and induces anginal chest pain, shortness of breath, or MI.
- Triggers irregular heartbeats (arrhythmias).
- Lowers HDL cholesterol, which is protects against CHD.
- Reduces the amount of oxygen that can be carried by red blood cells to all tissues and to the heart.
- Produces constriction of the systemic arteries and causes hypertension.
- Increases fibrinogen in the blood, making it thicker and more likely to clot.

Smoking also affects those around you. Secondhand smoke increases other people's risk of CHD. Smoking cessation reduces CHD and overall cardiovascular morbidity and mortality rates rapidly after you stop smoking.

TOP 25 MODIFIABLE CHD RISK FACTORS

Measuring only the top five CHD risk factors is not sufficient to determine your CHD and MI risk. It is mandatory that your top 25 modifiable CHD risk factors be properly evaluated and effectively treated. In addition, new definitions and testing of the top five risk factors is also now recommended. All of these are reviewed below.

Following are the top five CHD risk factors:

1. **Hypertension**. In addition to office and home BP monitoring, the new gold standard to evaluate BP and CHD risk is the 24-hour ambulatory blood pressure monitoring (ABM) discussed in the Hypertension section of this chapter.
2. **Dyslipidemia**. An advanced blood lipid analysis that measures all the different lipid particle numbers and sizes as well a test to determine the function of HDL is much more accurate than the old standard lipid profiles. See the Dyslipidemia section in this chapter.

3. **Hyperglycemia, metabolic syndrome, insulin resistance, and diabetes mellitus**. Evaluate all blood tests to determine the fasting glucose, two-hour GTT, insulin levels, and HbA1C. See this section in this chapter.
4. **Obesity**. Measure the total body fat, regional fat, visceral fat, and lean muscle with the BIA. See the Obesity section in this chapter.
5. **Smoking**. Must be stopped immediately. Also stop all e-cigarettes and vaping. See the Smoking section in this chapter.

Following are the other top 20 CHD risk factors:

Hyperuricemia

Hyperuricemia (high uric acid) increases risk of hypertension, endothelial dysfunction, metabolic syndrome, stroke, heart failure, high coronary artery calcium, CHD, MI, thrombosis, fat storage, insulin resistance, metabolic syndrome, diabetes mellitus, dyslipidemia, sympathetic nervous system overactivity, and chronic kidney disease. The risk for CHD, MI, and hypertension is increased three to five times proportional to the blood uric acid level. Hyperuricemia is present in 25–47% of hypertensive patients, and 50% of these are on diuretics. Hyperuricemia causes resistant hypertension in the elderly. The risk of hypertension in all patients starts at 6 mg/dL in men and 5 mg/dL in women. Each increase in 1 mg/dL in blood uric acid increases the risk of hypertension by 13%. Humans lack an important enzyme called uricase that converts uric acid to a soluble form. The uric acid level should be kept below 5 mg/dL. Uric acid can be both antioxidant and pro-oxidant depending on the blood level and the location in or outside of the cell. At high levels, uric acid will increase oxidative stress, inflammation, immune dysfunction of blood vessels, increase artery stiffness, lower nitric oxide, and increase sodium sensitivity and blood pressure. Hyperuricemia may be genetic, drug-related (diuretics, beta blockers, aspirin), or due to obesity, alcohol consumption, increased red meat in the diet, fructose intake, insulin resistance, metabolic syndrome, diabetes mellitus, obesity, dehydration, chronic kidney disease, some types of seafood, small bowel disease, and cancer. There are many ways to reduce uric acid in addition to looking at all of these causes and removing or treating them. For example:

- Allopurinol lowers BP 3.3/1.3 mm Hg, and improves endothelial dysfunction, arterial stiffness, CHD, and heart failure.
- Estrogen increases the urinary excretion of uric acid.
- Folate also lowers uric acid.
- Losartan lowers uric acid. This is a drug used for hypertension.

Renal Disease

Chronic kidney disease (CKD) is associated with CHD risk that is equal to or greater than other established very high-risk conditions. In fact, CKD is 2 times the risk compared to diabetes, metabolic syndrome, and smoking. The glomerular filtration rate (GFR), blood creatinine and cystatin C, the urinary protein (proteinuria), and microalbumin in the urine (microalbuminuria) all correlate as independent risk factors for CHD.

Albumin in the urine (microalbuminuria) is one of the earliest abnormalities in the vascular system and kidney that reflects endothelial dysfunction and increased vascular permeability. Microalbuminuria has a high correlation with progression to more proteinuria, renal disease, CHD, MI, heart failure, and stroke. A single urine sample with a blood creatinine can measure an important lab called the albumin creatinine ratio (ACR). An increased ACR predicts a linear relationship between coronary microvascular dysfunction and CHD. There is a linear relationship of ACR starting at 5 mg/gm. There are many BP drugs that can improve both CKD and microalbuminuria, such as angiotensin-converting enzyme inhibitors (ACEI) and angiotensin receptor blockers (ARB). In addition, omega-3 FA and R lipoic acid may reduce microalbuminuria.

Elevated Fibrinogen
An elevated blood fibrinogen is directly associated with increased thrombotic rate and increases the risk of CHD, MI, stroke, and overall mortality by 81%. For each 50 mg/dL increase in blood fibrinogen, there is a 30% increase in CHD. Plasma viscosity, or the thickness of the blood, correlates with fibrinogen and is an independent CHD risk factor. Fibrinogen is an acute and chronic phase reactant that is made in the liver. It increases plaque growth, platelet adhesion, inflammation, atherosclerosis, and cholesterol synthesis. The blood levels should be below 380 mg/dL but the lowest risk is below 288 mg /dL. See Figure 18.3.

Elevated Serum Iron
Iron can be measured in the blood by serum iron and a ferritin which is the liver storage form of iron. Both are important and predict increased risk for CHD and other medical problems. The blood iron is bound to proteins that transport it and is called the total iron-binding capacity (TIBC). So, a high serum iron and a high saturation are bad. Below are some of the problems that are associated with an increase in iron levels.

1. Increased risk of CHD, MI, and carotid artery disease.
2. Increased iron levels may be genetic due to the HFE C282Y mutation and is called hemochromatosis.
3. Enhanced iron mediates oxidative stress, LDL oxidation, contributes to dyslipidemia, endothelial dysfunction, atherosclerosis, and CHD.
4. High iron is related to the severity of perfusion, blood flow, and functional abnormalities of coronary arteries. Although there may be an obstruction with plaque, most of the time iron causes coronary microvascular disease with angina and endothelial dysfunction of the coronary arteries.
5. Ferritin measures iron stores and CHD risk. Ferritin levels over 200 µg/L increase CHD and MI risk by 2 times.
6. Insulin resistance and hyperglycemia and DM are increased.
7. Increases LDL and decreases HDL.
8. Nonalcoholic fatty liver disease (NAFLD).

FIBRINOGEN

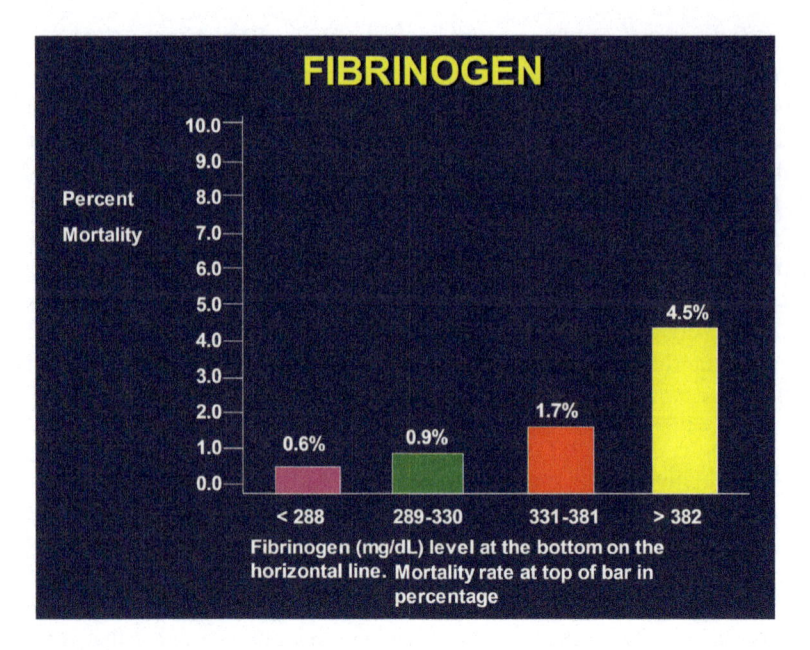

Fibrinogen (mg/dL)	Percent Mortality
< 288	0.6%
289-330	0.9%
331-381	1.7%
> 382	4.5%

Fibrinogen (mg/dL) level at the bottom on the horizontal line. Mortality rate at top of bar in percentage

FIGURE 18.3 Fibrinogen increases total mortality depending on the blood level. The higher the level the greater the death rate. Levels should be below 380 mg/dL, but the lowest risk would be below 288 mg/dL.

9. Increased mitochondrial dysfunction.
10. Oxidative stress is increased.
11. Inflammation is increased.

Lowering the iron burden with professional chronic phlebotomy improves cardiovascular outcomes of CHD, MI, stroke, and life expectancy. A serum ferritin level of 76.5 ng/ml has the lowest event rate for CHD.

Trans fatty acids and refined carbohydrates. Avoid TFA and decrease refined carbohydrates to less than 25 grams per day. See this section in Chapter 12 Part I on Nutrition.

Low dietary omega 3 fatty acids. Increase dietary omega 3 FA until the blood omega 3 index is over 8. See this section in Chapter 12 Part I on Nutrition.

Low dietary potassium and magnesium with high sodium intake. Increase dietary potassium and magnesium and decrease dietary sodium. See this section in Chapter 12 Part I on Nutrition.

Inflammation in the arteries. Decrease all inflammation blood markers to normal. See Chapter 6.

Increased vascular oxidative stress and decreased oxidative defense. Reduce oxidative stress and increase oxidative defense. See Chapter 6.

Increased vascular immune dysfunction. Lower vascular immune markers to normal. See Chapter 6.

Coronary Heart Disease Risk Factors

Lack of Sleep

Short sleep duration of less than six hours is an independent risk factor for CHD, MI, hypertension, heart failure, diabetes mellitus, obesity, metabolic syndrome, silent cerebral infarcts and future strokes, and increased mortality. Prolonged sleep if over ten hours increases CVA risk also. Eight hours appears to be the perfect sleep duration for most people to prevent CHD, MI, and these other events.

Lack of Exercise

Physical inactivity will increase CHD risk. The exercise recommendations are discussed in detail in Chapter 21 on aerobic and resistance training with the ABCT exercise program. Please read that chapter for details.

- 4200 kilo joules (17,566 KC) per week of aerobic exercise to achieve maximal CHD reduction.
- Do not exercise for more than one hour per day. 40 minutes resistance and 20 minutes aerobic training is recommended.
- Exercise at least four days per week.
- Sitting for long periods of time is equivalent to smoking. A study in the British Journal of Sports Medicine found that those who sit (for example, watching TV) for an hour reduce their lifespan by 22 minutes, whereas smokers shorten their lives by 11 minutes on average per cigarette. One study in Diabetologia, the journal of the European Association for the Study of Diabetes, confirmed the danger by comparing those who spent the most time sitting versus those who sat the least. Those who sat the most faced a 112% increase in diabetes, a 147% increase in death from cardiovascular events, a 90% increase in death from cardiovascular causes, and a 49% increase in death from all causes.

Stress, Anxiety, and Depression

Chronic stress, anxiety, and depression increase the risk of CHD and MI though an imbalance of the autonomic nervous system (ANS) with increased sympathetic nervous system (SNS) activity. See Chapter 14.

Homocysteinemia

Homocysteine is a sulfur-based amino acid derived from the plant and animal-based amino acid called methionine. The metabolism is very complex and involves many biochemical pathways and many vitamins via methylation (50%) and transulfuration (50%). I will try to simplify the relationship of elevated homocysteine levels to CHD below. See also Figure 18.4.

- There is a continuum of risk for CHD starting at levels of 5 micromoles/L, but the greatest risk starts at 12 mm/L or more.
- Elevated homocysteine blood levels are associated with oxidative stress, inflammation, inhibition of the important enzyme glutathione peroxidase (GPx), arterial damage, endothelial dysfunction, thrombosis, platelet aggregation, stroke, neurodegenerative disease, CHD, and renal disease.

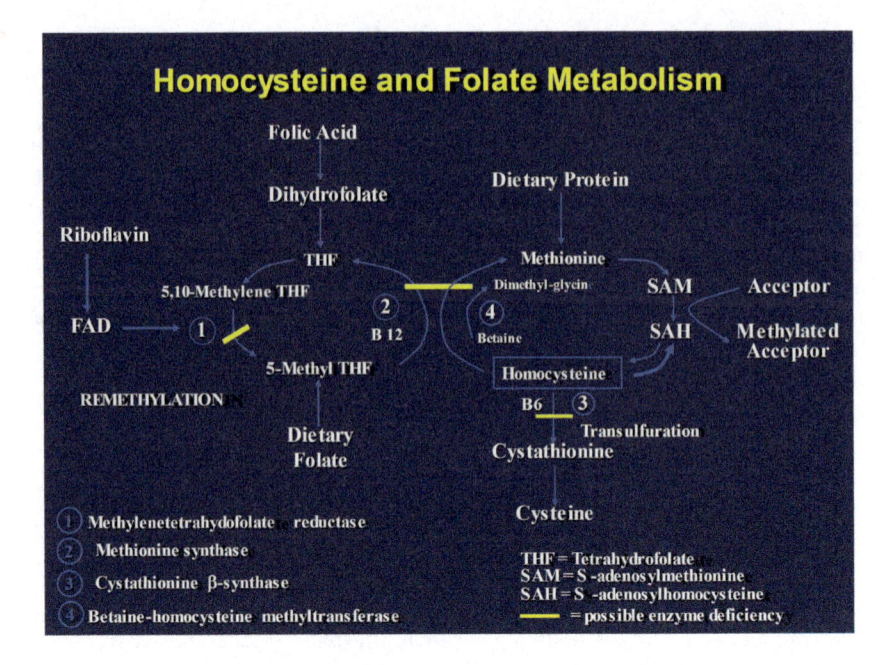

FIGURE 18.4 Homocysteine and folate metabolism.

- Many factors are thought to raise levels of homocysteine. These include a poor diet, smoking, high coffee consumption, increased alcohol intake, some prescription drugs, such as cholestyramine, metformin, nicotinic acid (niacin), and fibric acid derivatives (used to lower triglycerides and cholesterol), and diseases, such as diabetes, rheumatoid arthritis, kidney disease, and poor thyroid function.
- The treatment of elevated homocysteine involves B vitamins (vitamins B6, B2, B12), methylated folic acid, and several amino acids (S adenosyl methionine SAME, betaine, serine, and glycine)—and possibly vitamin C, vitamin E, and N acetyl cysteine (see Figure 18.4).
- Other treatments include genetic evaluation for enzyme defects and removing or treating any causes as stated above.

Subclinical Hypothyroidism
The thyroid gland, which resides in the lower neck, produces the hormones thyroxine (or T4) and tri-idothyronine (or T3) which control many metabolic processes in the body. It is one of the most important endocrine glands. An underactive thyroid (hypothyroidism) that does not produce enough T4 or T3 is associated with CHD. Blood tests will identify thyroid problems and include total T4, free T4, free T3, thyroid stimulating hormone (TSH) (from the pituitary gland in the brain), thyroid-binding proteins, such as thyroid binding globulin (TBG), and thyroid antibodies. Once identified, there are many supplemental thyroid medications that can

be administered to correct the thyroid function back to normal. Hypothyroidism is associated with these conditions:

- CHD, MI, heart failure, and general atherosclerosis.
- Heart rhythm problems, such as ventricular tachycardia, ventricular fibrillation, and atrial fibrillation.
- Sudden death.
- Hypertension and dyslipidemia.
- Homocysteinemia.
- Endothelial dysfunction.
- Obesity.
- Metabolic syndrome, insulin resistance, and diabetes.
- Increased carotid disease.
- Arterial stiffness.

The TSH is one of the earliest blood abnormalities seen with primary hypothyroidism of the thyroid gland. A TSH over 2.5 mIU/ml is considered abnormal. There are normal values in the lab for each of the thyroid tests mentioned and all should be measured.

Hormonal Imbalances in Both Genders

There is evidence that declining sex hormones in both genders will increase the risk of CHD and MI. There is also evidence that the appropriate and optimal replacement of sex hormones in the correct clinical setting will reduce the risk of CHD and MI. The lab tests for men should include total testosterone, free testosterone, and sex hormone-binding globulin (SHBG). The lab tests for women should include estrone, estradiol, estriol, progesterone, follicle-stimulating hormone (FSH), and luteinizing hormone (LH). If treatment is required, then bioidentical hormone replacement (BIHRT) is preferred. You should consult with a specialist in hormone replacement or an endocrinologist. A gynecologist is also suggested for women.

Chronic Clinical or Subclinical Infections

All types of organisms including bacteria, viruses, parasites, tuberculosis, and fungi (commonly called microorganisms) can increase the risk of CHD and MI. This association is true for both active, previous, or inactive infections. This is called the lifetime pathogenic burden of infections. There is clearly a link between infectious burden, history of infections, active infections, and atherosclerosis. The greater the number of microorganisms to which one displays a previous immune experience (reactive T cells or IgG antibody levels), the greater is the CHD. This pathogenic burden of various microorganisms has a significant correlation with endothelial dysfunction, coronary artery spasm, the presence coronary artery plaque and the severity of CHD as defined by coronary artery calcification (CAC) scan and coronary arteriograms. I will now outline the present relationships and findings of all of these infections and the risk of CHD and MI.

- Individual microorganisms also have significant correlations with CHD, including periodontal microbes, herpes simplex virus (HSV), cytomegalovirus (CMV), H. Pylori, Chlamydia Pneumoniae, Hepatitis A, B, C, HIV, and Epstein Barr virus (EBV), and all bacteria and others as defined by the production of IgG and IgM antibodies. *H. pylori*-positive patients with endothelial dysfunction (ED) had resolution of the ED after treatment and eradication of the *H. pylori* infection.
- The DNA of many viruses and bacteria can be detected in the coronary arteries and in the plaque at autopsy.
- In the presence of a chronic bacterial infection (periodontal or gum disease being the most common), mononuclear cells, such as T cells, infiltrate the gum tissue, gobble up the corresponding microbes, and digest them. If that mononuclear cell happens to traverse a coronary artery containing an active plaque, it may be nonspecifically pulled into the lesion. There it may display gum microbes which induce inflammation and an immune response that activates the plaque to an unstable form that is prone to rupture.
- There is a strong association between an acute myocardial infarction and acute bacterial and viral infections. This risk is highest the first few weeks after the infection but remains elevated for a year after severe infection. This is mediated by trained immunity.

Micronutrient Deficiencies

Almost any micronutrient deficiency can increase the risk of CHD and MI through a wide variety of mechanisms. The list is extensive and includes all vitamins, antioxidants, minerals, various enzymes, amino acids, and other nutrients (Table 18.1). Various supplements such as vitamin K2 MK 7, omega-3 fatty acids, niacin, luteolin, (celery, green pepper, rosemary, carrots, oregano, oranges, olives), curcumin, quercetin, magnesium, grapeseed extract, NAC (N acetyl cysteine), and aged garlic reduce CHD risk, plaque formation, and CAC. In addition, foods such as pomegranate, broccoli, green tea, or green tea extract (EGCG) will be effective. Two other natural proprietary compounds called NEO 40 and Arterosil will improve nitric oxide levels, dilate the coronary arteries, stabilize plaque, and reduce plaque (Arterosil) The combination of many of these will stabilize plaque, reduce plaque volume, reduce the lipid core, and decrease the coronary artery calcium (CAC).

Heavy Metals

All heavy metals such as mercury, lead, arsenic, cadmium, and iron increase the risk of CHD and MI through many mechanisms, such as inflammation, oxidative stress, and mitochondrial damage. These should be measured in both blood and urine in a baseline state and the repeated in a provoked method using chelators. If any are increased, then the source must by identified and removed with oral or intravenous chelation therapy.

Environmental Pollutants

Any type of environmental pollutant in the water, air, and food, such as pesticides, organocides, particulate matter, or other sources will increase the risk of CHD and

TABLE 18.1
List of Micronutrients

Trace minerals
- Zinc
- Calcium
- Copper
- Chromium
- Selenium
- Manganese
- Molybdenum
- Cobalt
- Iron
- Fluorine
- Iodine
- Magnesium
- Phosphorous
- Boron
- Potassium
- Sodium
- Chloride

Vitamins
- **Vitamin B complex**
  - **Vitamin B$_1$ (thiamin)**
  - **Vitamin B$_2$ (riboflavin)**
  - **Vitamin B$_3$ (niacin)**
  - **Vitamin B$_5$ (pantothenic acid)**
  - **Vitamin B$_6$ group:**
    - **Pyridoxine**
    - **Pyridoxal-5-Phosphate**
    - **Pyridoxamine**
  - **Vitamin B$_7$ (biotin)**
  - **Vitamin B$_9$ (folate)**
  - **Vitamin B$_{12}$ (cobalamin)**
  - **Choline**
- **Vitamin A**
- **Vitamin D**
- **Vitamin C (ascorbic acid)**
- **Vitamin K**
  - **Vitamin K$_1$ (phylloquinone)**
  - **Vitamin K$_2$ (menaquinone)**
  - **Vitamin K$_3$ (menadione)**
- **Vitamin E (tocopherols and tocotrienols)**
- **Bioflavonoids**
- **Lycopene**
- **Luteolin**
- **Vitamin A (e.g., retinol (see also–provitamin A carotenoids))**
- **Vitamin C (Ascorbic acid)**
- **Carotenoids (not accepted as essential nutrients)**
  - **Alpha carotene**
  - **Beta carotene**

(Continued)

TABLE 18.1 (CONTINUED)
List of Micronutrients

- Cryptoxanthin
- Lutein
- Lycopene
- Zeaxanthin

Amino acids
- Lysine
- Histidine
- Methionine
- Phenylalanine
- Threonine
- Tryptophan
- Asparagine
- Aspartic acid/Aspartate
- Alanine
- Arginine
- Cysteine/Cystine
- Glutamine
- Glutamic acid/Glutamate
- Glycine
- Proline
- Serine
- Tyrosine
- BCAAs (Isoleucine, Leucine, Valine)

Others
EGCG
Omega-3 fatty acids
Lipoic acid
Co enzyme Q 10
Inositol
Curcumin
Quercetin
Sulfur
Carnitine

MI. These can be measured by specialty labs with blood and urine samples. These are more difficult to treat once found.

THE OTHER 400 CHD RISK FACTORS

In addition to the top 5 and the top 25 CHD risk factors there are about 400 new and emerging CHD risk factors that should be evaluated and measured in certain cases based on history, clinical exam, family history, known CHD risk factors, known CHD or previous MI and other labs. This list is very extensive so I will only list these in Table 18.2. Due to the limited space in this book, it is not possible to discuss these in detail.

TABLE 18.2
Partial List of Some of the Additional 400 CHD Risk Factors

Hypertension
- Dyslipidemia
- Diabetes mellitus
- Smoking
- Insulin resistance and hyperinsulinemia
- LVH and diastolic dysfunction
- Stress, anxiety and depression
- Sleep Apnea
- Chronic lack of sleep or too much sleep
- Type A personality and aggressive behavior
- Male gender
- Hyperuricemia over 5.0
- Caffeine use depending on CYP 1 A 2 genotypes
- Age
- Elevated HS-CRP (high sensitivity C-reactive protein) and other markers of vascular inflammation
- Any cause for inflammation increases CVD
- Excessive intake of alcohol: may increase risk; however, small quantities of alcohol increase HDL and have anti-inflammatory or other effects that are associated with decreased risk of CHD. Alcohol has a U-shaped curve for CV risk. 20 grams per day for men and 10 grams per day for women.
- Obesity
- Diet soft drinks and regular soft drinks
- Renin and angiotensin II hypertension hormones are vasculotoxic and induce LVH
- Sympathetic nervous system overactivity, elevated catecholamines, vasculotoxic (NE and epinephrine [EPI] levels.
- Elevated blood viscosity.
- Increased oxidative stress and /or decreased oxidative defense.
- Increased vascular immune function
- Cystatin C predicts CKD, CVD and total mortality better than serum Cr or GFR. Not effected by muscle mass, gender, age, or other factors
- Hyperfibrinogenemia especially gamma fibrinogen, increased VWF, and increased clotting factors (Factor V Leiden mutation, VIII, IX, X, XII), protein C_1S.
- Elevated homocysteine and SAM levels
- Shortened telomeres
- Increased ADMA (asymmetric dimethyl arginine)
- Chronic renal disease, elevated creatinine, proteinuria and microalbuminuria (MAU)
- Metalloproteinases MMP 2 and MMP 9 and TIMPs (inhibitors)
- Reduced intake of PUFA, omega-3 fatty acids, ALA, and MUFA in diet
- Trans fatty acids
- Increased intake of omega 6 fatty acids
- Reduced dietary antioxidants like vitamin E, C, lycopene, etc.
- Low intake of fruits and vegetables
- Leukocytosis
- Corneal arcus

(*Continued*)

TABLE 18.2 (CONTINUED)
Partial List of Some of the Additional 400 CHD Risk Factors

- Diagonal earlobe crease
- Hairy earlobes
- Short stature or very high stature
- Low levels of dehydroepiandrosterone sulfate (DHEAS)
- Nonspecific ST-T wave changes on ECG
- Prolonged PR interval even within the normal range
- Short or long QT interval even within normal range
- Abnormal heart rate variability
- Elevated TSH and hypothyroidism both clinical and subclinical
- Lower socioeconomic status
- Elevated serum estradiol in men
- Chromium deficiency
- Lean hypertensive men (bottom 20% of IBW)
- Low muscle strength and low lean muscle mass
- Abnormal heart rate variability
- Increased nocturnal heart rate
- Elevated serum iron and ferritin
- Chronic infections (HSV, CMV, EBV, H. flu, chlamydia pneumonia, mycoplasma pneumonia, H. pylori)
- Chronic periodontal infection
- Osteoporosis at menopause
- Neopterin (activated macrophages) and high levels of CD3, CD4, and CD28 lymphocytes: all increase inflammation
- Chronic cough or chronic inflammatory lung disease
- Hypochloremia
- Low vitamin K levels
- Increased serum microparticles
- Increased myeloperoxidase levels (MPO)
- Low lean muscle mass in men
- Polycythemia
- Low serum copper
- Hypomagnesemia
- Male pattern baldness
- Chronic tachycardia and slow heart rate recovery after TMT
- I Haplotype of Y chromosome increases CHD 55-67% In 15% in UK and 60% in Scandinavia and Eastern Europe. Which genes and gene expression are involved are not known
- Metallothioneins MT1A +125 and +647 Haplotype related to zinc homeostasis in the heart and CHD risk
- Genetic increase in certain SNP (single nucleotide polymorphisms) such as TSP-1 and TSP-4 (thrombospondins), MAT1A(methylation), 9p21 DRB1, D6589, TNF alpha, chromosome 2,3,6,9,10 and 12 (SH2B3) and Chromosome 1, etc.
- ACE polymorphisms (ACE–DD)
- Alpha adducin (ADD 1) GLY 460 Trp SNP
- 242 T variant of CYBA SNP that encodes for p22 phox, a component of NADPH oxidases

(Continued)

TABLE 18.2 (CONTINUED)
Partial List of Some of the Additional 400 CHD Risk Factors

- YKL-40 serum levels (human cartilage glycoprotein 39)
- Taq IB CETP polymorphism
- Anti-apoprotein A-I IgG
- Mitochondrial DNA variant 16189 T>C
- KIF 6 polymorphism of Trp719Arg (Kinesin 9 family)
- NT5E mutation for CD73 encoding nucleotidase that converts AMP to adenosine
- Osteoprotegerin (OPG): Circulating glycoprotein of TNF superfamily. Increased in MAU, multivessel CHD. Independent risk factor for atherosclerosis
- Increased platelet adhesion or aggregation (or both) and abnormal thrombogenic potential (TPA/PAI-I ratio)
- Hemodynamic effects that alter arterial flow disturbances and induce endothelial damage, which enhances atherosclerosis (blood velocity, surfaces) such as increased pulse pressure, oscillatory flow issues, bifurcation areas
- PAPP-A: pregnancy associated plasma protein
- Lp-PLA2
- Increased plasma leptin
- Hostility
- Aortic calcification
- High myeloid related protein 8/14
- Low serum bilirubin
- Elevated HSP (heat shock protein) and HSP antibodies in serum
- Low serum calcitonin gene-related peptide (CGRP)
- Low sex hormone binding globulin (SHBG).
- Low free testosterone in men
- Low estradiol, estriol, and progesterone in women
- Low plasma adiponectin
- High levels of PAI-I
- Increased CD-14 monocytes and NK T cells
- Toll-like receptors 2 and 4 expression
- Increased phospholipase A_2
- Increased serum leptin
- Increased desaturated lecithin
- Uncarboxylated matrix Gla Protein
- Fibroblast growth factor 23. especially CHF
- Low heme oxygenase I (HOI): Fe metabolism
- Elevated hepcidin (inhibits ferroportin): Fe
- Low RBC glutathione peroxidase (GPx) SNP and low selenium with 1–4 GPx.
- Increased serum amyloid A (SAA)
- Low levels of serum coenzyme Q-10
- Increased aldosterone
- N-terminal pro-brain natriuretic peptide
- Sensitive troponin
- Galectin 3

(Continued)

TABLE 18.2 (CONTINUED)
Partial List of Some of the Additional 400 CHD Risk Factors

- TPA/PAI-I ratio (tissue plasminogen activator/plasminogen activator inhibitor)
- Heavy metals and other toxins, pesticides and organocides: mercury, lead, cadmium, arsenic, iron
- Low vitamin D levels
- Increased sodium intake
- Metabolic syndrome is CHD risk equivalent
- Abnormal ABI
- Abnormal CAPWA
- Abnormal endothelial function
- Abnormal carotid IMT and plaque
- Abnormal coronary artery calcium score (CAC)
- Increased oxidative stress
- Reduced oxidative defense
- Autoimmune dysfunction and markers
- Growth differentiation factor 15: stress responsive cytokine
- Air pollution
- Water pollution
- Cytokines
  1. Proinflammatory interleukins: IL 1, IL 6 and IL 8, and TNF alpha
  2. Colony stimulating factors: G-CSF, M-GSF, and GM-GSF
  3. Chemotactic factors: MCP-1, MIP-IB, PAF, L-B4, complement components, N –formal peptides, and Gro-alpha
  4. Interleukin I cluster
- Cell adhesion molecules (CAMs)
  1. Selectins: P-selectin, E-selectin, L-selectin, and CD 34
  2. Immunoglobulin family: ICAM 1-5, VCAM, MADCAM, and PECAM-1
  3. Cadherins: N, P R, B and E, desmogleins 1 and 3, desmocollins
  4. Integrins: B-1-B-4 for leukocytes and platelets, VLA
- Serum ATHERO-ELAMS: Endothelial-Leukocyte Adhesion Molecules
- ESPP: Endothelial Soluble Proatherogenic Products
- Cluster of Differentiation molecules: CD 30
- HSP: Heat Shock Proteins: HSP 60,65,70,72,
- HSP-Ab-Heat Shock Protein Antibodies: Anti-HSP 60, 65,70,72.
- Salusin beta
- Environmental pollutants: POPs (persistent organic pollutants): PCBs, dioxin, BDE47, pesticides
- C-terminal pro-vasopressin
- C-terminal pro-endothelin-1
- Mid-regional pro-adrenomedullin
- Mid-regional pro-atrial natriuretic peptide
- Tissue inhibitor metalloproteinase-1
- Atrial natriuretic peptides
- Under carboxylation of MGP (matrix Gla P)
- Low apelin levels
- Copeptin

(Continued)

TABLE 18.2 (CONTINUED)
Partial List of Some of the Additional 400 CHD Risk Factors

- Myeloid related protein 8/14
- Choline
- Heart type FA binding protein
- Ischemia modified albumin
- Placental growth factor
- Soluble tyrosine kinase-1
- Fatty acid binding proteins 4 and 5
- Haptoglobin phenotype
- Coupling factor 6
- Low serum sRAGE and high AGE/sRAGE ratio
- Serum P53
- Low intraplatelet melatonin (cyclooxygenase 1,2 mRNA)
- High levels of cardiac troponin T (cTnT)
- Low levels of serum osteocalcin
- Urinary proteomic risk factors to identify disease specific diagnostic polypeptides (over 238 identified)
- Serum metabolomics: TBD
- Soluble Lox -1 receptor (sLOX-1 receptor) increases CHD, plaque burden, complexity, ACS, and plaque rupture
- Epicardial fat volume increases CHD
- Serum metabolomics several 100 under review
- Soluble Lox -1 receptor (sLOX-1 receptor) increases CHD, plaque burden, complexity, ACS, and plaque rupture

SUMMARY AND KEY TAKE AWAY POINTS

1. The top five CHD risk factors must be redefined and evaluated accurately and treated in all patients.
2. The top modifiable 25 CHD risk factors should now be included in the evaluation of patients with any risk of CHD or known CHD.
3. Other CHD risk factors, metabolomics, and proteomics in blood and urine are now emerging and will be important in the future evaluation of CHD risk.
4. Early measurement of the top 25 modifiable CHD risk factors and early, aggressive, and appropriate prevention and treatment will help prevent and reduce CHD and MI.

BIBLIOGRAPHY

1. Houston MC. *Handbook of Hypertension*. Oxford, UK: Wiley, Blackwell; 2009.
2. O'Brien E, White WB, Parati G, Dolan E. Ambulatory blood pressure monitoring in the 21st century. *J Clin Hypertens* 2018; 20:1108–11.

3. Houston M. The role of nutraceutical supplements in the treatment of dyslipidemia. *J Clin Hypertens* 2012; 14:121–32.
4. Petrie JR, Guzik TJ, Touyz RM. Diabetes, hypertension, and cardiovascular disease: Clinical insights and vascular mechanisms. *Can J Cardiol* 2018; 34:575–84.
5. Mah E, Bruno RS. Postprandial hyperglycemia on vascular endothelial function: Mechanisms and consequences. *Nutr Res* 2012; 32:727–40.
6. Fazio S, Linton MF. High-density lipoprotein therapeutics and cardiovascular prevention. *J Clin Lipidol* 2010; 4:411–9.
7. van der Steeg WA, Holme I, Boekholdt SM, et al. High-density lipoprotein cholesterol, high-density lipoprotein particle size, and apolipoprotein A-I: Significance for cardiovascular risk: The IDEAL and EPIC-Norfolk studies. *J Am Coll Cardiol* 2008; 51:634–42.
8. Houston M. et al. Clinical roundup: Selected treatment options for hyperlipidemia. *Alternative and Complementary Medicine* 2018;24(5):1–4.
9. Houston M. Treatment of hypertension with nutrition and nutraceutical supplement: Part 1. *Alternative and Complimentary Medicine.* 2019;24: 260–275.
10. Houston M. Treatment of hypertension with nutrition and nutraceutical supplement: Part 2. *Alternative and Complementary Medicine.* 2019;25: 23–36.
11. Houston M. Nutrition and nutraceutical supplements in the management of dyslipidemia: Part 1. *Alternative and Complementary Medicine* 2019;25(2):77–84.
12. Grundy SM. Metabolic syndrome update. *Trends Cardiovasc Med.* 2016 May;26(4):364–73.
13. Houston M. Nutrition and nutraceutical supplements in the management of dyslipidemia: Part 2. *Alternative and Complementary Medicine* 2019;25(1):23–36.
14. Houston M. Three finite vascular responses that cause coronary heart disease and cardiovascular disease: Implications for diagnosis and treatment. *Alternative and Complimentary Medicine.* 2019;25(4):181.
15. Harreiter J, Roden M. Diabetes mellitus-Definition, classification, diagnosis, screening and prevention (Update 2019)]. *Wien Klin Wochenschr.* 2019 May;131(Suppl 1):6–15.
16. Houston M. *Hypertension and Dyslipidemia Chapters in Nutrition and Integrative Strategies in Cardiovascular Medicine.* Houston and Sinatra, Editors. Boca Raton: CRC Press; 2015.
17. Sinatra S, Houston M, Editors. *Nutrition and Integrative Strategies in Cardiovascular Medicine.* Boca Raton: CRC Press; 2015.
18. Houston M, Lamb J and Hays, A. *Vascular Biology in Clinical Practice.* Denver: Outskirts Press; November 2019.
19. Houston MC, Editor. *Personalized and Precision Integrative Cardiovascular Medicine.* Philedelphia: Wolters Kluwer Publishers; December 2019.
20. Houston M. *Revolutionary Concepts in the Prevention and Treatment of Cardiovascular Disease, in Integrative and Functional Medical Nutrition Therapy.* New Jersey: Humana Press. Noland and Drisko Editors; 2019, pp. 823–41.
21. Houston M, Bell L. *Controlling High Blood Pressure Though Nutrition, Nutritional Supplement, Lifestyle and drugs.* Boca Raton and Oxford: CRC Press; 2021.
22. Landmark K. [Smoking and coronary heart disease].*Tidsskr Nor Laegeforen.* 2001 May 30;121(14):1710–2.
23. Kachur S, Lavie CJ, de Schutter A, Milani RV, Ventura HO *Obesity and cardiovascular diseases.* Torino, Italy: Minerva Med2017.

19 Testing
Labs, Noninvasive, and Invasive Testing

CLINICAL SIGNS, BLOOD TESTS, AND NONINVASIVE VASCULAR TESTING FOR HYPERTENSION AND CARDIOVASCULAR DISEASE

There are many clinical signs, blood tests, and other noninvasive vascular tests that your doctor can order to determine if you have developed any damage to your arteries, endothelium, heart, brain, kidneys, eyes, or aorta. These tests will accurately define your present clinical cardiovascular damage that is related to all of your CHD risk factors and also predict your risk of future cardiovascular, such as a CHD, MI, stroke, heart failure, kidney failure, retinal damage, or aortic aneurysm. Let us take a look at each of these organs that can be damaged, what tests are available, and discuss what they mean and how to treat the findings (1–19).

THE ARTERIES

The Computerized Arterial Pulse Wave Analysis (CAPWA) is a machine (CV Profiler—see the Sources section on testing equipment) that can measure the stiffness or elasticity of the large, medium-size, and small resistance arteries throughout the body in a noninvasive manner using computer analysis (20–24). The results are standardized, based on your age and gender, to give you a score called "compliance of the arteries", which is a measure of their elasticity. The scores for compliance for the big and medium arteries are called "C-1 compliance", and for the small resistance arteries, they are called the "C -2 compliance". The higher the compliance number the more elastic the artery and the lower is your risk of a future CHD or a cardiovascular event. The stiffer the artery the lower the compliance score and the more likely you already have damage from CHD risk factors. This gives you a point in time that accurately estimates your vascular age and your future risk for a CHD, MI, stroke, heart failure, and possibly kidney disease. In many cases, the smaller resistance arteries are abnormal decades before a CHD event occurs. This is an excellent clinical tool to detect and then treat you early before more arterial damage might occur (20–24).

ENDOTHELIAL FUNCTION AND DYSFUNCTION

One of the best validated early detection tests for functional abnormalities of the endothelium is the EndoPAT (see the Sources section) which determines endothelial function and dysfunction (25–29). The EndoPAT measures the blood flow in your arm after a five-minute occlusion of the brachial (arm) artery which is an excellent indirect measure of nitric oxide (NO) bioavailability. Endothelial dysfunction in the arteries may predict future CHD, MI, and hypertension or be directly related to a disease that now exists and is causing arterial damage (Figure 19.1). If the EndoPAT shows endothelial dysfunction, it predicts accurately the future risk for hypertension, CHD, angina, MI, congestive heart failure, cardiac death, hospitalization, coronary artery bypass graft, stent restenosis, the presence of plaque in the coronary arteries that are rupture prone, peripheral arterial disease (PAD), and stroke (25–29). Endothelial dysfunction is defined as a score of less than 1.67. A really good score is over 2.1.

CAROTID ARTERY ULTRASOUND

The carotid arteries are located on both sides of the neck and supply blood to the brain. They can become thick, stiff, or blocked with plaque. This will reduce the blood supply to the brain, or a piece of the plaque could break off and go to the brain and cause a stroke (embolic stroke) (Figures 19.2 and 19.3) (28–30). Hypertension is a

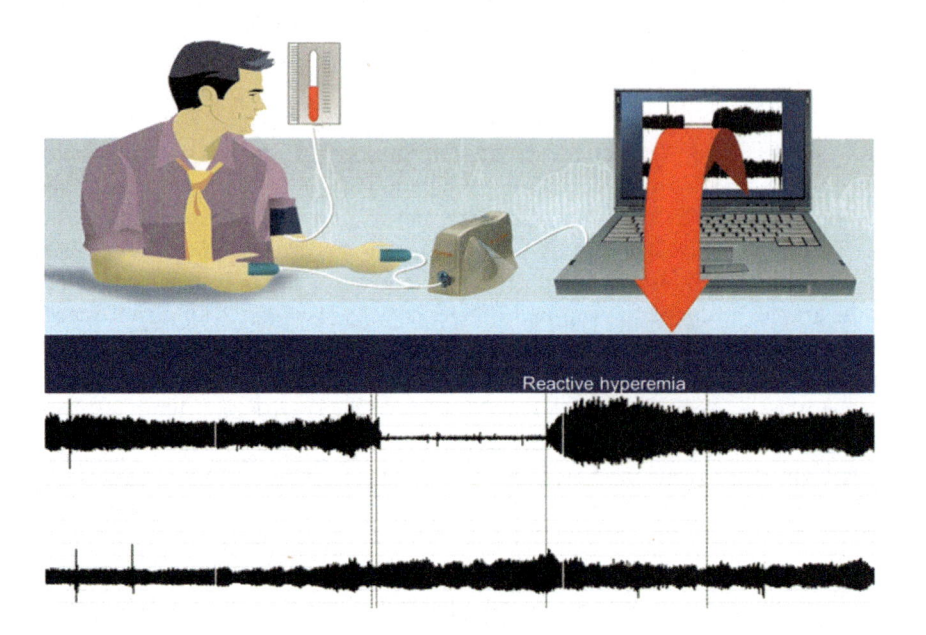

FIGURE 19.1 The measurement of endothelial function and dysfunction with the EndoPAT. The top curve shows a normal artery with proper dilation after a brief occlusion of the brachial artery. Note the height of the black curve, called "reactive hyperemia", which is increased compared to the other arm at the bottom. The ratio of the two curves determines the score for the EndoPAT.

Testing 197

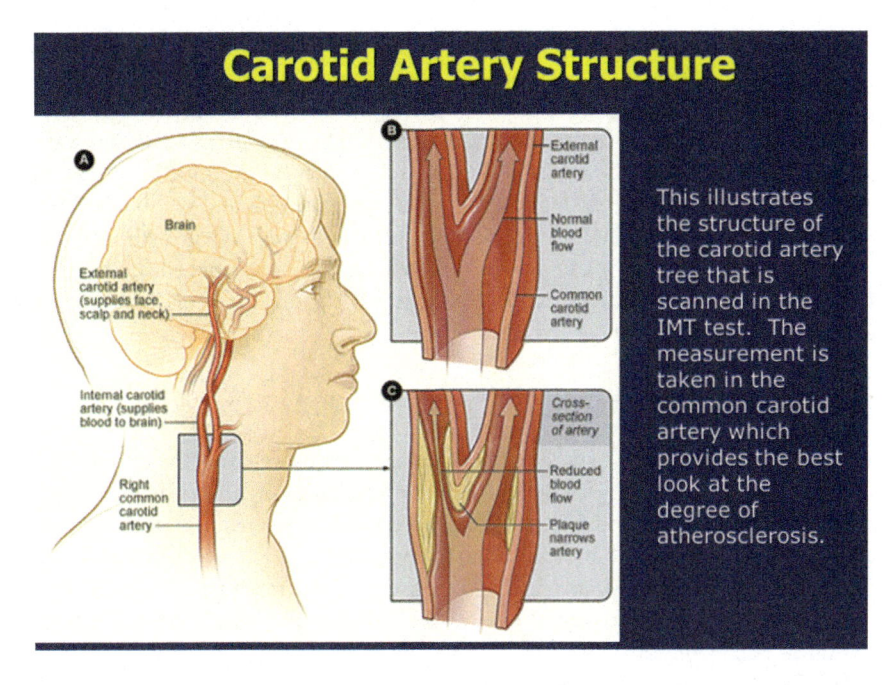

FIGURE 19.2 Carotid artery structure showing the IMT and plaque.

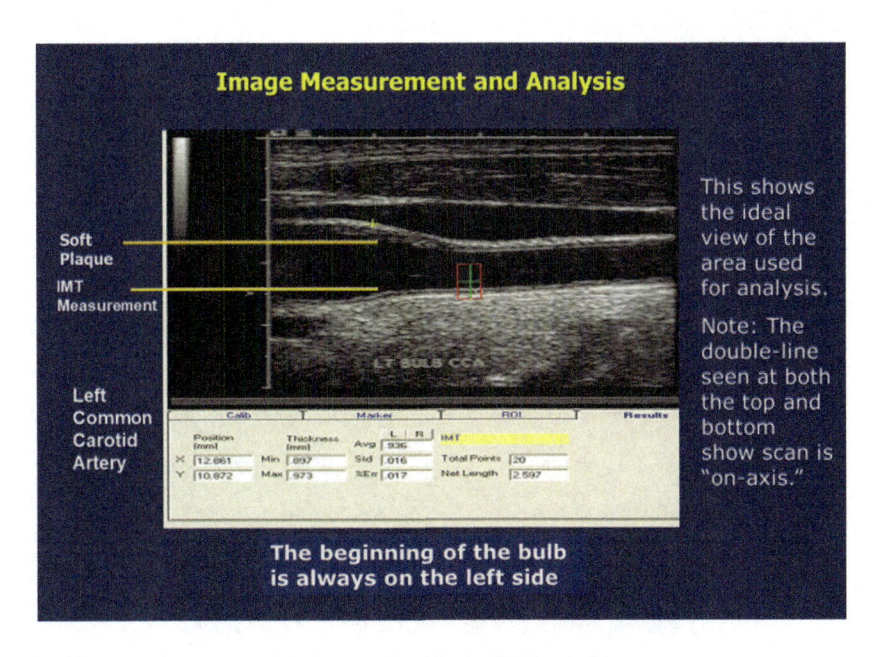

FIGURE 19.3 Actual carotid ultrasound with the IMT and plaque measurement.

common cause of plaque and stroke, especially if there are other risk factors such as diabetes mellitus, high cholesterol, obesity, or smoking. The thickness of the endothelium and the muscle of the carotid artery is called the carotid intimal media thickness (IMT). This IMT predicts future risk of CHD, MI, and stroke (30–32). Normal values for the carotid IMT must be adjusted for age and gender. A carotid IMT of less than 0.6 mm is normal to low risk, 0.6 to 0.7 mm is moderate risk and 0.7 to 0.95 mm is high risk for future CHD, MI, or stroke. The normal carotid IMT growth rate is less than 0.016 mm/year. The risk for a CHD and MI is 26% greater per 0.10 mm growth of the carotid artery IMT difference over five years. The risk for stroke is 32% higher per 0.10 mm growth of the carotid artery IMT over five years (30–32).

THE EYE AND THE RETINA

The retina or fundus examination (back of the eye) which contains the retinal arterioles is a window to the arteries in the brain and correlates highly with hypertension, diabetes mellitus, dyslipidemia, smoking, CHD, atherosclerosis and with disease in the small arteries in other organs throughout the body (33–36). Retinal damage indicates microvascular (small vessels) disease. Retinal microvascular endothelial dysfunction assessed with special eye testing shows that retinal artery health and dilation is dependent on nitric oxide and predicts hypertension, stroke, CHD, and MI (33–36). Endothelial dysfunction of the coronary arteries by EndoPAT has a very high correlation with endothelial dysfunction of the retina. In addition, hemorrhages, exudates, artery thickening, and damage may be seen (Figure 19.4).

CORONARY ARTERY CALCIFICATION

Coronary artery calcification (CAC) happens as the coronary arteries age and calcium is deposited in the muscular wall of the artery or a plaque inside the artery lumen, thus obstructing the blood flow (37–40) (Figure 19.5). The calcium in the four primary larger coronary arteries increases as you age and is related to all risk factors, such as hypertension, high cholesterol, diabetes mellitus, obesity, or smoking. The higher the CAC score, based on age and gender, the greater the risk for CHD and MI. A normal

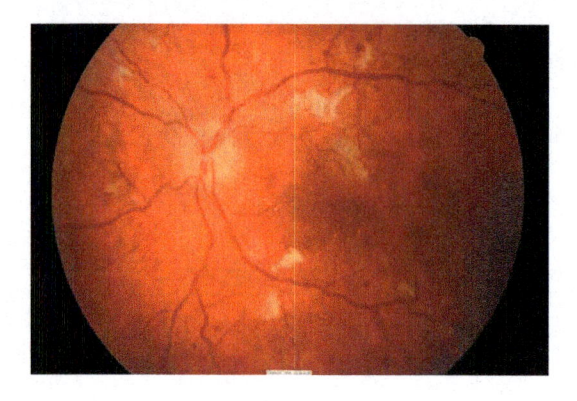

FIGURE 19.4 Hypertension retinopathy with AV nicking, hemorrhages, and exudates.

Testing

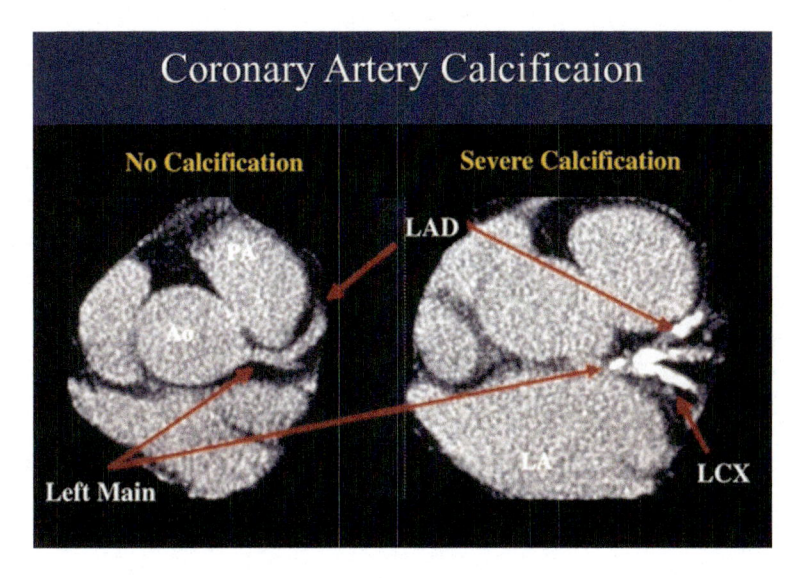

FIGURE 19.5 Coronary artery calcium in the main coronary arteries, including the left anterior descending (LAD), left main, and left circumflex arteries (LCX).

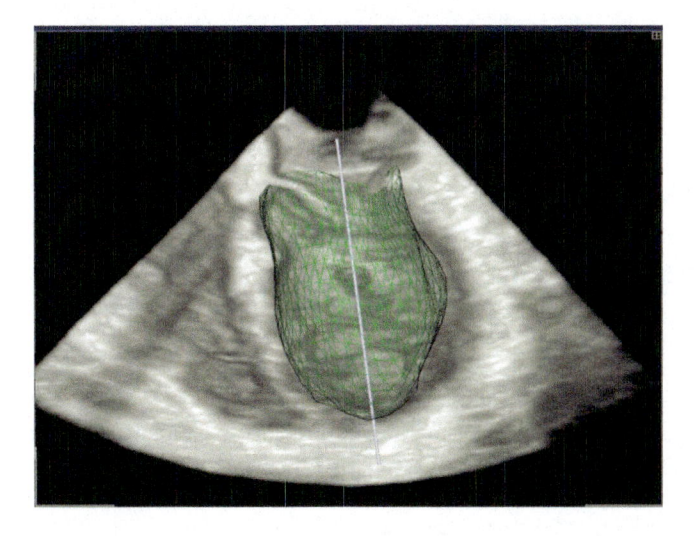

FIGURE 19.6 ECHO of the heart, showing the left ventricle muscle in white and the left ventricle chamber in green.

score is zero. A baseline CAC score predicts CHD and MI risk beyond traditional risk factors. A CAC score of over 300 has a risk of a MI that is increased ten-fold (37–40).

ECHOCARDIOGRAPHY (HEART ULTRASOUND)

Echocardiography (ECHO) or 2D echocardiography is an ultrasound of the heart (Figure 19.6) (41). This noninvasive test can determine the size (dilation) of the atria

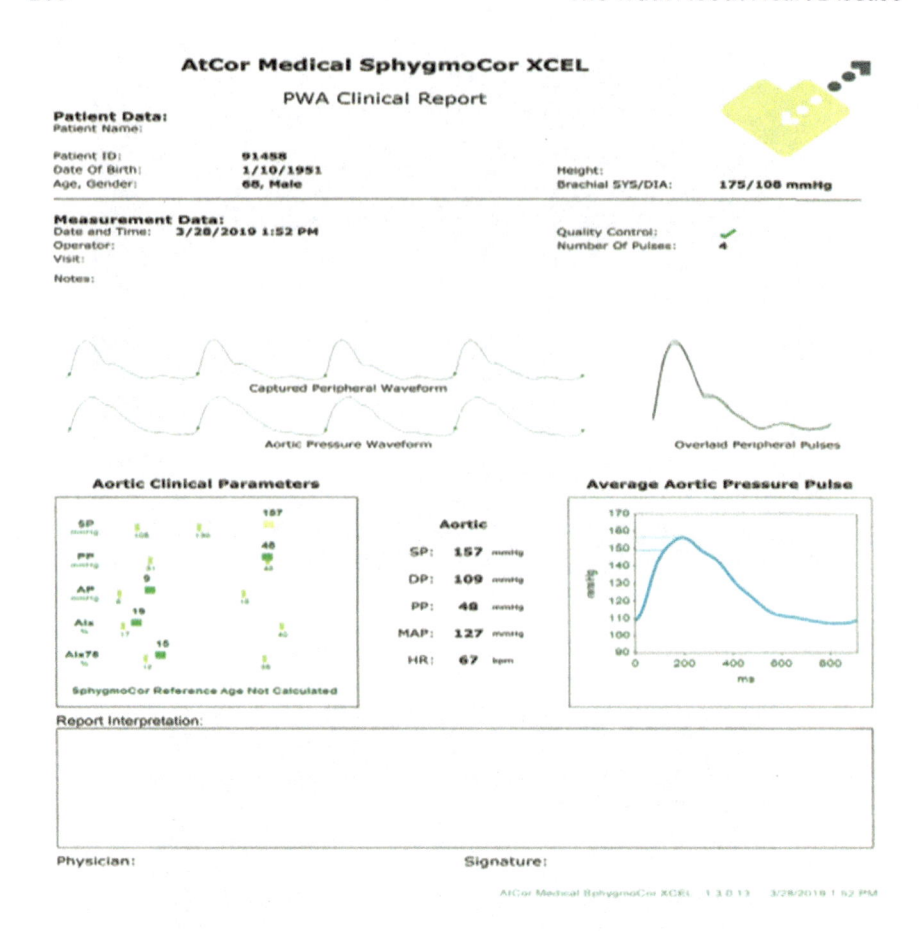

FIGURE 19.7 Central blood pressure and aortic stiffness. This AtCor test measures the central systolic and diastolic blood pressure, the pulse pressure, the heart rate, and the augmentation index.

and ventricular chambers, thickening, stiffness, and enlargement of the heart muscle (hypertrophy), pressures in the heart and heart valve damage (insufficiency or regurgitation), and stenosis. The ECHO also shows the heart function and pumping action (ejection fraction), stiffness of the ventricles (diastolic dysfunction), and congestive heart failure. Hypertension and CHD with blockage of the blood flow to the heart muscle can dilate the various chambers of the heart, make them stiff, cause enlargement, hypertrophy (LVH), and heart failure (41). Also, if a valve is leaking or stenotic, the ECHO will show it (Figure 19.6). Finally, the ECHO may show an aneurysm of the thoracic aorta (big artery in the chest as it comes out of the heart).

AORTIC ULTRASOUND FOR ANEURYSMS AND KIDNEY SIZE

An ultrasound of the abdomen can show the size of the aorta, if an aneurysm is present, or if it is dilated or has plaque present. An aneurysm is a weakened artery wall that enlarges, and when a certain size is reached, it may rupture. In addition a renal ultrasound will show the size of the kidneys. Small kidneys indicate damage from hypertension and atherosclerosis with declining function and progression to chronic renal disease (CKD). As noted previously, chronic kidney disease is a major risk factor for CHD and MI.

CENTRAL BLOOD PRESSURE AND AORTIC STIFFNESS (AUGMENTATION INDEX)

It is possible to measure the blood pressure centrally in the aorta just as it comes out of the heart (42). This is a more accurate blood pressure reading than the one that we take in the arm. In addition, the **AtCor** machine measures the stiffness or elasticity of the aorta and the distal artery branches to predict the augmentation index (Figure 19.7). Hypertension causes the arteries to become stiff so that a wave of pressure in the artery wall bounces back to the aorta near the heart that further increases the systolic blood pressure and the risk for stroke, heart failure, CHD, and MI. This is called the "augmentation index" (42).

ELECTROCARDIOGRAM AND CARDIOPULMONARY EXERCISE TESTING

An electrocardiogram shows heart rate and rhythm and possible heart damage from a previous MI. Types of electrical heart blocks, extra heartbeats (from the atria, premature atrial beats or PACs and from the ventricle, premature ventricular beats or PVCs), other electrical issues, heart chamber enlargement, heart stress, and other problems due to CHD. Cardiopulmonary Exercise Testing (CPET) will determine the risk for CHD with blockage in the arteries, spasm of the coronary arteries, extra heartbeats with exercise, lung function, blood pressure, and heart rate response to exercise and recovery during exercise. These and other findings help to predict cardiovascular events.

COMPLETE AND ADVANCED CARDIOVASCULAR LABORATORY TESTING

A complete cardiovascular blood panel with advanced testing should be done to evaluate cardiovascular, CHD, and kidney damage. These are listed in Table 19.1.

SUMMARY AND KEY TAKE AWAY POINTS

1. Several noninvasive vascular tests will predict the risk for CHD or MI decades before an event occurs.
2. These noninvasive vascular tests will predict the presence and severity of CHD and the risk for MI.
3. Numerous supporting blood tests also predict CHD and MI.
4. All of these tests are available clinically.
5. Immediate prevention and treatment programs can be started to correct these abnormal tests and prevent the progression of CHD or an MI.

TABLE 19.1
Complete List of Advanced Cardiovascular Lab Tests

- Complete blood count (CBC) with differential
- Urinalysis
- Complete metabolic profile (CMP 12) which includes kidney function, creatinine, and calculates effective glomerular filtration (eGFR) and electrolytes.
- Cystatin C and SDMA are measures of kidney function
- Advanced lipid profile for particle size and particle number of all classes of lipids
- Complete thyroid panel, including a free T4, T3, TSH, RT3, and thyroid antibodies
- Magnesium
- Iron, total iron-binding capacity (TIBC), and Ferritin
- Fibrinogen
- HS-CRP (C reactive protein)
- Homocysteine
- Uric acid
- Microalbuminuria
- Gammaglutamyl transpeptidase (GGTP) and hepatic profile
- Myeloperoxidase (MPO)
- Cardiovascular genomics from Vibrant Labs
- Toxicology and heavy metal screen: spot or 24-hr urine or blood
- Vitamin D3
- Fasting C peptide, hemoglobin A1C, insulin, proinsulin, 2-hr glucose tolerance test (GTT)
- Plasma renin activity (PRA) and aldosterone
- Free testosterone, sex hormone-binding globulin (SHBG), estradiol, estriol, progesterone, dehydroepiandrosterone (DHEA and DHEAS)
- Omega-3 index

REFERENCES

1. Whelton PK et al. ACC/AHA/AAPA/ABC/ACPM/AGS/APhA/ASH/ASPC/NMA/PCNA Guideline for the prevention, detection, evaluation, and management of high blood pressure in adults: A report of the American College of Cardiology/American Heart Association Task Force on Clinical Practice Guidelines. *Hypertension.* 2018 Jun;71(6):e13–e115.
2. Houston M. The role of nutrition and nutraceutical supplements in the treatment of hypertension. *World Journal of Cardiology* 2014;6(2): 38–66.
3. Houston M. Nutrition and nutraceutical supplements for the treatment of hypertension: Part 1. *J. Clinical Hypertension* 2013; 15:752–757.
4. Houston M. Nutrition and nutraceutical supplements for the treatment of hypertension: Part II. *J. Clinical Hypertension* 2013; 15:845–851.
5. Houston M. Nutrition and nutraceutical supplements for the treatment of hypertension: Part III *J Clinical Hypertension* 2013; 15:931–7.
6. Borghi C, Cicero AF Nutraceuticals with a clinically detectable blood pressure-lowering effect: A review of available randomized clinical trials and their meta-analyses. *Br J Clin Pharmacol.* 2017;83(1):163–171.
7. Sirtori CR, Arnoldi A, Cicero AF. Nutraceuticals for blood pressure control. Review. *Ann Med.* 2015;47(6):447–56.

8. Cicero AF, Colletti A. Nutraceuticals and blood pressure control: Results from clinical trials and meta-analyses. *High Blood Press Cardiovasc Prev.* 2015;22(3):203–13.
9. Turner JM, Spatz ES. Nutritional supplements for the treatment of hypertension: A practical guide for clinicians. *Curr Cardiol Rep.* 2016;18(12):126. Review
10. Caligiuri SP, Pierce GN. A review of the relative efficacy of dietary, nutritional supplements, lifestyle and drug therapies in the management of hypertension. *Crit Rev Food Sci Nutr.* 2016 Aug 5:0.
11. Houston MC, Fox B, Taylor N. *What Your Doctor May Not Tell You About Hypertension. The Revolutionary Nutrition and Lifestyle Program to Help Fight High Blood Pressure.* New York: AOL Time Warner, Warner Books; September 2003.
12. Houston M. Treatment of hypertension with nutrition and nutraceutical supplement: Part 1. *Alternative and Complimentary Medicine.* 2019;24: 260–275
13. Houston M. Treatment of hypertension with nutrition and nutraceutical supplement: Part 2. *Alternative and Complimentary Medicine.* 2019;25: 23–36.
14. Sinatra S and Houston M, Editors. *Nutrition and Integrative Strategies in Cardiovascular Medicine.* Boca Raton: CRC Press; 2015.
15. The seventh report of the joint national committee on prevention, detection, evaluation, and treatment of high blood pressure (JNC-7). *JAMA* 2003; 289:2560–2572.
16. Thomopoulos C, Parati G, Zanchetti A. Effects of blood pressure lowering on outcome incidence in hypertension: 7. Effects of more vs. less intensive blood pressure lowering and different achieved blood pressure levels: Updated overview and meta-analyses of randomized trials. *J Hypertens.* 2016;34(4):613–22.
17. Ettehad D, Emdin CA, Kiran A, Anderson SG, Callender T, Emberson J, Chalmers J, Rodgers A, Rahimi K. Blood pressure lowering for prevention of cardiovascular disease and death: A systematic review and meta-analysis. *Lancet.* 2016;387(10022):957–67.
18. ESH/ESC Task Force for the Management of Arterial Hypertension. 2013 Practice guidelines for the management of arterial hypertension of the European Society of Hypertension (ESH) and the European Society of Cardiology (ESC): ESH/ESC Task Force for the Management of Arterial Hypertension. *J Hypertens* 2013; 31: 1925–38.
19. Flack JM, Calhoun D, Schiffrin EL. The New ACC/AHA Hypertension Guidelines for the Prevention, Detection, Evaluation, and Management of High Blood Pressure in Adults. *Am J Hypertens.* 2018;31(2):133–135.
20. Matsuzawa Y, Sugiyama S, Sumida H, Sugamura K, Nozaki T, Ohba K, Matsubara J, Kurokawa H, Fujisue K, Konishi M, Akiyama E, Suzuki H, Nagayoshi Y, Yamamuro M, Sakamoto K, Iwashita S, Jinnouchi H, Taguri M, Morita S, Matsui K, Kimura K, Umemura S, Ogawa H. Peripheral endothelial function and cardiovascular events in high-risk patients J Am Heart Assoc. 2013 Nov 25;2(6):e000426
21. Prisant LM, Pasi M, Jupin D, Prisant ME Assessment of repeatability and correlates of arterial compliance.*Blood Press Monit.* 2002;7(4):231–5.
22. Cohn JN, Hoke L, Whitwam W, Sommers PA, Taylor AL, Duprez D, Roessler R, Florea NScreening for early detection of cardiovascular disease in asymptomatic individuals. *Am Heart J.* 2003;146(4):679–85.
23. Nelson MR Stepanek J, Cevette M, Covalciuc M, Hurst RT, Tajik AJ. Noninvasive measurement of central vascular pressures with arterial tonometry: Clinical revival of the pulse pressure waveform? *Mayo Clin Proc.* 2010;85(5):460–72.
24. Hashimoto J, Ito SSome mechanical aspects of arterial aging: Physiological overview based on pulse wave analysis. *Ther Adv Cardiovasc Dis.* 2009;3(5):367–78.
25. Matsuzawa Y, Sugiyama S, Sugamura K, Nozaki T, Ohba K, Konishi M, Matsubara J, Sumida H, Kaikita K, Kojima S, Nagayoshi Y, Yamamuro M, Izumiya Y, Iwashita S, Matsui K, Jinnouchi H, Kimura K, Umemura S, Ogawa H Digital assessment of endothelial function and ischemic heart disease in women. *J Am Coll Cardiol.* 2010;55(16):1688–96.

26. Bonetti PO, Pumper GM, Higano ST, Holmes DR Jr, Kuvin JT, Lerman A Noninvasive identification of patients with early coronary atherosclerosis by assessment of digital reactive hyperemia. *J Am Coll Cardiol.* 2004;44(11):2137–41.
27. Hamburg NM, Keyes MJ, Larson MG, Vasan RS, Schnabel R, Pryde MM, Mitchell GF, Sheffy J, Vita JA, Benjamin EJ Cross-sectional relations of digital vascular function to cardiovascular risk factors in the Framingham Heart Study. *Circulation.* 2008;117(19):2467–74.
28. Schoenenberger AW, Urbanek N, Bergner M, Toggweiler S, Resink TJ, Erne P Associations of reactive hyperemia index and intravascular ultrasound-assessed coronary plaque morphology in patients with coronary artery disease. *Am J Cardiol.* 2012;109(12):1711–6.
29. Matsuzawa Y, Sugiyama S, Sumida H, Sugamura K, Nozaki T, Ohba K, Matsubara J, Kurokawa H, Fujisue K, Konishi M, Akiyama E, Suzuki H, Nagayoshi Y, Yamamuro M, Sakamoto K, Iwashita S, Jinnouchi H, Taguri M, Morita S, Matsui K, Kimura K, Umemura S, Ogawa H. Peripheral endothelial function and cardiovascular events in high-risk patients J Am Heart Assoc. 2013 Nov 25;2(6):e000426
30. Johnsen SH, Mathiesen EB. Carotid plaque compared with intima-media thickness as a predictor of coronary and cerebrovascular disease. *Curr Cardiol Rep.* 2009;11(1):21–7.
31. Bots ML, Taylor AJ, Kastelein JJ, Peters SA, den Ruijter HM, Tegeler CH, Baldassarre D, Stein JH, O'Leary DH, Revkin JH, Grobbee DE.. Rate of change in carotid intima-media thickness and vascular events: Meta-analyses cannot solve all the issues. A point of view. *J Hypertens.* 2012;30(9):1690–6.
32. Lorenz MW, Markus HS, Bots ML, Rosvall M, Sitzer M..Prediction of clinical cardiovascular events with carotid intima-media thickness: A systematic review and meta-analysis. *Circulation.* 2007;115(4):459–67.
33. Rizzoni D, Porteri E, Duse S, De Ciuceis C, Rosei CA, La Boria E, Semeraro F, Costagliola C, Sebastiani A, Danzi P, Tiberio GA, Giulini SM, Docchio F, Sansoni G, Sarkar A, Rosei EA. Relationship between media-to-lumen ratio of subcutaneous small arteries and wall-to-lumen ratio of retinal arterioles evaluated noninvasively by scanning laser Doppler flowmetry. *J Hypertens.* 2012;30(6):1169–75.
34. Ying GS, Maguire M, Pistilli M, Daniel E, Alexander J, Whittock-Martin R, Parker C, Mohler E, Lo JC, Townsend R, Gadegbeku CA, Lash JP, Fink JC, Rahman M, Feldman H, Kusek JW, Xie D, Coleman M, Keane MG; Chronic Renal Insufficiency Cohort (CRIC) Study Group Association between retinopathy and cardiovascular disease in patients with chronic kidney disease (from the Chronic Renal Insufficiency Cohort [CRIC] Study). *Am J Cardiol.* 2012;110(2):246–53.
35. Virdis A, Savoia C, Grassi G, Lembo G, Vecchione C, Seravalle G, Taddei S, Volpe M, Rosei EA, Rizzoni D Evaluation of microvascular structure in humans: A 'state-of-the-art' document of the Working Group on Macrovascular and Microvascular Alterations of the Italian Society of Arterial Hypertension. *J Hypertens.* 2014;32(11):2120–9.
36. Al-Fiadh AH, Wong TY, Kawasaki R, Clark DJ, Patel SK, Freeman M, Wilson A, Burrell LM, Farouque O Usefulness of retinal microvascular endothelial dysfunction as a predictor of coronary artery disease. *Am J Cardiol.* 2015;115(5):609–13.
37. Choi Y, Chang Y, Ryu S Cho J, Kim MK, Ahn Y, Lee JE, Sung E, Kim B9 Ahn J, Kim CW, Rampal S, Zhao D Zhang Y, Pastor-Barriuso R Lima JA, Chung EC Shin H, Guallar E Relation of Dietary Glycemic Index and Glycemic Load to Coronary Artery Calcium in Asymptomatic Korean Adults. *Am J Cardiol.* 2015;116(4):520–6.
38. Ahmadi N, Tsimikas S, Hajsadeghi F, Saeed A, Nabavi V, Bevinal MA, Kadakia J, Flores F, Ebrahimi R, Budoff MJ.Relation of oxidative biomarkers, vascular dysfunction, and progression of coronary artery calcium. *Am J Cardiol.* 2010;105(4):459–66.

39. Raggi P, Callister TQ, Shaw LJ Progression of coronary artery calcium and risk of first myocardial infarction in patients receiving cholesterol-lowering therapy *Arterioscler Thromb Vasc Biol*. 2004;24(7):1272–7.
40. Criqui MH, Denenberg JO, Ix JH, McClelland RL, Wassel CL, Rifkin DE, Carr JJ, Budoff MJ, Allison MA Calcium density of coronary artery plaque and risk of incident cardiovascular events. *JAMA*. 2014;311(3):271–8.
41. Cheitlin MD, AWACC/AHA/ASE 2003 guideline update for the clinical application of echocardiography-summary article. *J Am Coll Cardiol*. 2003;42:954–70.
42. Park CM, Korolkova O, Davies JE, et al. Arterial pressure: Agreement between a brachial cuff-based device and radial tonometry. *J Hypertens*. 2014 Apr;32(4):865–72.

20 How to Assess Your Risk Using CHD Scoring Tests

Once you have obtained the results of all of your history, physical exam, blood tests, genetics, and noninvasive cardiovascular disease tests, your doctor can calculate your risk for coronary heart disease (CHD) and myocardial infarction (MI) using several different validated CHD risk-scoring systems. These will be discussed in this chapter. I recommend that these scoring systems be calculated and interpreted by your physician, as they are complicated. I would not recommend that you attempt to do these on your own but ask your doctor to obtain the tests for each one, calculate the score for you and interpret the results, and then institute your prevention and treatment programs.

FRAMINGHAM RISK SCORE

The Framingham Risk Score (FRS) estimates the risk of developing CHD within a ten-year time period. It is now a common practice to double the FRS if there is a family history of premature CHD in a first-degree relative (men <55years, women <65years). Patients with a ten-year risk of CHD ≥5% are considered high risk. Please note that the FRS is an older scoring system and, although still used, is not as accurate as others that are discussed in this chapter. Also note the FRS measures only the RISK of developing CHD and not mortality. Secondly, it is a ten-year estimate.

These are the criteria that are measured:

1. Gender.
2. Age.
3. Total cholesterol.
4. HDL cholesterol.
5. Systolic blood pressure.
6. On medication for hypertension.
7. Smoker.
8. Diabetes mellitus.
9. Known vascular disease, such as CHD or stroke.

If the family history is positive for CHD, then the score is doubled.

Framingham Risk Score for Women Based on Age with Points for Each Age Group:

> **Age.** 20–34 years: Minus 7 points. 35–39 years: Minus 3 points. 40–44 years: 0 points. 45–49 years: 3 points. 50–54 years: 6 points. 55–59 years: 8 points. 60–64 years: 10 points. 65–69 years: 12 points. 70–74 years: 14 points. 75–79 years: 16 points.

DOI: 10.1201/b22808-21

Total cholesterol, mg/dL. Age 20–39 years: Under 160: 0 points. 160–199: 4 points. 200–239: 8 points. 240–279: 11 points. 280 or higher: 13 points. • Age 40–49 years: Under 160: 0 points. 160–199: 3 points. 200–239: 6 points. 240–279: 8 points. 280 or higher: 10 points. • Age 50–59 years: Under 160: 0 points. 160–199: 2 points. 200–239: 4 points. 240–279: 5 points. 280 or higher: 7 points. • Age 60–69 years: Under 160: 0 points. 160–199: 1 point. 200–239: 2 points. 240–279: 3 points. 280 or higher: 4 points. • Age 70–79 years: Under 160: 0 points. 160–199: 1 point. 200–239: 1 point. 240–279: 2 points. 280 or higher: 2 points.

If cigarette smoker. Age 20–39 years: 9 points. • Age 40–49 years: 7 points. • Age 50–59 years: 4 points. • Age 60–69 years: 2 points. • Age 70–79 years: 1 point.

All nonsmokers. 0 points.

HDL cholesterol, mg/dL. 60 or higher: Minus 1 point. 50–59: 0 points. 40–49: 1 point. Under 40: 2 points.

Systolic blood pressure, mm Hg. Untreated: Under 120: 0 points. 120–129: 1 point. 130–139: 2 points. 140–159: 3 points. 160 or higher: 4 points. • Treated: Under 120: 0 points. 120–129: 3 points. 130–139: 4 points. 140–159: 5 points. 160 or higher: 6 points.

10-year risk in %: Points total: Under 9 points: <1%. 9–12 points: 1%. 13–14 points: 2%. 15 points: 3%. 16 points: 4%. 17 points: 5%. 18 points: 6%. 19 points: 8%. 20 points: 11%. 21=14%, 22=17%, 23=22%, 24=27%, >25= Over 30%.

Framingham Risk Score for Men Based on Age with Points for Each Age Group:

Age. 20–34 years: Minus 9 points. 35–39 years: Minus 4 points. 40–44 years: 0 points. 45–49 years: 3 points. 50–54 years: 6 points. 55–59 years: 8 points. 60–64 years: 10 points. 65–69 years: 11 points. 70–74 years: 12 points. 75–79 years: 13 points.

Total cholesterol, mg/dL. Age 20–39 years: Under 160: 0 points. 160–199: 4 points. 200–239: 7 points. 240–279: 9 points. 280 or higher: 11 points. • Age 40–49 years: Under 160: 0 points. 160–199: 3 points. 200–239: 5 points. 240–279: 6 points. 280 or higher: 8 points. • Age 50–59 years: Under 160: 0 points. 160–199: 2 points. 200–239: 3 points. 240–279: 4 points. 280 or higher: 5 points. • Age 60–69 years: Under 160: 0 points. 160–199: 1 point. 200–239: 1 point. 240–279: 2 points. 280 or higher: 3 points. • Age 70–79 years: Under 160: 0 points. 160–199: 0 points. 200–239: 0 points. 240–279: 1 point. 280 or higher: 1 point.

If cigarette smoker. Age 20–39 years: 8 points. • Age 40–49 years: 5 points. • Age 50–59 years: 3 points. • Age 60–69 years: 1 point. • Age 70–79 years: 1 point.

All non-smokers. 0 points.

HDL cholesterol, mg/dL. 60 or higher: Minus 1 point. 50–59: 0 points. 40–49: 1 point. Under 40: 2 points.

Systolic blood pressure, mm Hg. Untreated: Under 120: 0 points. 120–129: 0 points. 130–139: 1 point. 140–159: 1 point. 160 or higher: 2 points.
- Treated: Under 120: 0 points. 120–129: 1 point. 130–139: 2 points. 140–159: 2 points. 160 or higher: 3 points.

10-year risk in %: Points total: 0 point: <1%. 1–4 points: 1%. 5–6 points: 2%. 7 points: 3%. 8 points: 4%. 9 points: 5%. 10 points: 6%. 11 points: 8%. 12 points: 10%. 13 points: 12%. 14 points: 16%. 15 points: 20%. 16 points: 25%. 17 points or more: Over 30%.

COSEHC GLOBAL CARDIOVASCULAR RISK CALCULATION

The Consortium of Southeastern Hypertension Centers (COSEHC) score estimates the risk of death from CHD or MI within five years with both absolute and relative risk. Please note that this score has more variables, estimates CHD death as absolute and relative risk, and also is a five-year (not ten-year) risk, like the FRS.

- **Absolute Risk**: probability of an adverse event occurring in an individual within a defined period of time.
- **Relative Risk:** probability of an adverse event happening in an individual compared to average or normal individuals sharing similar demographics other than the risk factor.
- High CHD risk is defined as follows:
  Relative Risk ≥ 60th percentile (see the table below)
  Absolute Risk: Risk score ≥ 40
     ≥ 2.3% risk of CHD death in 5 years

Approximate 60th Percentile Relative Risk Scores for Men and Women Based on Age

Men	Age Range	Women
29	35–39	18
32	40–44	21
<u>36</u>	45–49	27
40	50–54	31
44	55–59	<u>36</u>
48	60–64	41
53	65–69	45
57	70–74	49

Men at age 50 = relative risk of > 60th percentile
Women at age 60 = relative risk of > 60th percentile

The absolute risk scores for men and women are different. There are only 12 validated measures for women and 17 for men.

COSEHC Risk Score for Women: 12 Risk Factors

- Age (years).
- Extra for cigarette smoking.
- Systolic blood pressure (mm Hg).
- Total cholesterol conc. (mg/dL). — convert to mg/dL - UK mmol/L
- Height (inches). 38.66
- Creatinine conc. (mg/dL).
- Homocysteine (µmol/L).
- Prior MI.
- Prior stroke.
- LVH.
- Diabetes.
- Nondiabetic, FBS (fasting blood sugar) (mg/dL).

COSEHC Risk Score for Men: 17 Risk Factors

- Being male.
- Age (years).
- Extra for cigarette smoking.
- Systolic blood pressure (mm Hg).
- Total cholesterol conc. (mg/dL).
- LDL cholesterol (mg/dL).
- HDL cholesterol (mg/dL).
- Triglyceride (mg/dL).
- Height (inches).
- Creatinine concentration (mg/dL).
- Homocysteine (µmol/L).
- Prior MI.
- Family history of MI pre age 60.
- Prior stroke.
- LVH.
- Diabetes.
- Nondiabetic, FBS (fasting blood sugar) (mg/dL).

Figure 20.1 shows the COSEHC scoring by points for each of the measured lab tests or risk factors

Figure 20.2 indicates the absolute risk score on the left and the risk of death from CHD within five years

RASMUSSEN CENTER CHD SCORING

The variables measured in the Rasmussen CHD scoring system include computerized arterial pulse wave analysis (CAPWA), blood pressure at rest and exercise, left ventricular mass by ECHO, microalbuminuria, brain natriuretic peptide (BNP), retinal score, Carotid IMT (intimal medial thickness), carotid ultrasound and the electrocardiogram (EKG) (see Figure 20.3). See Chapter 19 for detailed discussion of

Assess Your Risk Using CHD Scoring Tests

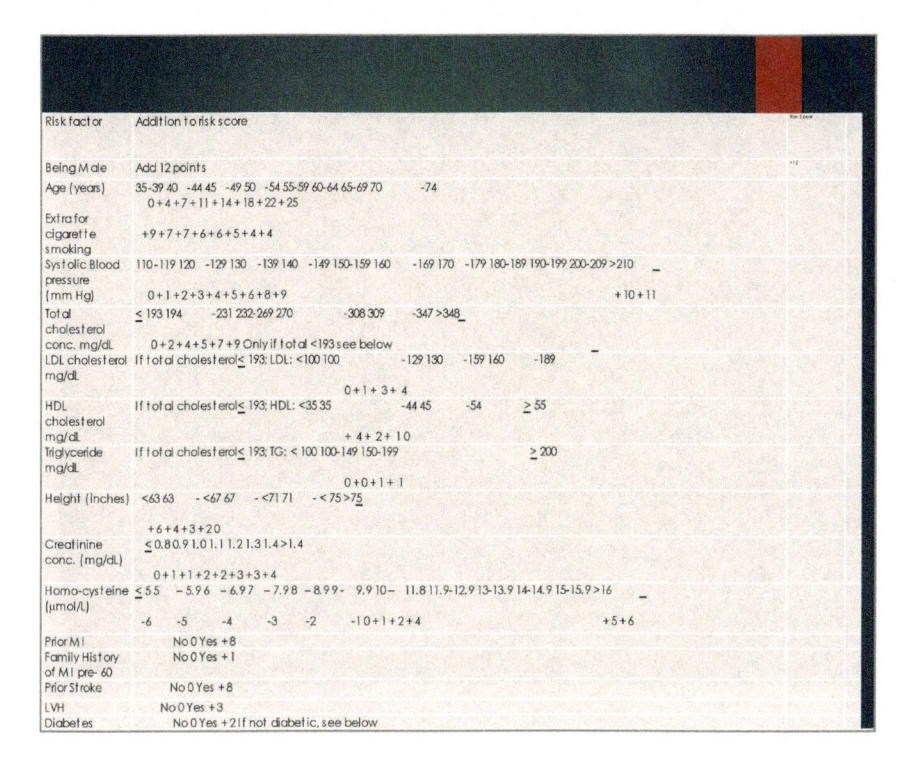

Risk factor	Addition to risk score
Being Male	Add 12 points
Age (years)	35-39 40 -44 45 -49 50 -54 55-59 60-64 65-69 70 -74 0 +4 +7 +11 +14 +18 +22 +25
Extra for cigarette smoking	+9 +7 +7 +6 +6 +5 +4 +4
Systolic Blood pressure (mm Hg)	110-119 120 -129 130 -139 140 -149 150-159 160 -169 170 -179 180-189 190-199 200-209 >210 0 +1 +2 +3 +4 +5 +6 +8 +9 +10 +11
Total cholesterol conc. mg/dL	≤ 193 194 -231 232-269 270 -308 309 -347 >348 0 +2 +4 +5 +7 +9 Only if total <193 see below
LDL cholesterol mg/dL	If total cholesterol ≤ 193; LDL: <100 100 -129 130 -159 160 -189 0 +1 +3 +4
HDL cholesterol mg/dL	If total cholesterol ≤ 193; HDL: <35 35 -44 45 -54 ≥ 55 + 4 + 2 + 1 0
Triglyceride mg/dL	If total cholesterol ≤ 193; TG: < 100 100-149 150-199 ≥ 200 0 +0 +1 +1
Height (inches)	<63 63 - <67 67 - <71 71 - <75 >75 +6 +4 +3 +2 0
Creatinine conc. (mg/dL)	≤0.8 0.9 1.0 1.1 1.2 1.3 1.4 >1.4 0 +1 +1 +2 +2 +3 +3 +4
Homo-cysteine (µmol/L)	≤5 5 -5.9 6 - 6.9 7 -7.9 8 -8.9 9- 9.9 10- 11.8 11.9-12.9 13-13.9 14-14.9 15-15.9 >16 -6 -5 -4 -3 -2 -1 0 +1 +2 +4 +5 +6
Prior MI	No 0 Yes +8
Family History of MI pre- 60	No 0 Yes +1
Prior Stroke	No 0 Yes +8
LVH	No 0 Yes +3
Diabetes	No 0 Yes +2 If not diabetic, see below

FIGURE 20.1 COSHEC point score for each CHD risk factor.

these tests. For each of these tests a numeric score is assigned as shown in the table below. The total score is calculated and the six-year risk of having a CHD event is shown below. The Rasmussen scoring is superior to the FRS.

- Disease score 0–2: no CHD events in six years.
- Disease score 3–5: 5% chance of CHD events in six years.
- Disease score over 6: 15% chance of CHD events in six years.

The higher the score the greater the risk of CHD within six years (see Figure 20.4).

CHAN2T 3 CHD SCORING SYSTEM

The CHAN2T 3 CHD scoring system predicts the risk of a CHD event in ten years as shown below. A score of 5 indicates a 25% risk of a CHD event during this time period of ten years (see Figure 20.5).

PULS CARDIAC TEST (CHL)

The PULS is a blood test predicts the five-year risk of having an MI and estimates your heart age. Figures 20.6–20.9 and the following tables illustrate the test and how this is calculated.

Scores Exceeding 40 are HIGH ABSOLUTE RISK category

Risk Score	% dying from cardiovascular disease in 5 years
0	0.04
5	0.07
10	0.11
15	0.19
20	0.31
25	0.51
30	0.84
35	1.4
40	2.3
45	3.7
50	6.1
55	9.8
60	15.6
65	24.5

FIGURE 20.2 COSHEC absolute risk of CHD death in 5 years.

Rasmussen Center CV Scoring
J Am Society of Hypertension 2011;5:401

Test	Normal	Borderline	Abnormal
Score for each test	0	1	2
Large artery elasticity		(age- and gender dependent)	
Small artery elasticity		(age- and gender dependent)	
Resting BP (mm Hg)	SBP <130 and DBP <85	SBP 130–139 or DBP 85–89	SBP ≥140 or DBP ≥90
Treadmill exercise BP (mm Hg)	SBP increase <30 and SBP ≤169	SBP increase 30–39 or SBP 170–179	SBP increase ≥40 or SBP ≥180
Optic fundus photography retinal vasculature	A/V ratio >3:5	A/V ratio ≤3:5 or mild A/V crossing changes	A/V ratio ≤1.2 or A/V nicking
Carotid IMT		(age- and gender dependent)	
Microalbuminuria (mg/mmol)	≤0.6	0.61–0.99	≥1.00
Electrocardiogram	No abnormalities	Nonspecific abnormality	Diagnostic abnormality
LV ultrasound LVMI (g/m^2)	<120	120–129	≥130
NT-proBNP (pg/dl)	<150	150–250	>250

FIGURE 20.3 Rasmussen Center scoring for CHD event in 6 years.

Assess Your Risk Using CHD Scoring Tests   213

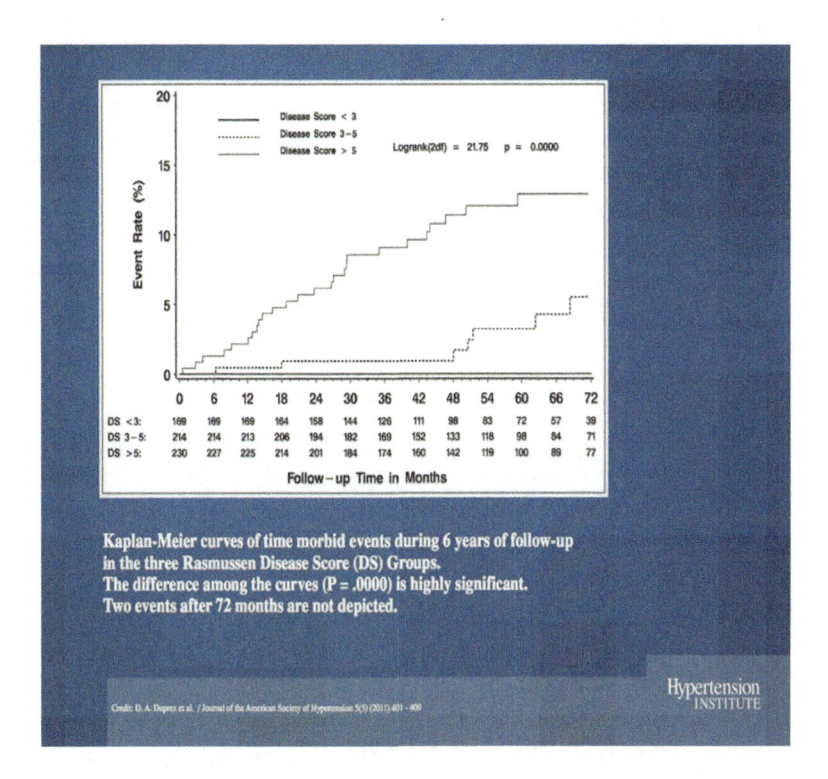

FIGURE 20.4 Kaplan-Meir curve for Rasmussen CHD event in 6 years.

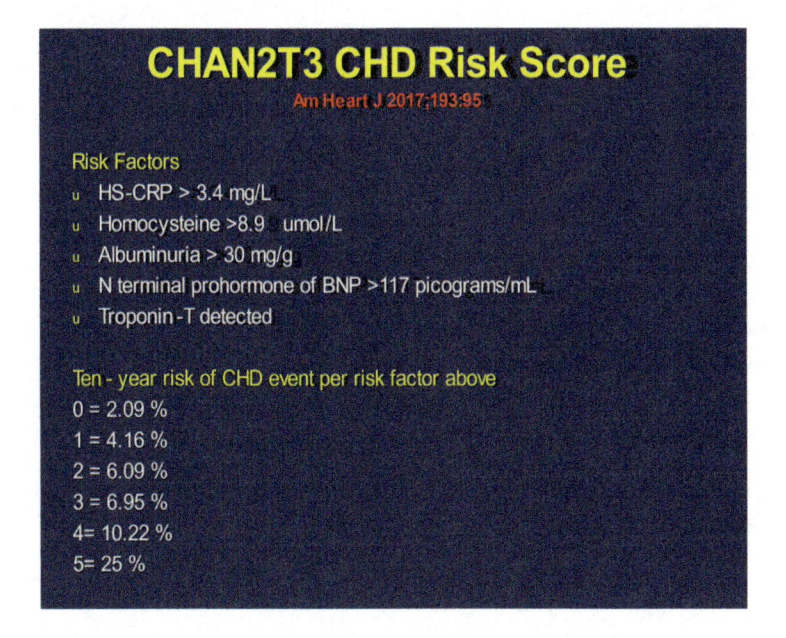

CHAN2T3 CHD Risk Score
Am Heart J 2017;193:95

Risk Factors
- HS-CRP > 3.4 mg/L
- Homocysteine >8.9 umol/L
- Albuminuria > 30 mg/g
- N terminal prohormone of BNP >117 picograms/mL
- Troponin-T detected

Ten-year risk of CHD event per risk factor above
0 = 2.09 %
1 = 4.16 %
2 = 6.09 %
3 = 6.95 %
4 = 10.22 %
5 = 25 %

FIGURE 20.5 CHANT risk score for 10 years risk for CHD event.

Elevated score related to:

- CHD development.
- Presence of unstable or vulnerable arterial plaque.
- Increased near-term risk of myocardial infarction.

Biomarkers:

- MCP-3: immune cell direction and activity.
- sFas: prevents apoptosis (cell death).
- Fas ligand: initiates cell recycling and death.
- Eotaxin: activates immune cells in areas of injury.
- CTACK: helps to clean up damaged cells.
- IL-16: recruits and activates immune cells and indicates inflammation.
- HGF: stimulates tissue repair.

Normal < 3.5. Borderline 3.5–7.49. Elevated > 7.5

GENE EXPRESSION TESTING

GES (Gene Expression Score) Corus CHD

CORUS is a blood test that measures the risk of having plaque or obstruction in your coronary arteries. The higher the score the higher the risk.

- Gene expression test that measures changes in WBC/RNA (white blood cell/ribonucleic acid) levels that are sensitive to the presence of coronary plaque.
- Measures expression levels of 23 genes grouped into six categories.
- Highly correlated with quantitative coronary artery angiogram (QCA) and degree of stenosis.
- Correlated QCA with Corus gene expression.
- The higher the Corus score (0–40) the greater the chance of a 50% or greater stenosis in one major coronary artery and greater risk of future major adverse CV event (MACE).
- Score < 15 = low risk for CHD.
- Score 28 = 50% chance of major CHD.
- Score 40 = 68% chance of major CHD.

MEASUREMENTS

The score is divided into three groups including a low, moderate, and high risk of CHD, MI, stroke, or cardiovascular death. Low risk is 0–3; moderate is 4–7; high risk is greater than 7.

Assess Your Risk Using CHD Scoring Tests

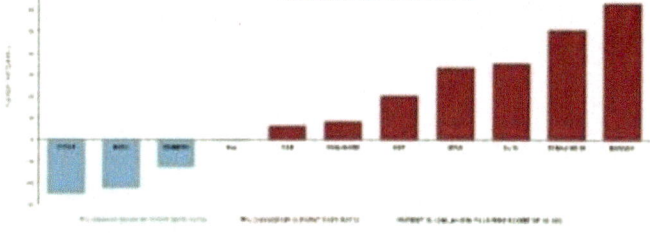

FIGURE 20.6 PULS score for MI risk in 5 years.

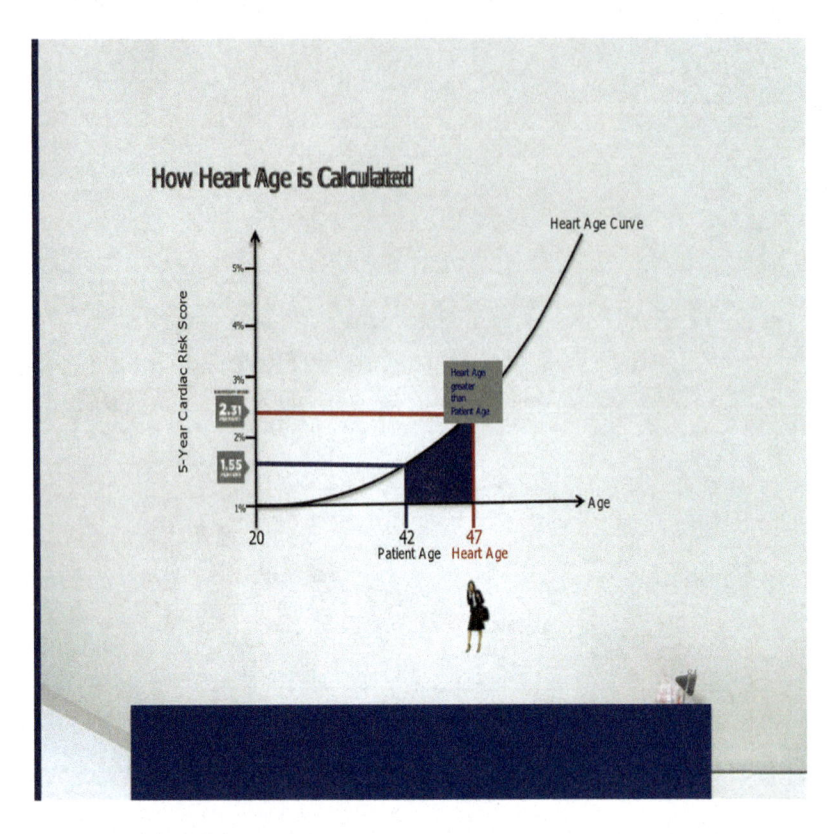

FIGURE 20.7 PULS heart age score based on age.

Tissue inhibitor of matrix metalloproteinase (TIMP) regulates the extra cellular matrix (ECM) degradation in vascular walls by inhibiting the activity of matrix metalloproteinases (MMPs).

SUMMARY AND KEY TAKE AWAY POINTS

1. There are many validated scoring systems that will predict your risk for CHD and MI.
2. Each measures different variables such as blood tests, genetics, gender, vascular tests, and predicts a CHD event, CHD death, MI, or obstructive CHD.
3. It is important to obtain all of the tests for each of the scoring systems and have your doctor calculate your risk for CHD and MI and interpret the results for you.
4. Based on these scoring systems, additional diagnostic testing and treatment will be indicated.

Assess Your Risk Using CHD Scoring Tests

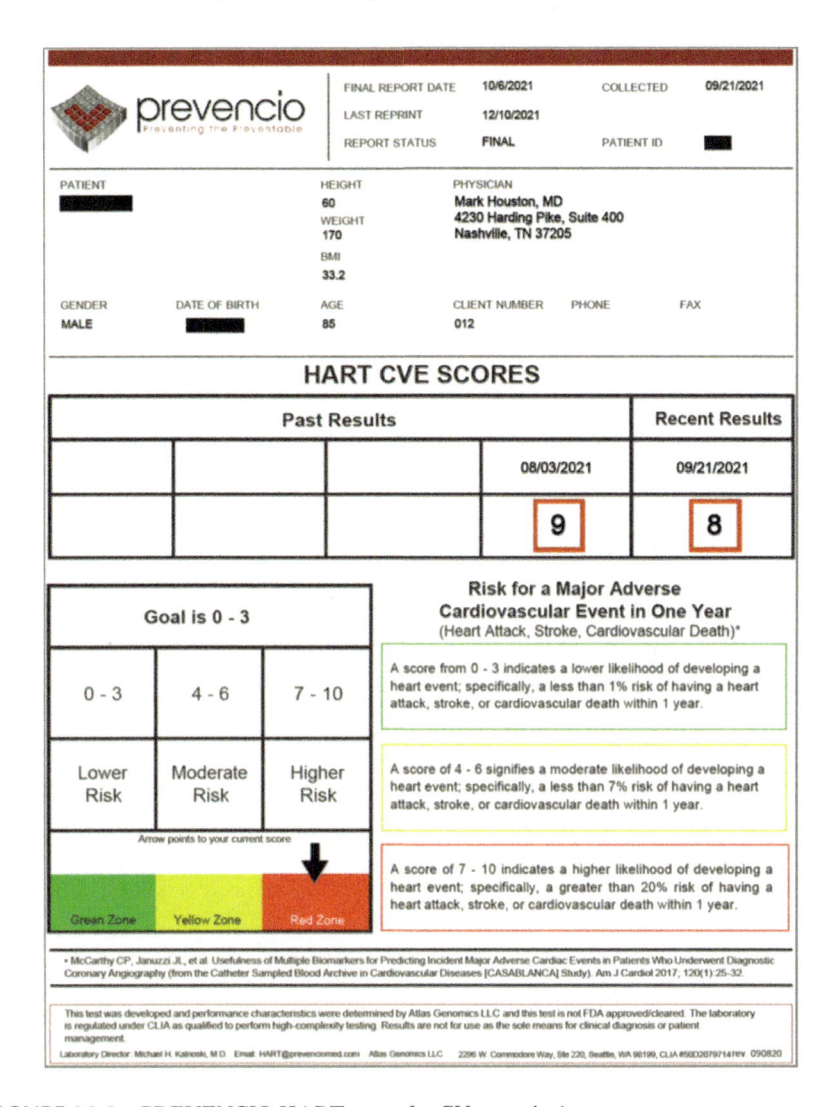

FIGURE 20.8 PREVENCIO HART score for CV event in 1 year.

Significance of the four proteins included in the HART CVE blood test:
- **NT-proBNP:** associated with stress in your heart
- **Kidney Injury Molecule-1 (KIM-1):** associated with dysfunction of your kidneys and heart and with vascular inflammation
- **Osteopontin:** associated with calcification and plaque obstruction in your heart arteries
- **TIMP-1:** associated with risk of heart artery plaque rupture

FIGURE 20.9 PREVENCIO proteins measured in HART CV event score.

BIBLIOGRAPHY

1. Houston MC, Editor. *Personalized and Precision Integrative Cardiovascular Medicine.* Philelephia: Wolters Kluwer Publishers; December 2019.
2. Houston M. Revolutionary concepts in the prevention and treatment of cardiovascular disease. In *Integrative and Functional Medical Nutrition Therapy.* New Jersey: Humana Press Noland and Drisko Editors; 2019, pp. 823–41.
3. Houston M, Bell L. *Controlling High Blood Pressure Through Nutrition, Nutritional Supplement, Lifestyle and Drugs.* Boca Raton and Oxford: CRC Press; 2021.

21 The Integrative Coronary Heart Disease (CHD) Prevention Program

NUTRITION

Nutrition is the foundation for the prevention and treatment of coronary heart disease (CHD). Nutritional studies provide definitive proof that CHD can be prevented or reduced with a weighted plant-based diet that includes 12 servings of fruits and vegetables per day (4 fruits and 8 vegetables). You should consume more monounsaturated fats (MUFA), omega-3 fatty acids, nuts, whole grains and coldwater fish. The MUFA should include 25 grams per day of olive products, such as olives and olive oil (especially extra virgin olive oil (EVOO)), and mixed nuts. The omega-3 fats should be about 3–5 grams per day with coldwater fish and supplemental omega-3 fatty acids. Reduce all of the starches, sweets, refined carbohydrates, and sugars to less than 25 grams per day. Sucrose, sugar substitutes, high fructose corn syrup, and processed foods should be stopped. One should completely eliminate all trans fats (TFA). Eggs and dairy products are not associated with CHD. Long chain SFAs should be reduced to less than 10 percent of fat intake. Organic grass-fed beef and wild game may reduce CHD. High intake of potassium of about 5 grams per day and magnesium to 1000 mg per day are recommended in conjunction with sodium restriction of 1500 mg per day. The best overall nutritional programs include the modified and updated DASH 1 and DASH 2 diets and the PREDIMED-TMD diet (Mediterranean diet). See Chapter 12 on nutrition for a detailed discussion.

NUTRITIONAL SUPPLEMENTS

Many nutritional supplements, vitamins, antioxidants, and minerals will help prevent and treat CHD by improving coronary artery elasticity and vasodilation, endothelial dysfunction, and glycocalyx dysfunction; increasing blood flow in the coronary arteries; reducing elevated blood pressure; stabilizing or reducing plaque formation; decreasing or stabilizing coronary artery calcium; and lowering vascular inflammation, vascular oxidative stress, and vascular immune dysfunction. All patients should have blood tests to measure all the micronutrients (micronutrient testing or MNT) to determine any deficiencies, and you should then take the proper dose of the micronutrient, vitamin, mineral, or antioxidant. The recommended MNT testing companies are Vibrant America Labs and Spectracell Labs. See the Sources section at the end of this book. In addition to correcting micronutrient deficiencies, you should take the

CHD, cardiac, and vascular supplements listed in this section to prevent and treat CHD and other cardiac and vascular problems. The list of all these supplements, their mechanisms of action, and clinical use in CHD, cardiac, and vascular health are discussed in this section.

NEO 40

NEO 40 is a concentrated beet supplement that provides nitrites to the blood to make nitric oxide. It also contains vitamin C, vitamin B12, methyl folate, and a proprietary nitric oxide blend (beetroot powder, Hawthorn berry extract, L-citrulline, and sodium nitrite). The dose is one wafer twice per day. It is available from Human labs as Superbeet chewables and Superbeet powder. There are no side effects. See the Sources section at the end of this book.

NEO 40 lowers blood pressure in hypertensive patients, increases blood flow, improves endothelial dysfunction, dilates both the small and large arteries to improve arterial elasticity (C1 and C2 compliance of the arterial wall); and improves the augmentation index, pulse wave velocity, central pressure, and forward and backward pulse waves. In addition, NEO 40 improves anginal chest pain, reduces coronary artery spasm and risk of clotting, decreases platelet stickiness, increases energy and overall cardiovascular function, and improves exercise function. NEO 40 improves vascular inflammation, vascular oxidative stress, and vascular immune dysfunction. Another recommended supplement for nitric oxide is Berkley Life Nitric Oxide support. Be cautious of any other nitric oxide-boosting supplements, as they have not undergone the scientific clinical studies as done with NEO 4 (see Chapters 5 and 19).

Arterosil

Arterosil is a compound containing a marine-based specialized sulfated polysaccharide (SSP), which has been verified to regenerate the glycocalyx. Arterosil contains rhamnan sulfate, a specialized sulfated polysaccharide derived from the green seaweed *Monostroma nitidum*. Rhamnan sulfate is a glycocalyx regenerating compound (GRC) and has been reported to possess anticoagulant and antithrombotic activity. GRCs are regenerating glycocalyx compounds from marine polysaccharides with a similar structure to the glycosaminoglycans found in the human endothelial glycocalyx. The Arterosil ingredients include 900 mg green seaweed (Monostroma sp.) extract, grape (seed and skin) extract, green tea (leaf) extract, grape pomace (fruit) extract, tomato (fruit), carrot (root) juice, bilberry (fruit), broccoli (aerial parts), green cabbage (leaf), onion (bulb), garlic (bulb), grapefruit (fruit), asparagus (stalk), papaya (fruit), pineapple (fruit), strawberry (fruit), apple (fruit), apricot (fruit), cherry (fruit), orange (fruit), blackcurrant (fruit), olive (fruit), and cucumber (fruit). The dose is one capsule twice per day with food. It is produced by Calroy Health Sciences. See the Sources section at the end of this book.

Arterosil increases nitric oxide levels, reduces blood pressure in hypertensive patients, increases blood flow, improves glycocalyx and endothelial dysfunction, pulse wave velocity, central pressure, forward and backward pulse waves. It

The Integrative CHD Prevention Program

stabilizes and reduces vulnerable arterial plaque (patent approved for this indication), decreases soft plaque in the carotid arteries, reduces the size of the lipid-rich core in plaques; reduces clotting, leukocyte adhesion, and cell adhesion molecules; decreases atherosclerosis; and improves large and small arterial stiffness (C1 and C2 arterial compliance); and improves the augmentation index. And finally, it improves vascular inflammation, vascular oxidative stress, and vascular immune dysfunction. There are no side effects. See Chapters 5 and 19.

D Ribose Powder

D-Ribose is a five-carbon sugar with excellent absorption and proven benefits in heart disease. It does not increase blood sugar, and, in some patients, it will lower blood sugar. In fact, it can be used as a healthy sweetener for beverages. The dose is 5 grams three times per day in the water. It is produced by Designs for Health Labs (DFH), Biotics Research Labs, and others. See the Sources section at the end of this book.

D Ribose (part of the structure of adenosine triphosphate (ATP)) increases energy in the heart muscle. Clinical results show that it improves heart function and contractility, reduces anginal chest pain, shortness of breath, and fatigue; improves coronary artery blood flow, heart muscle stiffness, and ejection fraction; decreases heart failure and rhythm abnormalities of the heart; and increases energy and exercise capability. Rare adverse effects include low blood glucose, gout, and diarrhea. See Chapters 5 and 19.

Vitamin K2 MK 7

There are two forms of vitamin K—called K1 and K2—and there are subtypes of each of these. Vitamin K is a fat-soluble vitamin. Vitamin K1 (phylloquinone) is found in plant foods, like leafy greens. Vitamin K2 (menaquinone) is found in animal and fermented foods. However, supplementing with K2 MK 7 is recommended to obtain an adequate blood level in order to prevent CHD, coronary artery calcification, and heart disease. While vitamin K1 is absorbed and used primarily by the liver, vitamin K2 MK 7 is available, and measurable, in the blood. As such, unlike vitamin K1, vitamin K2 does not get sequestered in the liver for only the liver's function. Rather, because it is available in serum blood, it can travel and go to work in other areas of the body.

The most important vitamin K in relation to CHD is vitamin K2 MK 7. Vitamin K2 MK 4 is also good but is less potent and requires 5–10 times the dose of vitamin K2 MK 7. Vitamin K2 MK 7 also has better absorption and a longer half-life of 3 days vs hours compared to vitamin K2 MK 4. Menaquinones are vitamin K2 (gut bacteria) (75% of K) are more effective than phylloquinone or vitamin K1 (dietary intake) (25% of K) to prevent vascular disease and vascular calcification. Vitamin K2MK7 is important for the production of a protein called gamma carboxyglutamate (Gla matrix protein-MGP) through a process called "carboxylation" that is important in the prevention of calcification of the coronary arteries, calcification of coronary

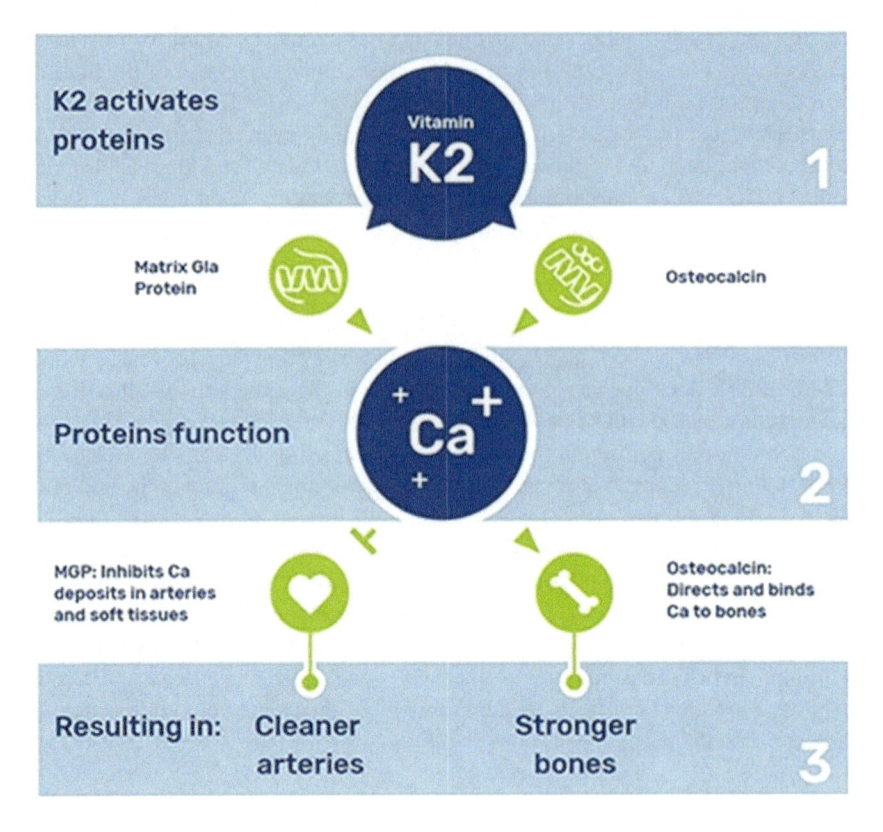

FIGURE 21.1 Mechanism of Vitamin K with Gla matrix protein and reduction in arterial calcification.

artery plaque, CHD, and MI. (Figure 21.1) The dose is 360 micrograms per day or two caps per day. K2 MK 7 is produced by Ortho Molecular labs. See the Sources section at the end of this book.

Vitamin K is also important in blood coagulation, bone formation and bone density, osteopenia and osteoporosis, soft tissue calcification, atherosclerosis, cell growth and normal cell death (apoptosis), cell signaling, cancer prevention, and brain cell protection. Vitamin K decreases HS-CRP, carotid IMT, and plaque; increases arterial elasticity; lowers blood pressure; reduces the risk of type 2 diabetes mellitus; and reduces CAC, aortic calcification, CHD, and total mortality. It improves pulse wave velocity, central pressure, forward and backward pulse waves (see Chapters 5 and 19). Vitamin K2 supplementation at 360 micrograms per day increases maximal cardiac output by 12%. Vitamin K2 MK 7 activates proteins C and S, which inhibit clotting and reduce inflammation. In addition, it increases mitochondrial ATP and induces plaque stabilization. Other effects of vitamin K include:

- Enhancing athletic performance by increasing cardiac output when serum stores are optimal.

- Supporting joint health.
- Supporting GI microbiome health.
- Protecting against vascular dementia.
- In diabetes it helps in preventing and reducing neurodegenerative diseases/neuropathies.
- Playing a role in magnesium and vitamin D metabolism.
- Supporting healthy aging through its protective actions in mitochondrial health.

Vitamin K effects are enhanced with the co-administration of vitamin D and vitamin A. At doses of K2 MK 7 below 360 micrograms there is no alteration in the risk of clotting as measured by the lab test called PT-INR in patients taking the blood thinner Warfarin. Vitamin K is safe to take with any other blood thinners. It does not increase the risk of clotting. Vitamin K1 converts to K2. Vitamin K deficiency occurs in seven days on a vitamin K deficient diet. Vitamin K2 MK 7 is safe without adverse effects and has many uses to improve CHD and overall heart health.

Clinical studies in CHD and coronary artery calcification are very positive regarding vitamin K2. The Nurses Health Study (NHS) of 72,874 female nurses showed a 16% reduction in CHD from lowest to highest quintile of vitamin K1 intake. In the Health Professional Study (HPFS) of 40,087 men, there was a 13–16% reduction in CHD. In a population-based study of 4807 subjects, the incident risk for CHD was reduced by vitamin K1 (phylloquinone) and vitamin K2 (menaquinone). Vitamin K2 MK 7 reduced CHD by 57% in upper vs lower tertile and reduced all-cause mortality by 26% in upper vs lower tertile. K2 reduced aortic calcification by 52% in upper vs lower tertile and reduced total mortality by 26%. Evidence linking vitamin K2 intakes to cardiovascular benefits started to come to light in 2004 with the landmark Rotterdam Study, which showed that high dietary intake of vitamin K2—but not vitamin K1—has a strong protective effect on cardiovascular health. Findings from this ten-year population-based study indicated that eating foods rich in natural vitamin K2 (at least 32 mcg/day) resulted in 50% reduction of arterial calcification, 50% reduction of cardiovascular risk, and 25% reduction of all-cause mortality. In 2009, these findings were confirmed by another population-based study with 16,000 subjects (ranging in age from 49 to 70) from the Prospect-EPIC cohort population. After following female participants for up to eight years, the researchers found that for every 10 mcg vitamin K2 (MK-7, MK-8, and MK-9) consumed—again, not vitamin K1—the risk of coronary heart disease was reduced by 9%.

The American Health Association published a population study that evaluated vitamin K status in 835 randomly recruited Flemish individuals. Researchers observed that higher vitamin K was associated with lower pulse wave velocity, central pressure, forward pulse wave, and backward pulse wave. Stiffening and calcification of the large arteries are forerunners of cardiovascular complications. The 2015 Knapen study provides solid evidence that, in addition to prevention, vitamin K2 can reverse existing levels of calcification and restore arterial flexibility. Among participants with an elevated arterial stiffness at baseline, this study demonstrated a significant decrease in stiffness among the MK-7 test group after three years (at doses of 180 μg/day), compared with a slight increase in stiffness for the control group. The MK-7 group demonstrated a restoration of arterial elasticity and flexibility—essentially the return of the circularity system to a previous state or degree of health (see Chapters 5 and 19 and Figure 21.1).

AGED GARLIC: KYOLIC GARLIC CARDIOVASCULAR FORMULATION

Aged garlic (Kyolic garlic) extract has been proven to have beneficial effects in the prevention of CHD, as well as lowering blood pressure and serum lipids and cholesterol. Aged garlic reduces coronary calcium and plaque progression and plaque volume in humans in those on statins and in those not on statins. In a trial of 23 patients over one year, consuming aged garlic at 4 grams per day, the CAC was slowed significantly. The garlic-treated patients increased CAC only 7.5% but the placebo-treated patients increased CAC by 22.2%. There was a high correlation with oxidative stress, ox LDL, Lp (a), and vascular dysfunction with the change in the CAC. Also aged garlic with B vitamins, folate, and arginine retarded coronary atherosclerosis reducing progression CAC of 6.8% vs 26.5% with placebo that was significant at p=0.005.

In the FAITH Trial, garlic and Coenzyme Q10 (CoQ10) evaluated the effects on vascular elasticity and endothelial function. Sixty-five firefighters in Los Angeles were given aged garlic at 300 mg a day with CoQ10 at 30 mg per day vs a placebo for 12 months. The pulse wave velocity (PWV) significantly decreased by 1.21 m/s (vascular elasticity) ($p < 0.005$), and endothelial function improved in the treated group significantly ($p < 0.01$).

In a recent study in 2020, 104 patients were randomized with 46 in the active group given aged garlic at 2400 mg daily for one year and 47 in the placebo group. There was a significant ($p < 0.05$) change in CAC progression. blood glucose and IL-6 in favor of the active group. There was also a significant ($p = 0.027$) decrease in systolic blood pressure in the aged garlic group. See Chapters 5 and 19.

There are many active ingredients in garlic, such as Allicin, Ajoene, S allyl cysteine which provide the following mechanisms of action:

- Reduces vascular muscle enlargement and vascular stiffness and augmentation index.
- Reduces oxidative stress and inflammation, HS-CRP, and TNF alpha which are blood markers for inflammation.
- Lowers blood pressure.
- Improves lipids, lowers LDL, increases HDL, and decreases ox LDL.
- Prevents entry of lipids into arterial wall and macrophages to help reduce plaque formation especially low-attenuation plaque (LAP) volumes.
- Decreases endothelial damage and improves endothelial dysfunction.
- Increases nitric oxide.
- Improves the microcirculation.
- Reduces platelet aggregation and adhesion to reduce clotting.
- Increases antioxidant defense with glutathione superoxide dismutase and (SOD).

The dose is 600 mg twice per day of Kyolic garlic cardiovascular formula and can be obtained from most drug stores, grocery stores or health food stores. See Chapters 5, and 19.

OMEGA-3 FATTY ACIDS

Omega-3 fatty acids are a type of fat the human body cannot make on its own. However, they are an *essential* fat, which means they are needed to survive and assure good health. We get omega-3 fatty acids from our foods. The human body can make most of the types of fats it needs from other fats or raw materials. That is not the case for omega-3 fatty acids. These are essential fats that come from fish, vegetable oils, nuts (especially walnuts), flax seeds, flaxseed oil, algae, and leafy vegetables. They are an integral part of all cell membranes throughout the body and affect the function of the cell receptors in these membranes. They provide the starting point for making hormones that regulate blood clotting, contraction and relaxation of artery walls, and inflammation. They also bind to receptors in cells that regulate genetic function. Likely due to these effects, omega-3 fats have been shown to help prevent and treat all types of heart disease, especially CHD and stroke, and other health conditions.

Omega-3 fats are a key family of polyunsaturated fats. There are three main omega-3s:

- Eicosapentaenoic acids (EPA) come mainly from fish, so they are sometimes called marine omega-3s.
- Docosahexaenoic acids (DHA) come mainly from fish, so they are sometimes called marine omega-3s.
- Alpha-linolenic acid (ALA), the most common omega-3 fatty acid in most Western diets, is found in vegetable oils and nuts (especially walnuts), flax seeds and flaxseed oil, leafy vegetables, and some animal fat, especially in grass-fed animals. The human body generally uses ALA for energy, and conversion into EPA and DHA is very limited. The conversion from ALA to EPA and DHA may be only 5%. This conversion requires several enzymes and many nutrients. Cofactor deficiencies in magnesium or vitamin B6 or interfering substances or conditions such as alcohol, trans fats, saturated fats, high omega-6 FA, obesity, insulin resistance, and DM will decrease the conversion to EPA and DHA. It is important to increase the anti-inflammatory omega-3 fatty acids in the diet and reduce the inflammatory omega-6 fatty acids. (See Figure 21.2 below.)

Numerous published clinical research shows that omega-3 fatty acids can improve your cardiovascular health and especially reduce CHD and MI. Most of this research involves EPA + DHA, but ALA may also have some cardiovascular benefits. In 25 studies of 280,000 patients, there is an inverse association between fish consumption, omega-3 FA, and morbidity and mortality from CHD. They also reduce vascular inflammation.

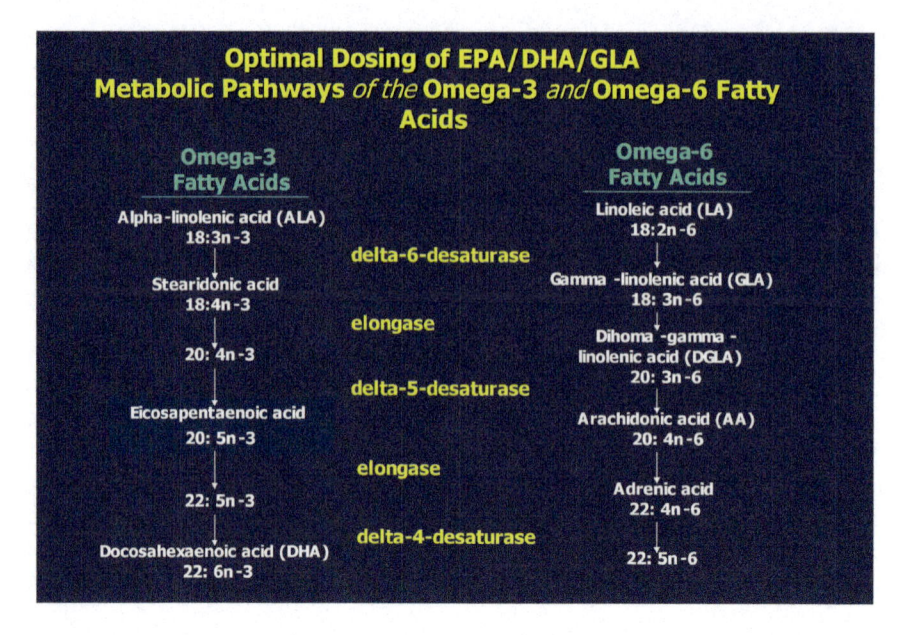

Optimal Dosing of EPA/DHA/GLA
Metabolic Pathways *of the* Omega-3 *and* Omega-6 Fatty Acids

Omega-3 Fatty Acids		Omega-6 Fatty Acids
Alpha-linolenic acid (ALA) 18:3n-3		Linoleic acid (LA) 18:2n-6
↓	delta-6-desaturase	↓
Stearidonic acid 18:4n-3		Gamma-linolenic acid (GLA) 18:3n-6
↓	elongase	↓
20:4n-3		Dihoma-gamma-linolenic acid (DGLA) 20:3n-6
↓	delta-5-desaturase	↓
Eicosapentaenoic acid 20:5n-3		Arachidonic acid (AA) 20:4n-6
↓	elongase	↓
22:5n-3		Adrenic acid 22:4n-6
↓	delta-4-desaturase	↓
Docosahexaenoic acid (DHA) 22:6n-3		22:5n-6

FIGURE 21.2 Omega-3 pathway and the conversion of ALA to EPA and DHA. Linoleic acid: 18:2 n-6 found in nuts, seeds, vegetable oils, like corn, sunflower, safflower, canola, and soybean oil. Alpha linolenic acid: 18:3 n-3 found in seeds of flax, rape, perilla, walnuts, chia, and chloroplasts of leafy green vegetables. It takes five enzymes for conversion from ALA or linoleic acid to EPA or AA: 3 desaturases (6, 5, and 4) and 2 elongases. Enzymes are the same for both n-3 and n-6 conversion.

The Cardiovascular Benefits of Omega-3 Fatty Acids

The benefits of omega-3 fatty acids are:

- Reduced risk of all cardiovascular diseases and atherosclerosis.
- Improves angina.
- Prevention and treatment of CHD and MI.
- Reduced risk of death if you have cardiovascular disease.
- Reduced risk of sudden cardiac death caused by an abnormal heart rhythm such as ventricular tachycardia or ventricular fibrillation and improved heart rate variability.
- Improves parasympathetic tone and reduces sympathetic nervous system activity.
- Reduced risk of blood clots because omega-3 fatty acids help prevent blood platelets from clumping together.
- Improves endothelial dysfunction and increases nitric oxide.
- Improves arterial elasticity, augmentation index, and pulse wave velocity.
- Reduces plaque from forming in the coronary arteries. Reduces plaque rupture and reduces the size of the lipid-rich core.
- Reduces coronary artery calcium (CAC) score.

- Reduces stenosis of the coronary artery after the placement of stents.
- Reduces restenosis of the coronary arteries after coronary artery bypass graft (CABG).
- Lowers triglyceride levels and raises HDL cholesterol.
- Decreases LDL particle number and increases LDL particle size.
- Lowers blood pressure.
- Lowers glucose and improves insulin resistance.
- Decreases inflammation and HS-CRP.

See Chapters 5 and 19 for more information.

Here are some of the best sources of omega-3 fatty acids and the content:

- **Mackerel**
  - Serving Size: 3 ounces (100 grams)
  - Amount of omega-3 fat: 2.5–2.6 grams
- **Salmon (wild)**
  - Serving Size: 3 ounces (100 grams)
  - Amount of omega-3 fat: 1.8 grams
- **Herring**
  - Serving Size: 3 ounces (100 grams)
  - Amount of omega-3 fat: 1.3–2 grams
- **Tuna (Blue fin)**
  - Serving Size: 3 ounces (100 grams)
  - Amount of omega-3 fat: 1.2 grams
- **Lake Trout**
  - Serving Size: 3 ounces (100 grams)
  - Amount of omega-3 fat: 2 grams
- **Anchovy**
  - Serving Size: 3 ounces (100 grams)
  - Amount of omega-3 fat: 1.4 grams
- **Tuna (Albacore)***
  - Serving Size: 3 ounces (100 grams)
  - Amount of omega-3 fat: 1.5 grams
- **Lake White fish (freshwater)**
  - Serving Size: 3 ounces (100 grams)
  - Amount of omega-3 fat: 1.5 grams
- **Bluefish**
  - Serving size: 3 ounces (100 grams)
  - Amount of omega-3 fat: 1.2 grams
- **Halibut**
  - Serving size: 3 ounces (100 grams)
  - Amount of omega-3 fat: 0.9 grams
- **Striped Bass**
  - Serving size: 3 ounces (100 grams)
  - Amount of omega-3 fat: 0.8 grams

- **Sea Bass (mixed species)**
  - Serving size: 3 ounces (100 grams)
  - Amount of omega-3 fat: 0.65 grams
- **Tuna, white meat canned**
  - Serving size: 3 ounces drained
  - Amount of omega-3 fat: 0.5 grams

Clinical Research Studies

In this meta-analysis, there were many randomized clinical trials (RCTs) (93,000 subjects) and prospective cohort studies (732,000 subjects) examining EPA+DHA from foods or supplements and CHD, including myocardial infarction, sudden cardiac death, coronary death, and angina in primary and secondary prevention.

Among RCTs, there was a 6% reduction in CHD risk with EPA+DHA treatment. A subgroup analysis of data from RCTs indicated a statistically 14–16% significant CHD risk reduction with EPA+DHA treatment among higher-risk populations, including participants with elevated triglyceride (TG) levels over 150 mg/dL and elevated low-density lipoprotein cholesterol above 130 mg/dL.

Meta-analysis of data from prospective cohort studies resulted in a statistically 18% significant reduction in CHD for higher intakes of EPA+DHA over one gram per day and risk of any CHD event. Sudden cardiac death (SCD) was reduced by 47%.

In patients with high TG over 150 mg/dL and a higher dose of omega-3 FA over one gram per day there was a reduction of CHD by 25%.

- **Dart Study**. 30% reduction in total and CHD mortality in men with previous MI over two years with fatty fish twice a week at 200–400 grams (500–800 mg n-3 FA).
- **Gissi Study.** Decrease in all-cause mortality (28%), CV mortality (30%), CHD mortality (28%), and sudden death (47%) within four months with one gram omega-3 FA in 11,000 patients with recent MI. No change in CVA or nonfatal MI.
- **Kupoio Heart Study**. Men in the highest quartile vs lowest quartile had a 44% reduction in fatal and nonfatal CHD.
- **Jelis Study**. 18% reduction in CHD events in patients already on a statin with or without preexisting CHD, especially in those with preexisting CHD with high TG, low HDL or hyperglycemia by 19% on 1.8 grams of EPA.
- Cohort study of over 50,000 patients 30% lowered risk of MI in men but not women over eight years in the highest quintile vs lowest quintile.
- **GISSI Heart Failure (CHF) study**: reduction in sudden death in ischemic and non-ischemic moderate to severe CHF by 9% to 20% on one gram omega-3 FA for four years.
- **Meta-analysis** of cohort studies and clinical trials with low dose EPA and DHA 250 mg per day reduced the risk of fatal CHD by 36%.
- **Meta-analysis** of 5 prospective cohort studies showed that the risk of fatal CHD was 21% lower among subjects with a high intake of ALA vs low intake of ALA.

- **Meta-analysis** of 17 prospective studies of 315,812 subjects over 16 years showed a 16% reduction in CHD and sudden death.
- **Reduce PCTA and stent re-stenosis** in low doses and reduce angina and MI.
- **Reduce CABG graft occlusion** after one year of 3–4 grams of omega-3 FA.
- **Japanese have less atherosclerotic plaque** burden consuming 8–15 times more omega-3 fatty acids than the US population.
- **Plaque stabilizing effect.** Incorporation of omega-3 FA into carotid plaque within 42 days reduced the number of macrophages and a plaque morphology that suggested increased stability and reduced plaque progression.
- **Coronary artery plaque** prevention and regression and CAC stabilization and reduction.

Safety of Omega-3 Fatty Acids

Omega-3 fatty acids are safe and have very few adverse effects from the recommended daily dose of 1000–5000 mg per day in divided doses twice per day. Supplementation of less than 4 grams per day when co-prescribed with antiplatelet and anticoagulants are not associated with an increased risk of major or minor bleeding episodes. Ten grams EPA equals 320 mg ASA for bleeding risk. The very low risk of possible atrial fibrillation remains controversial and may be limited to very special populations or poor-quality omega-3 fatty acids. These studies and analyses need more verification. This is not a reason to avoid omega-3 FA.

DOSE, INGREDIENTS, AND QUALITY

The ingredients and relative components of the omega-3 fatty acid supplement should be as follows:

The EPA to DHA ratio should be 3:2. The gamma linolenic acid (GLA) dose should be at 50% of the total dose of DHA and EPA (1:2 ratio). The gamma/delta tocopherol at 100 mg per 1000 mg DHA/EPA/GLA with no more than 20% as alpha tocopherol. GLA converts to DGLA which is anti-inflammatory. GLA depletes DHA and EPA. EPA and DHA deplete GLA.

The Omega-3 index of 8% or higher has been reported to be associated with the greatest cardioprotection. Omega-3 index of less than 4% gives the least cardioprotection.

The best supplement with these ingredients is EFA Sirt Supreme: 3–5 capsules twice per day with food (from Biotics Research). See the Sources section at the end of this book.

CURCUMIN

Turmeric is a spice that has received much interest from both the medical and scientific worlds as well as from the culinary world. It is a rhizomatous herbaceous perennial plant (*Curcuma longa*) of the ginger family. The medicinal properties of turmeric, the source of curcumin, have been known for thousands of years. It has antioxidant, anti-inflammatory, antimicrobial, and anticancer properties.

Curcumin, a polyphenol, has been shown to target multiple signaling molecules while also demonstrating activity at the cellular level, which has helped to support its multiple health benefits. It has been shown to benefit CHD, kidney function, inflammatory conditions, metabolic syndrome, pain, and arthritis. Most of these benefits are due to its antioxidant and anti-inflammatory effects. Despite its reported benefits via inflammatory and antioxidant mechanisms, one of the major problems with ingesting curcumin by itself is its poor bioavailability, which appears to be primarily due to poor absorption, rapid metabolism, and rapid elimination. Piperine, a known bioavailability enhancer, is the major active component of black pepper and is associated with an increase of 2000% in the bioavailability of curcumin. Therefore, the issue of poor bioavailability appears to be resolved by adding agents such as piperine that enhance bioavailability, thus creating a curcumin complex.

The cardiovascular and other health benefits of curcumin are as follows:

- Decreases proinflammatory cytokines during CABG procedures.
- Decreases cardiomyocytic apoptosis (death) after cardiac injury or MI.
- Reduces MI post-CABG from 30% to 13% at 4 grams per day given three days before and five days after CABG.
- The administration of curcumin 45 mg dose daily for seven days prior to percutaneous angioplasty PCA until 48 hours post-PCA is useful in reducing inflammatory response post-PCA with stable CHD.
- Decreases HS-CRP, inflammation, and oxidative stress markers.
- Reduces the expression and activity of metalloproteinases, such as MMP-2 and MMP-9, which can increase the risk of CHD and plaque rupture.
- Membrane stabilizing effect on the cardiac muscle.
- Inhibits platelet activation.
- Improves endothelial dysfunction.
- Improves vascular muscle stiffness and thickening.
- Decreases PWV.
- Increases adiponectin and lowers leptin, which are blood markers to improve diabetes.
- Improves insulin resistance.
- Lowers TG, LDL, and increases HDL.
- Lowers blood uric acid.
- Decreases visceral and total body fat.

See Chapters 5 and 19.

Curcumin is very safe with minimal adverse effects. There is no toxicity at doses of 10 grams per day.

The recommended and best-absorbed Curcumin supplements are Curcumin RX 2 twice per day (from Biotics Research), Curcumin Evail 2 twice per day (from Designs for Health) or Turvia 2 twice per day (from Ortho Molecular). See the Sources section at end of this book.

QUERCETIN

Quercetin, a flavonoid found in fruits and vegetables, has unique biological properties that may improve CHD and other cardiovascular problems. It also has other properties including anticarcinogenic, anti-inflammatory (lowers HS-CRP), antiviral, antioxidant, and has the ability to inhibit lipid peroxidation, platelet aggregation, and capillary permeability, and stimulate mitochondrial biogenesis. Quercetin (25 mg/kg) administered to rabbits on a high cholesterol diet had reduced progression and increased regression of atherosclerosis. It lowers the oxidative stress substance called myeloperoxidase (MPO), increases the removal of LDL cholesterol from cells called reverse cholesterol transport (RCT), and increases paraoxonase 2 (PON 2), which improves HDL function and decreases the oxidation of LDL. In addition, quercetin will improve endothelial function and increase nitric oxide, lower blood pressure, and decrease platelet aggregation and thrombosis. Exercise endurance is enhanced. There is an increase in muscle mitochondria in humans. Quercetin may slow aging by blocking senescence proteins.

Fruits and vegetables are the primary dietary sources of quercetin, particularly citrus fruits, apples, onions, parsley, sage, tea, and red wine. Olive oil, grapes, dark cherries, and dark berries such as blueberries, blackberries, and bilberries are also high in quercetin and other flavonoids. The recommended dose is 500 mg twice per day as Quercetin with nettles (from Designs for Health). **See the Sources section.**

COENZYME Q10

Ubiquinone (CoQ10) (oxidized form) and ubiquinol (CoQ10H2) (reduced form of Coenzyme Q10) are essential for optimal cellular function and production of ATP by the electron transport chain (ETC), a carrier of electrons and protons, in all human cells.

CoQ10 is a fat-soluble molecule (benzoquinone with 10 units in an isoprenoid side chain) that is found primarily in the mitochondria, but also in serum lipoproteins and cell membranes. The clinical use in cardiovascular disease includes hypertension, endothelial function, coronary heart disease (CHD), myocardial infarction (MI), and systolic and diastolic congestive heart failure (CHF).

Absorption, Pharmacokinetics, and Drug Depletions

CoQ10 is poorly absorbed unless combined in a lipid-soluble or emulsified environment or delivery system. It declines with age and is lower in patients that exercise or have high oxidative stress, cancer, diabetes mellitus (DM), hypertension, or CHF. Many drugs deplete CoQ10, such as statins, beta blockers, diuretics, and metformin.

The ability to convert to ubiquinone may also be impaired in DM, advanced CHF, and with certain genetic SNPs (NQ01 gene) . The optimal COQ 10 level is about 2–3 μmol/L, but in advanced CVD, levels of 3 μmol/L or greater are recommended. However, there may be a discrepancy between actual intracellular and mitochondrial levels and serum levels due to intracellular transport problems. Thus, a normal serum

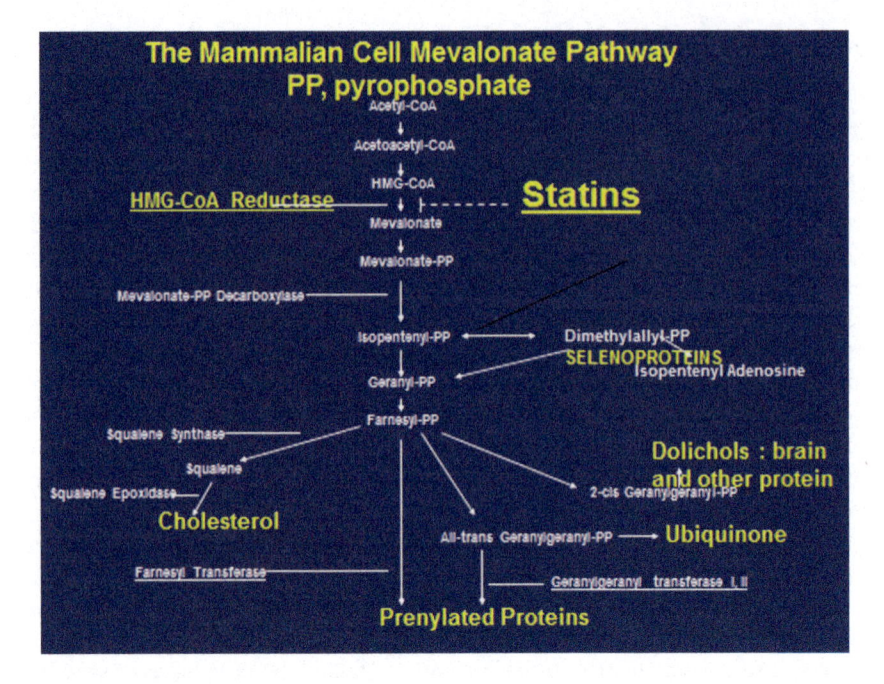

FIGURE 21.3 Production of CO Q-10 and how statins reduce the levels.

level may not reflect reduced intracellular or reduced mitochondrial levels. Intracellular levels of C0Q 10 should be measured in these cases using micronutrient testing.

CoQ10 is synthesized via acetyl CoA as shown below into the all trans geranyl-geranyl-pyrophosphate pathway (GGPP). Statins inhibit the HMG-COA reductase enzyme and reduce the downstream metabolites into the GGPP pathway and thus lower CoQ10. This may result in myalgias, myopathy, severe fatigue, oxidative stress, inflammation, vascular immune dysfunction, and clinical cardiovascular diseases, such as hypertension, endothelial dysfunction, CHD, and CHF. See Figure 21.3.

Clinical Physiology and Functions

CoQ10 is involved in numerous body functions. It is an antioxidant and is involved in DNA synthesis, lysosomal function, gene expression, mitochondrial protein uncoupling, mitochondrial permeability, mitochondrial ETC (electron transport chain) and ATP production, membrane function, reduction of lipid peroxidation and reduction of oxLDL, apoptosis, and recycling of other micronutrients especially tocopherols and vitamin C. The ETC complex 1–4 on the inner mitochondrial membrane produces ATP via electron transport with CoQ10 involvement particularly at Complex 1 and 2 (Figure 21.4).

Cardiovascular Disease: CHF, CHD, Angina, and MI

The heart uses about 6 million mg of ATP per day at normal heart rates of 60–70 beats per minute. This is more than any other organ and is about 20–30 times its own

The Integrative CHD Prevention Program

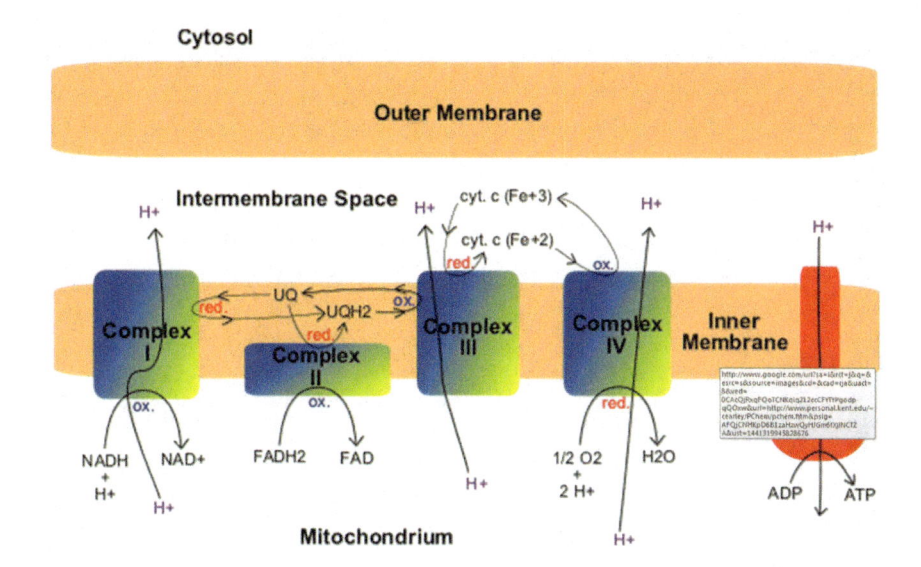

FIGURE 21.4 The electron transport chain of the mitochondria and the production of ATP.

weight. The loss of ATP results in diastolic dysfunction, followed by LAH, LVH, and systolic CHF, with low ejection fraction (EF). The primary energy sources for the heart are:

Free fatty acids: 60% or more
Carbohydrates: Glucose 35% or less
Amino acids and ketone bodies: 5%

The heart is different from most organs as it prefers FA to glucose for energy. Mitochondrial dysfunction is one of the most important events in the development of CHF, which explains the importance of CoQ10 in the mitochondrial ETC and ATP production. The metabolic pathways with glycolysis, TCA cycle, ETC, and final conversion to contractile energy are shown below with the importance of glucose, free fatty acids, the carnitine shuttle, beta oxidation, ETC production of ATP, then creatine kinase and the phosphocreatine shuttle for cardiac myocyte contractility. The three important parts of cardiac energy production are:

- Energy Substrate Utilization: Energy from food, uptake of glucose and FFA by cardiac myocyte then breakdown by glycolysis, Krebs cycle (TCA) and beta oxidation with the production of ATP.
- Oxidative Phosphorylation (ETC): Mitochondria and ATP production.
- ATP transfer and utilization: Creatine kinase and phosphocreatine energy shuttle, actomyosin ATPase in myofibril, Ca-ATPase in sarcoplasmic reticulum and Na-K ATPase in sarcolemma.

See Figure 21.5 (CHART 1).

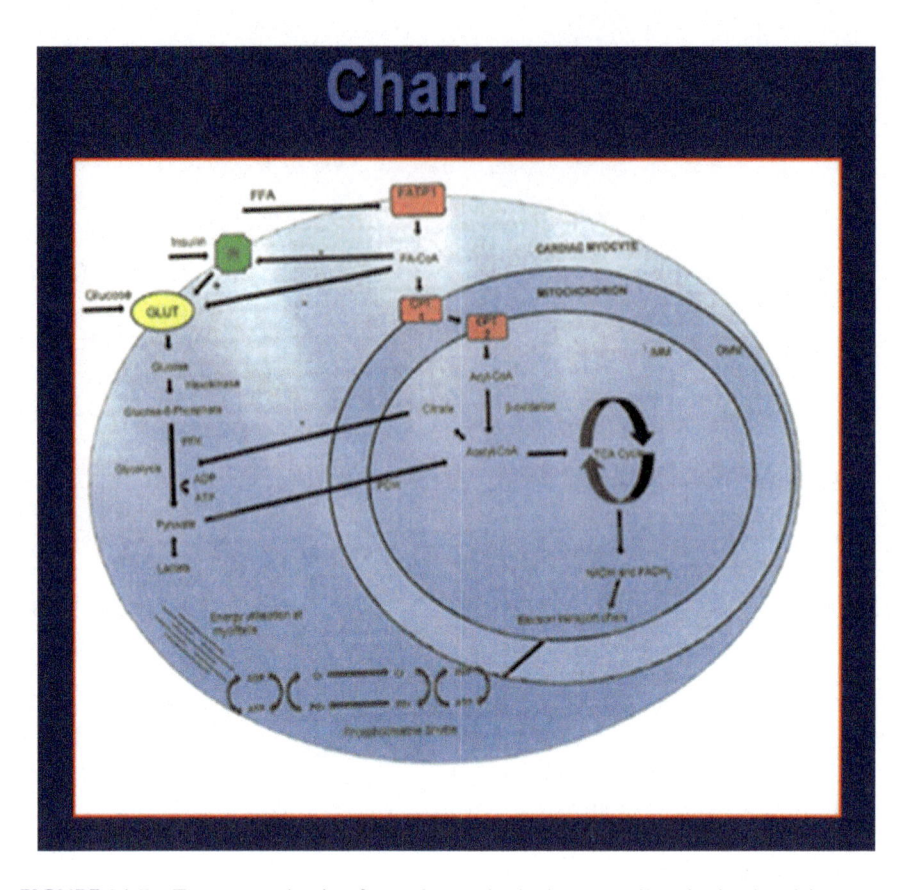

FIGURE 21.5 Energy production from glucose in the human cell and mitochondria.

Reduced levels of CoQ10 are an independent risk predictor of increased mortality in CHF, CHD, and MI patients. CoQ10 improves angina, CHD, diastolic dysfunction, CHF, and ventricular arrhythmia. CoQ10 improves endothelial dysfunction and mitochondrial function in patients with ischemic LV systolic CHF within eight weeks. CoQ10 improves functional abnormalities on ECHO, increases EF, improves clinical symptoms, and enhances overall cardiac function. . These clinical effects are related to its antioxidant activity, protection of membrane phospholipids and the mitochondrial membrane protein, reduction in oxidative stress increases in oxidative defenses with SOD and catalase, lowering of LDL oxidation NT-proBNP, cell adhesion molecules, and IL-6 and increasing in HDL-C.

CoQ10 reduced post-MI reperfusion ventricular arrhythmias, improved LV function and total cardiac death. In a DBRPC trial of 144 subjects with acute MI, CoQ10 at 120 mg per day administered within the first three days of an MI resulted in significant improvements in the treated group in all parameters ($p < 0.05$).

- Angina (9.5% vs 25.3%).
- Arrhythmias (9.5% vs 25.3%).

The Integrative CHD Prevention Program

- LVF improved (8.2 vs 22.5%).
- Total cardiac events and death reduced at 15% vs 30.9% (P < 0.02).

In patients with acute MI, COQ10 administration at 200 mg per day over 12 weeks had higher HDL, lower LDL, lower fibrinogen, and lower BP compared to the placebo group. These risk factor improvements would reduce future risk for MI. CoQ10 pretreatment with 100–300 mg per day for 7–14 days prior to CABG improved short-term outcome measures. Several placebo-controlled trials in patients with chronic stable angina demonstrated improved exercise tolerance and reduced or delayed EKG changes of myocardial ischemia in the CoQ10 treated group.

A meta-analysis of eight and 11 randomized trials over —one to six months showed significantly increased LVEF and CO . In a study over four weeks, in patients with moderate to severe CHF, administration of COQ 10 at 300 mg/day improved significantly the LV contractility and peak VO2. CoQ10 also increased peak VO2 vs placebo ($p < 0.05$).

CoQ10 reverses statin-induced diastolic dysfunction and improves systolic and diastolic dysfunction, endothelial function, and mitochondrial function at 300 mg per day within eight weeks. In a DBPC study, 641 patients with CHF, CoQ10 reduced hospitalization by 20% vs placebo.

Published clinical trials with 395 patients, from 1985 to 2005 in which 79% of the subjects were men, ages 50–68, conducted over 4–28 weeks demonstrated impressive results.

- Ejection fraction (EF) increased from 22% to 46% and NYHA class from 2.3 to 3.4.
- Pooled net change of 3.67% in EF (–3.0% to 17.8%) and improved NYHA functional class (CI 1.60 to 5.74%)
- Best increase in EF was in those with EF over 30%.
- Baseline Coq10 was .61–1.01 µg/ml which increased 1.4 µg/ml in most of the studies at doses of CoQ10 of 60–300 mg/d. The ideal blood level is 2–3ug/ml but this was achieved in only three studies. If all studies had achieved the optimal serum levels, the results would have likely been much better.

The Q–SYMBIO Trial of CoQ10 and CHF is the most important of all the studies yet published on CoQ10 and heart failure.

A summary of the results of this study are shown below:

- Randomized trial of 420 patients with moderate to severe CHF defined as NYHA class 3 and 4 over ten years.
- 2 mg/kg CoQ10 per day (100 mg tid) vs placebo plus standard therapy. Increased serum COQ 10 levels three times.
- Primary short-term endpoints: NYHA function class, six-minute walk test, NT-pro BNP. No difference between groups. No significant change in EF.
- Reduced MACE (major adverse cardiac events) by 43% ($p = 0.003$ CI: 32–80) and all-cause mortality 42% (major adverse cardiovascular events:

hospitalization or death due to CHF, CV/MI death, SCD, cardiac transplant, mechanical circulatory support.)
- Reduced all-cause mortality by 50% (p = 0.01) and CHF.

Hypertension and Endothelial Function

CoQ10 is very effective in lowering blood pressure and improving endothelial function. It reduces SVR and BP which correlates with an increase in serum CoQ-10 level (↑ .97 µg/ml) (p <0.02) compared to pretreatment CoQ10 levels. Office BP reductions average 17/10 mm Hg and the 24-hour ABM decreases 18/10 mm Hg (p < 0.001).

In a meta-analysis of 12 trials with 362 patients, the range in BP reduction was 11–17/8–10 mm Hg. Therapeutic plasma levels are 3.0 ng/ml. The heart rate decreased five beats per minute. The BP effects occur at 4–12 weeks and are gone at 7–10 days after discontinuation.

There is a deficiency of Co Enzyme Q-10 in 39% of hypertensive patients vs 6% of control patients. CoQ10 is reduced with age, disease, oxidative stress, statins, HLP, CHD, HBP, DM, aerobic exercise, and atherosclerosis. The administration of CoQ10 reduces the dose and number of BP medications. The best BP reduction occurs with the lowest initial CO-Q-10.

In a 12-week R, DB placebo-controlled trial of 83 men and women with systolic hypertension (165/81–82 mm Hg) were given 60 MG CO-Q-10 bid. The serum levels increased by 2.2 µg/ml. The SBP fell 18 mm Hg (P < 0.01) and the DBP fell 2.6 mm Hg.

Patients administered Co10 with enalapril improved 24-hour BP better than enalapril monotherapy and normalized endothelial dysfunction.

In older mice, CoQ10 improved NO bioavailability normalized superoxide production and oxidative stress (nitrotyrosine abundance), as well as with increases in markers of vascular mitochondrial health, including antioxidant status. It also reversed the age-related increase in endothelial susceptibility to acute mitochondrial damage. This study and others indicate that preservation of vascular endothelial function with advancing age may prevent of age-related CVD. In a placebo-controlled trial in diabetic patients over 12 weeks, CoQ10 at 200 mg per day improved EDV (endothelial-dependent vasodilation). Patients on statins also have improved EDV when given CoQ10 at 200 mg per day for 12 weeks. A meta-analysis of five randomized controlled trials with 194 subjects found that CoQ10 at 150 to 300 mg per day for 4–12 weeks improved EDV by 1.7%.

SUMMARY

CoQ10 improves endothelial function, lowers BP, reduces CHF morbidity and mortality, improves CHD, angina, and MI. It may be used in conjunction with other drugs and supplements to enhance efficacy. Doses of CoQ10 should be adjusted to both serum levels and intra-cellular levels to achieve optimal cardiovascular improvement. CoQ10 is safe even at high doses.

The recommended dose is 200 mg per day of COQNOL (from Designs for Health). The optimal COQ 10 level is about 3 µmol/L. See the Sources section at the end of this book.

The Integrative CHD Prevention Program

MITOQUINONE: MITO Q

Mito Q is an enhanced form of COQ10 that has been engineered to enter the heart mitochondria at significant levels. No other antioxidant, CO Q 10, or form of CO Q 10 can do this as effectively. Once inside the mitochondria, the production of ATP is increased, heart muscle functions better and there is improvement in CHD and CHF. MitoQ improves cardiac enlargement and remodeling, fibrosis, LV dysfunction, and dysregulation of RNAs in pressure-overloaded hearts. I have seen patients with a marked reduction in anginal chest pain, improvement in the cardiac ejection fraction, improved CHF symptoms, reduction in diastolic dysfunction, and improved energy and exercise levels.

MULTI-NUTRIENT CARDIOVASCULAR PROTECTION SUPPLEMENT

There are many micronutrients that are additive to the individual supplements discussed in this section to prevent and treat CHD and other cardiovascular problems.

VasculoSirt is made by Biotics Research and is a revolutionary nutritional supplement designed to slow vascular aging, promote vascular health, and provide healthy support for blood pressure, cholesterol, glucose, and insulin levels within the normal range.

Dosage
Five (5) capsules two (2) times each day as a dietary supplement or as otherwise directed by a healthcare professional.

Active Ingredients
Vitamin A (as mixed carotenoids), vitamin C (as ascorbic acid), vitamin D3 (as cholecalciferol), vitamin K (as Mena Q7®) menaquinone-7 (extract of Bacillus subtilis natto) and as phytonadione}, Thiamin (B1) (as thiamin mononitrate), Riboflavin (B2), Niacin, Vitamin B6 (as pyridoxine HCl), Folate (as calcium folinate) Vitamin B12 (as methylcobalamin), Biotin, Pantothenic Acid (as calcium pantothenate), Magnesium (as magnesium glycinate**), Zinc (as zinc picolinate), Selenium (as selenomethionine), Copper (as copper citrate), Coenzyme Q10 (emulsified), Trans-Resveratrol (from fermentation), R-Alpha Lipoic Acid (from stabilized sodium salt), Green Tea Extract (50% EGCG) (leaf), Acetyl-L-Carnitine hydrochloride, Olive Extract (Olea europaea) (fruit), Quercetin (Dimorphandra mollis), Ginkgo Extract (Ginkgo biloba) (leaf), Phytolens® (Lens esculenta extract), Lutien (from Aztec Marigold flower), Lycopene (from Tomato).

EXERCISE, CARDIOVASCULAR DISEASE, AND CHD

THE AEROBICS, BUILD, CONTOUR, AND TONE EXERCISE PROGRAM

The Aerobics, Build, Contour, and Tone (ABCT) exercise program has numerous health benefits such as reducing CHD, MI, CHF, and reduction in blood pressure. But, prior to starting any exercise program, you should consult with your physician for medical clearance with a history, physical exam, and cardiopulmonary exercise testing. This ensures that you do not have heart disease, CHD, angina-marked elevations in blood pressure with exercise, or other contraindications to an exercise program. Then the exercise should be slowly progressive until training is achieved. You should do a combination of aerobic training (AT) and resistance training (RT) for at least 60 minutes per day at least six days per week to reduce cardiovascular risk and lower blood pressure.

The aerobic exercises should be for 20 minutes per day up to 60–80% of maximal aerobic capacity (MAC) for age: MAC = MHR (maximal heart rate) = (220−age). Resistance training should be for 40 minutes per day alternating muscle groups each day such as upper or lower body exercises. Resistance training should be progressive, initially with lighter weights to avoid elevations in blood pressure, if not done under supervision. Progressive resistance training under supervision will reduce CHD risk, CHD risk factors, improve lean muscle mass, insulin sensitivity, lowers glucose, improves serum lipids and cholesterol, and reduces blood pressure.

Exercise increases nitric oxide, improves endothelial function, increases coronary artery blood flow, improves angina with time and training, dilates arteries, and reduces inflammation. Table 21.1 shows various exercises that you can do and the number of kilocalories burned (kilocalorie = 1000 calories).

Skeletal muscle (all of the muscles of your body) is a secretory/endocrine organ and exercise increases the metabolic and secretory/endocrine capacity of muscle (1). Specific kinds of exercise can alter the ways that genes function and how they interact with cells (1). By triggering the right exercise–gene interactions, inflammation, oxidative stress, and immune dysfunction are improved. (1) The slow physical deterioration of the cardiovascular system and body in general that is seen with age

TABLE 21.1
Exercise Activities and Kilocalories Used

Energy values in kilocalories per hour of selected activities

Weight (lbs.)	95	125	155	185	215	245
Slow walking	86	114	140	168	196	222
Fast walking	172	228	280	336	392	555
Hiking	285	342	420	504	588	666
Jogging	430	570	700	840	980	1110
Running	480	770	945	1134	1323	1499
Heavy work	194	256	315	378	441	500
Sweeping	108	142	175	210	245	278
Scrubbing	237	313	385	462	539	611
Tennis	301	399	490	588	686	777
Golf (walk)	237	313	385	462	539	611
Golf (in cart)	151	200	245	294	343	389
Swimming (light laps)	344	456	560	672	784	888
Swimming	430	570	700	840	980	111

The Integrative CHD Prevention Program

is not inevitable. It is largely the result of diet and movement—or the lack thereof. Movement is one of the primary keys to overall health and especially cardiovascular health. The movement required is the same kind of natural movement that has kept humans in robust physical health for millennia. This is not the kind of exercise that most personal trainers, fitness enthusiasts, or doctors recommend. As a matter of fact, most doctors and trainers recommend the exact opposite approach to movement and exercise, one that may actually accelerate the deterioration of health, cardiovascular benefits, and overall aging. There is an optimal type and duration of exercise that is required to improve health. It is important to avoid the overtraining syndrome which actually increases muscle breakdown, elevates cortisol levels and increases sympathetic tone, increases oxidative stress and inflammation, and results in the opposite effects that one desires. The power of exercise relates to the numerous hormones, mediators, and signaling molecules that lower blood pressure, blood sugar, and cholesterol which are released with the proper type of exercise that influences genes, inflammation, oxidative stress, and immune function and reduces the risk of CHD and MI.

The ABCT exercise program is a modern way of exercising the way our ancestors did. The program is specifically designed to get the muscles and body moving in short burns of intense activity, mixing anaerobic with just enough aerobic exercise to improve cardiovascular health, CHD, overall health, and conditioning (1). In addition, proper nutrition is emphasized before, during, and after exercise training. The ABCT exercise program has numerous positive effects on the body and mind. The ABCT exercise program is simple, effective, scientifically proven, and adaptable to everyone's exercise needs. It allows for optimal training benefits in a shorter period of time to build and tone muscle, reduce body and visceral fat, lose weight, improve hormone levels, lower inflammation and oxidation, decrease blood sugar, reduce blood pressure, improve the lipid profile, decrease CHD, improve one's quality of life, increase life expectancy and slow the aging process. In addition, this type of exercise avoids overtraining associated with increased cortisol levels and catabolic effects.

THE ABCS OF EXERCISE WITH A TWIST

The most efficient and effective means of achieving all the health benefits of exercise is to combine interval aerobic training (AE) with anaerobic or resistance exercise or training (RT), in a way that causes body restoration and proper muscle growth and efficiency. The ABCT (aerobics, build, contour, and tone) exercise program is designed for both genders and allows for specific outcomes. It has additional meanings that help define its goals:

- **A = Aerobics** plus action and adaptation. The program focuses on the types of action best suited for muscle and cardiovascular conditioning to adapt to new exercises, so muscles do not accommodate to the same daily training.
- **B = Build**, plus bulk, burn, and breathe. The program builds and increases muscle strength more than any other exercise regimen while learning to

use muscle burn to the best advantage with proper breathing to increase oxygen consumption and eliminate carbon dioxide to improve cardiovascular and muscle conditioning and function while reducing muscle or chronic fatigue.
- **C = Contour**, plus core and controlling your genes. Muscular exercise regulates the expression of more than 400 genes that mediate the beneficial effects of physical activity. In addition to aerobics and resistance exercise, this program will combine core exercises that improve abdominal and back muscle strength as well as exercises for flexibility and balance.
- **T = Tone**, plus trim and tight. Total body fat, visceral body fat, and body weight are decreased while lean muscle mass increases.

The ABCT exercise program emphasizes interval aerobic and anaerobic resistance movements. Proper warm-up and stretching before beginning every exercise session and cooling down and stretching again when finished are mandatory to avoid muscle, tendon, and ligament injuries while promoting the flexibility that is necessary for exercising.

Before embarking on an exercise program of any kind or changing the exercise regimen you already have a complete history, physical exam, labs, cardiovascular risk factor analysis, and cardiopulmonary exercise test are suggested. Interval training with rapid bursts of activity may precipitate plaque rupture in a coronary artery in some predisposed individuals and result in myocardial infarction.

The Elements of ABCT

The main elements of ABCT, each of which will be discussed in more detail below, are:

- Resistance Training. Weight-lifting modified properly encourages optimal muscle physiology and the release of hormones, mediators, and interleukins. ABCT uses graduated weights and variable repetitions. In brief, initially lift the heaviest weight possible twelve times to get the muscle burn, then decrease the weight with each subsequent set, but keep increasing the number of times that weight is lifted. This maximizes post-exercise oxygen consumption, depletes glycogen, and increases the production of lactic acid to achieve all the muscle-, hormone-, cytokine-, and interleukin-stimulating effects that lead to the health benefits of exercise.
- Aerobic Training in Intervals. Jogging, swimming, biking, and other forms of continual movement should be done at specific levels of submaximal and maximal aerobic capacity (MAC) or estimated heart rate for age and level of exercise (MHR). The best technique is aerobic interval training, which consists of short periods ranging from twenty seconds to two minutes of "burst" aerobic training of varying intensities, depending on one's present level of exercise conditioning. This more closely mimics the natural activities we evolved to perform and benefit from and strings together several

The Integrative CHD Prevention Program 241

periods of intense and semi-intense activity into a single, longer exercise period that still burns calories and builds endurance.
- Proper Ratio of Aerobic Training to Resistance Training. The optimal ratio of resistance to interval aerobic training should be 2:1. For example, during a 60-minute workout, you would perform 40 minutes of resistance training and 20 minutes of interval aerobics, with the aerobics coming after the resistance training.
- Core Exercises. Exercises designed to improve abdominal and back strength while increasing flexibility. These exercises are important for the core (abdomen and lower back), which is often neglected.
- Time-Intense Exercise. Rather than methodically working one muscle or muscle group after another, then doing the aerobic exercises—or even saving the aerobics for the next day—ABCT challenges the body by combining exercises as much as possible. For example, instead of doing leg squats followed by shoulder presses, with ABCT they are done at the same time, mimicking real-life movements.
- Busy Rest Periods. These are used to insert small bursts of aerobic exercises into the resistance training period.
- Water and the ABCT Energy Shake. Drinking plenty of water while working out is vital. (If you get thirsty during the workout, you have waited too long to drink.) You must drink water before beginning to exercise, at set intervals during exercise, and afterward. Your water should be of high quality and not from plastic containers, due to the risk of certain chemical compounds like PCBs that get into the water from the plastic. In addition, about ten minutes after starting your workout, you begin consuming an energy drink consisting of fresh orange juice (4 ounces) and water (6 ounces), d-ribose (5 grams), and whey protein (30 grams) to provide ATP and energy as well as nutritional substrates to maximize exercise performance and increase muscle strength and performance as well as lean muscle mass (1, 14–24).
- Exercise in the Morning. Exercising in the morning after a 12-hour fast is best for numerous reasons, including the fact that an empty stomach optimizes fat burning and interleukin 10 (IL-10), which lowers inflammation but increases the myokine surges. This results in an increase in muscle strength, bulk, tone, and contour, as well as weight and body fat loss, and improved energy level, focus, and concentration during the day.
- Exercise on an Empty Stomach. Begin exercising on an empty stomach after a 12-hour fast, except for water and whey protein consumed about ten minutes before exercise, and have nothing but more water and energy drink while exercising. This allows for the depletion of liver and muscle glycogen while generating maximal surges of IL-10. It also increases fat burning and accelerates weight and fat loss from both inside and outside the skeletal muscle.
- Push and Rest. Exercise to maximal effort during each set until "the burn" is significant. The burn should be severe and last for about four to

five seconds after you stop the exercise. Then rest for 60 seconds before beginning the next set of exercises. You may take three-second rest periods between repetitions, if necessary. Also performing supersets with minimal rest between sets of exercises or using the rest period for core exercises or alternative upper or lower body exercises will improve the time intensity of the exercise session.
- Exercise Daily, Utilizing Cross-Training. To achieve the best results, perform interval aerobic and resistance training at least six days per week or, if desired, every day but alternate the muscle groups for the upper and lower body as well as the type of aerobic exercise performed. It is important to change the exercise routine every few weeks to avoid "muscle accommodation" to the exercise regimen.
- Breathe. Mastering proper breathing techniques will ensure ample oxygenation for muscle performance, as well as prompt the removal of carbon dioxide.

THE ABCT ELEMENTS IN DETAIL

Resistance Training

ABCT resistance training takes a radically different approach, mixing heavier weights and lower repetitions with lower weights and greater repetitions to increase the lactic acid burn as well as to maximize muscle contractions and the release of muscle enzymes (myokines). With ABCT, the real goal is not only to bulk and contour the muscles but to use muscle movements to improve body biochemistry and improve blood pressure, cardiovascular health, and overall health.

ABCT resistance training is based on five sets, each with a different number of repetitions in each set. Here's the ABCT Five-Set Schedule that starts with the heaviest weight with low repetitions and advances to lighter weight with increasing repetitions.

- ABCT Set 1: 12 repetitions at maximum weight.
- ABCT Set 2: 18 repetitions at 75 percent of maximum weight.
- ABCT Set 3: 24 repetitions at 50 percent of maximum weight.
- ABCT Set 4: 50 repetitions at 25 percent of maximum weight.
- ABCT Set 5: 12 repetitions at maximum weight.

ABCT Five-Set schedule should be phased in slowly to avoid injury or excessive fatigue, depending on one's present level of physical conditioning.

- Beginners: ABCT 1, or ABCT 1 and 2.
- Intermediate: ABCT 1, 2, and 3
- Advanced: ABCT 1, 2, 3, and 4
- Professional: ABCT 1, 2, 3, 4, and 5

The Integrative CHD Prevention Program

This burn scale can be used as a guide to the exercise level. The idea is to attempt to score a 5 multiple times during your workouts, stopping only briefly (three seconds) to clear the burn before continuing again.

1—No burn in the muscle
2—Light burn
3—Moderate burn
4—Strong burn
5—Intense burn; must rest

What, How, and When to Lift

Descriptions of the key ABCT resistance exercises are found at the end of the chapter under the heading "Getting Started with ABCT: Training Session Schedules and Descriptions of the Lifts."

Upping the Intensity with Supersets, Hybrids, and Rapid Sets

Simply following the ABCT 5-Set Schedule will improve blood pressure, cardiovascular and general health. Incorporating hybrids, supersets, and rapid sets as one gains strength and endurance will improve exercise outcomes.

- Hybrids are two exercises performed at once; i.e., do a full leg squat while also performing an overhead press. Using more muscles simultaneously increases the burn, lactic acid, release of IL-10, and post-exercise oxygen consumption.
- Supersets are exercises done back-to-back, with almost no rest period between (fifteen seconds maximum). These can be the same exercise, such as biceps curls back-to-back, or different exercises, such as a biceps curl followed immediately by a triceps lift. Supersets dramatically increase the burn and other beneficial effects of exercise. Supersets should be done only after one has trained for some time to avoid overuse injury or excessive heart rate.
- Rapid sets are sets performed faster than normal to compress the workout time, enhance mechanical and metabolic burnout, and improve both resistance and aerobic conditioning. For example, with a biceps curl, increase the speed from one every second to two every three seconds.

ABCT Resistance Training Hints

1. Take three-second breaks along the way, if necessary.
2. Drink water and the ABCT energy drink after each set of exercises.
3. Even if the number of repetitions cannot be done in each set, attempt to push to the limit until the maximum burn occurs.

4. If the percentage reduction in weight is a fraction of a number, round up to the next highest whole number on the weight system you are using. This may be one pound or five pounds in most systems.

Aerobic Training in Intervals

Aerobic means "with oxygen" and refers to the use of oxygen in the body's metabolic processes. Aerobic training consists of continuous movements that demand more oxygen consumption and ultimately improve the body's oxygen use. Rapid walking, jogging, running, swimming, bicycling, dancing, and aerobics classes can all be aerobic exercises if they keep the body in moderate to intense motion for a moderately long period of time, with an elevated heart rate representing the body's heightened level of activity.

For best results, aerobic training should be broken up into periods of differing intensities or interval training with differing lengths and intensities. In general, the interval would be at 90% of maximum heart rate (MHR) for a period of time followed by 50% of MHR for one to three times that period depending on conditioning. For example, a 30-second sprint would be followed by a 30–90-second slow jog.

ONE TO THREE

The ideal ratio for aerobic interval training is 1:3. This means for every unit of time spent exercising at 90% of your maximal heart rate, spend an additional three units of time exercising at 50% of maximal heart rate (MHR). Then you repeat this 1:3 sequence about six times over about 20 minutes. The MHR calculation is shown below.

For best results, an interval aerobic exercise program should consist of a five-minute warm-up period, followed by moderate to intense interval training involving large and multiple muscle groups lasting about 20 minutes, followed by a cooling-down period of about 5 minutes at the end.

Aerobic sessions should utilize cross-training, by rotating through different aerobic activities to improve cardiovascular and muscle performance. For example, jog or run on Monday, Wednesday, and Friday; swim on Tuesday and Thursday; bike on Saturday and Sunday.

MAXIMUM HEART RATE AND MAXIMUM AEROBIC CAPACITY

- Maximal Heart Rate Calculation (MHR)—220 minus age gives the maximum aerobic capacity (MAC), and then multiply by the desired heart rate, which should range between 50% and 90% depending on the level of exercise and age.

For example, if you are age 40 the MAC would be 220-40 or 180. If you wanted to push to 80% of MAC then multiply 180 × .80 = 144 beats /minute as MHR.

The Integrative CHD Prevention Program

ALWAYS COMBINE RESISTANCE AND AEROBIC EXERCISES

The optimal ratio is 2 parts resistance training to one part interval aerobic training, with the resistance training first. Here's how the ratio works out with differing total exercise time frames.

- 15 minutes total = 10 minutes of resistance training, 5 minutes of aerobic training.
- 30 minutes total = 20 minutes of resistance training, 10 minutes of aerobic training.
- 45 minutes total = 30 minutes of resistance training, 15 minutes of aerobic training.
- 60 minutes total = 40 minutes of resistance training, 20 minutes of aerobic training.

Core Exercises

Exercising body core, such as the belly and lower back, increases abdominal and back strength while improving flexibility and balance. These can be done in sets of one to four per exercise, with the number of repetitions necessary to create the same burn that one gets with the resistance weight training program. Doing the core exercises during the sixty-second break periods while the upper or lower body muscles are resting will improve the time efficiency of the workout. Core exercises include sit-ups, abdominal crunches, leg lifts, and leg scissor crosses, etc.

Time-Intensive Exercise

Two additional steps are required to efficiently and effectively build muscle strength, tone, and contour while simultaneously improving cardiovascular conditioning and cardiovascular health: time-intensive resistance exercises and combined aerobic and resistance training.

5. Time-Intensive Resistance Exercises. Performing time-intense exercises requires using multiple and large muscle groups simultaneously, with minimal rest periods; for example, lifting light weights over the head while doing deep knee bends. This increases the release of IL-10 and other muscle cytokines, reduces inflammation, increases lactic acid "burn", enhances post-exercise oxygen consumption, builds muscle, optimizes metabolic and hormonal responses, and increases fat metabolism and fat and weight loss.
6. Combined Aerobic and Resistance Training. Instead of standing or sitting during the sixty-second between-set rest periods, another resistance or aerobic exercise should be done to maintain heart rate and respiratory rate. For example, on completing an upper body exercise, immediately start doing a lower body exercise or a core exercise that engages large muscle groups and requires "big" action. This technique maintains heart rate and provides more cardiovascular and muscular conditioning

Nutrition, Water, and ABCT Energy Drink

Nutritional macronutrient and micronutrient intake relative to exercise is important to optimize recovery, enhance subsequent performance, synthesize muscle (anabolism > catabolism), and provide optimal metabolic and nutrient-genetic interaction for muscle signaling (17). During exercise, inflammation and oxidative stress are linked via muscle metabolism and muscle damage (18). Consumption of a carbohydrate supplement immediately after exercise will improve insulin action and synthesize muscle glycogen significantly faster than when the same amount of carbohydrate is consumed two hours post-exercise (17). Protein intake and exercise have synergistic effects on increasing the rate of muscle protein synthesis leading to a more positive protein balance but vary depending on the timing of the protein intake and the exercise session (17). The timing of whole protein may not be as crucial as the timing of specific amino acid supplements and anabolism (17). Most studies suggest that consuming free amino acids immediately prior to exercise or within ten minutes of the initiation of exercise is more effective to increase muscle protein accretion compared to consumption after exercise (15, 17). In contrast, whey protein ingestion may be consumed before, during, or after exercise. The effects on muscle anabolism are enhanced with immediate and simultaneous consumption post-exercise of carbohydrates and fats as long as this occurs within less than two hours post-exercise (17). Hydration before, during, and after the exercise program is vital. Two types of hydration are necessary: plain water and the ABCT energy drink. Begin the exercise program well hydrated, drinking about 6 ounces of water.

The amounts of fluids and water consumed depend on body size, the ambient temperature, and the length and intensity of the workout. As a rule of thumb, 24–32 ounces or more of fluid and water are needed during the typical 60-minute workout.

Whey protein supplies glutathione precursors is anabolic with an increase in muscle mass and reduces oxidative stress and inflammation (1, 17). The ingredients in whey help maximize ATP (adenosine triphosphate) production, improve muscle performance, increase muscle mass, and reduce muscle fatigue (1,17).

Exercise on an Empty Stomach

The exercising program should begin in the morning on an empty stomach, following an eight- to twelve-hour fast, with the exception of drinking water and whey protein that is followed during exercise by the ABCT Energy Drink.

Consumption of carbohydrates before exercise decreases fat burning and weight loss. Exercising on an empty stomach burns more than twice as much fat as does exercising after consuming carbohydrates. Exercising in a fasting state may increase the utilization of muscle protein for energy slightly, but this effect is relatively small and is minimized when whey protein is consumed.

Optimal exercise benefits occur when skeletal muscle and liver glycogen are depleted. Glycogen depletion triggers the maximal release of IL-10 from muscle, increases muscle growth and fat burning for energy, accelerates fat loss from inside and outside muscle, and improves weight loss. All of this, in turn, reduces inflammation by increasing the levels of IL-10 while lowering other interleukins, such as IL-1,

The Integrative CHD Prevention Program

IL-6, and increasing the production of testosterone and growth hormone, improving insulin sensitivity, and lowering insulin and glucose levels.

The intramuscular triglycerides are metabolized better during glycogen-depleted exercise. The intramuscular triglycerides are far less responsive to insulin which slows the breakdown of stored fat. Exercising while fasting suppresses insulin levels and maximizes the hormonal and cytokine effects of the exercise program.

HORMONAL CHANGES

- Testosterone levels increase, enhancing muscle growth, mass, tone, and contour; improving insulin sensitivity, which lowers blood sugar and reduces the risk of diabetes and heart disease; elevating the energy level and libido; and slowing aging.
- Growth hormone levels increase, improving muscle growth, mass, tone, and contour; improving energy; increasing the sense of well-being; and slowing aging.
- Insulin levels decrease as a result of the improved insulin sensitivity that develops as lean muscle mass increases. (Lean muscle accounts for about 80 percent of insulin sensitivity or resistance in humans.) These changes help reduce intramuscular triglycerides and extramuscular fat tissue while reducing the risk of heart disease and inflammation.
- Cortisol levels drop, improving muscle growth, lowering cholesterol and TG, reducing blood sugar, and decreasing visceral fat which is associated with inflammation, diabetes, metabolic syndrome, insulin resistance, high blood pressure, elevated cholesterol, cancer, heart disease, and stroke.

NUTRITION BEFORE, DURING, AND AFTER THE EXERCISE SESSION

Nutrient availability serves as a potent modulator of many acute responses and chronic adaptations to both RT and AE. Changes in the macronutrient intake quickly change the concentration of substrates and hormones with alterations in the storage profile of skeletal muscle and other insulin-sensitive organs. This, in turn, regulates gene expression and cell signaling. Nutrient-exercise interactions activate or inhibit biochemical pathways during training. Proper nutrition after exercise with a breakfast containing fluids, high-quality protein, complex carbohydrates, omega-3 fatty acids, and monounsaturated fatty acids is essential to increase muscle mass and overall muscle performance and cardiovascular conditioning for each subsequent exercise session.

- A small bowl of whole-grain cereal or steel-cut oats with whole milk, rice milk, or almond milk.
- ½ cup of one of the following: fresh blueberries, raspberries, blackberries, or strawberries.

- 4 ounces fresh orange or vegetable juice or another fresh juice, such as pomegranate, grape, or grapefruit; or have an orange, grapefruit, grapes; or green vegetables like kale, spinach, or broccoli.
- 4 ounces of smoked salmon with lemon juice, capers, and perhaps add some hot sauce, and jalapeño peppers.
- Whole-wheat toast with omega-3 margarine and raw honey.
- One egg.

Instead of salmon, you might try tuna or other coldwater fish, lean organic meat (buffalo, elk, venison, beef), organic chicken, or organic turkey.

GETTING STARTED WITH ABCT: TRAINING SCHEDULES AND DESCRIPTIONS OF THE LIFTS

This section contains different training schedules, ranging from beginner to professional levels:

- Alternate days with the various resistance programs (numbers 1–4) listed below for each of the week's sessions.
- Vary the type of aerobic exercise: for example, running one day, swimming the next, and bicycling the third.
- Do the aerobic exercise *after* the resistance exercises.
- Always do the correct number of sets with each type of ABCT session for the upper body, lower body, core, flexibility, and balance exercises. For example, if you are doing ABCT 1, do only one set for each exercise. With ABCT 2, do two sets for each exercise, with ABCT 3, do three sets for each exercise, and so on.
- Customize the exercise program depending on your goals and time commitment. If you wish to build more muscle, do ABCT 1, 2, and 5; or ABCT 1, 2, 3, and 5. If your goal is to contour and tone, do ABCT 2, 3, and 4. If you wish to have bulk, contour, and tone, then do ABCT 1–5.

The ABCT Training Schedules

Week One: Beginning Session #1, with ABCT 1
1. Resistance training for ten minutes: Pick the maximum weight you can do for 12 repetitions, and do one set for each exercise.
   - 2 upper body exercises: 1 biceps, 1 triceps.
   - 2 lower body exercises: squat, lunges.
   - 1 core: 25–50 or more sit-ups until maximum burn.
2. Aerobic exercise for five minutes.

Week One: Beginning Session #2, with ABCT 1
1. Resistance training for ten minutes: Pick the maximum weight you can do for 12 repetitions, and do 1 set for each exercise.
   - 2 upper body exercises: one chest, one deltoid.
   - 2 lower body exercises: leg press, hamstring press.
   - 1 core: abdominal crunches until a maximum burn.

The Integrative CHD Prevention Program

   2. Aerobic exercise for five minutes.

Week One: Beginning Session #3, with ABCT 1
   1. Resistance training for ten minutes: Pick the maximum weight you can do for 12 repetitions, and do 1 set for each exercise.
      - 2 upper body exercises: 1 shoulder, 1 forearm.
      - 2 lower body exercises: squat with weights, lunges.
      - 1 core: leg lift.
   2. Aerobic exercise for five minutes.

Week One: Beginning Session #4, with ABCT 1
   1. Resistance training for ten minutes: Pick the maximum weight you can do for 12 repetitions, and do 1 set for each exercise.
      - 2 upper body exercises: 1 reverse biceps curl, 1 pull-down back exercise.
      - 2 lower body exercises: leg press, hamstring press.
      - 1 core: leg scissor crosses.
   2. Aerobic exercise for five minutes.

Week Two: Beginning Session #1, with ABCT 1 and 2
   1. Resistance training for 20 minutes: Do 12 repetitions at the maximum weight you can do, then 18 repetitions at 75 percent of original weight.
      - 3 upper body exercises: 1 chest press, 1 biceps, 1 triceps.
      - 2 lower body exercises: squats, lunges.
      - 1 core exercise: sit-ups.
   2. Aerobic exercise for ten minutes.

Week Two: Beginning Session #2, with ABCT 1 and 2
   1. Resistance training for 20 minutes: Do 12 repetitions at the maximum weight you can do, then 18 repetitions at 75 percent of the original weight.
      - 3 upper body exercises: 1 chest press, 1 biceps, 1 deltoid.
      - 2 lower body exercises: lunges, hamstring leg press.
      - 1 core exercise: abdominal crunches.
   2. Aerobic exercise for ten minutes.

Week Two: Beginning Session #3, with ABCT 1 and 2
   1. Resistance training for 20 minutes: Do 12 repetitions at the maximum weight you can do, then 18 repetitions at 75 percent of the weight.
      - 3 upper body exercises: 1 upper shoulder and trapezius, 1 biceps with reverse curl, 1 forearm.
      - 2 lower body exercises: squat with overhead press, quadriceps leg press.
      - 1 core: leg lifts at variable heights.
   2. Aerobic exercise for ten minutes.

Week Two: Beginning Session #4, with ABCT 1 and 2
   1. Resistance training for 20 minutes: Do 12 repetitions at the maximum weight you can do, then 18 repetitions at 75 percent of the weight.
      - 3 upper body exercises: 1 reverse bicep, 1 pull-down back exercise, 1 chest.
      - 2 lower body exercises: lunges with weights, squats.
      - 1 core: leg scissor crosses.

2. Aerobic exercise for ten minutes.
Week Three: Intermediate Session # 1 with ABCT 1 through 3
   1. Resistance exercise for 30 minutes with ABCT 1–3: Use maximum weight for 12 repetitions, 75-percent weight for 18 repetitions, 50-percent weight for 24 repetitions.
      – 3 upper body exercises: 1 biceps, 1 chest, 1 triceps.
      – 3 lower body exercises: squats, lunges, quadriceps leg press.
      – 2 core exercises: sit-ups, leg lifts
   2. Aerobic exercise for 15 minutes.
Week Three: Intermediate Session #2 with ABCT 1 through 3
   1. Resistance exercise for 30 minutes with ABCT 1–3: Use maximum weight for 12 repetitions, 75-percent weight for 18 repetitions, 50-percent weight for 24 repetitions.
      – 3 upper body exercises: 1 deltoid, 1 reverse biceps curl, 1 pull-down back exercise.
      – 3 lower body exercises: squats with weights, lunges, hamstring press.
      – 2 core exercises: leg scissor crosses, abdominal crunches.
   2. Aerobic exercise for 15 minutes.
Week Three: Intermediate Session #3 with ABCT 1 through 3
   1. Resistance exercise for 30 minutes with ABCT 1–3: Use maximum weight for 12 repetitions, 75-percent weight for 18 repetitions, 50-percent weight for 24 repetitions.
      – 3 upper body exercises: 1 forearm, 1 upper shoulder and trapezius, 1 chest.
      – 3 lower body exercises: lunges with weights, quadriceps leg press, hamstring press.
      – 2 core exercises: leg lifts to chest with floor extension, supine "bicycle" movement with elbows to opposite knees.
   2. Aerobic exercise for 15 minutes.
Week Three: Intermediate Session #4 with ABCT 1 through 3
   1. Resistance exercise for 30 minutes with ABCT 1–3: Use maximum weight for 12 repetitions, 75-percent weight for 18 repetitions, 50-percent weight for 24 repetitions.
      – 3 upper body exercises: 1 biceps, 1 triceps, one forearm.
      – 3 lower body exercises: leg quadriceps press, squats, hamstring press.
      – 2 core exercises: leg lifts, sit-ups.
   2. Aerobic exercise for 15 minutes.
Week Four: Advanced Session #1 with ABCT 1 through 4
   1. Resistance exercise for 40 minutes with ABCT 1–4: Use maximum weight for 12 repetitions, 75-percent weight for 18 repetitions, 50-percent weight for 24 repetitions, 25-percent weight for 50 repetitions.
      – 4 upper body exercises: 1 biceps, 1 triceps, 1 shoulder and trapezius, 1 deltoid.

The Integrative CHD Prevention Program

- 3 lower body exercises: leg quadriceps press, leg hamstring press, squats.
- 2 core exercises: sit-ups, leg lifts.
2. Aerobic exercise for 20 minutes.

Week Four: Advanced Session #2 with ABCT 1 through 4
1. Resistance exercise for 40 minutes with ABCT 1–4: Use maximum weight for 12 repetitions, 75-percent weight for 18 repetitions, 50-percent weight for 24 repetitions, 25-percent weight for 50 repetitions.
   - 4 upper body exercises: 1 pull-down back exercise, 1 reverse curl, 1 forearm, one chest.
   - 3 lower body exercises: lunges, leg quadriceps press, squats.
   - 2 core exercises: abdominal crunches, leg scissor crosses.
2. Aerobic exercise for 20 minutes.

Week Four: Advanced Session #3 with ABCT 1 through 4
1. Resistance exercise for 40 minutes with ABCT 1–4: Use maximum weight for 12 repetitions, 75-percent weight for 18 repetitions, 50-percent weight for 24 repetitions, 25-percent weight for 50 repetitions.
   - 4 upper body exercises: 1 biceps front curl with reverse curl, 1 chest, 1 deltoid, 1 triceps.
   - 3 lower body exercises: hamstring press, lunges with weights, squats with weights.
   - 2 core exercises: Lay on back and do leg extensions from chest, supine "bicycle" touching opposite elbow to knee.
2. Aerobic exercise for 20 minutes.

Week Four: Advanced Session #4 with ABCT 1 through 4
1. Resistance exercise for 40 minutes with ABCT 1–4: Use maximum weight for 12 repetitions, 75 percent weight for 18 repetitions, 50 percent weight for 24 repetitions, 25 percent weight for 50 repetitions.
   - 4 upper body exercises: 1 forearm, 1 biceps, 1 upper shoulder trapezius, 1 pull-down back exercise.
   - 3 lower body exercises: lunges with weights, squats with weights, quadriceps leg press.
   - 2 core exercises: leg lifts, sit-ups.
2. Aerobic exercise for 20 minutes.

Week Five and Beyond: Professional Sessions with ABCT 1 through 5
1. Resistance exercise for 40 minutes with ABCT 1–5: Use maximum weight for 12 repetitions, 75-percent for 18 repetitions, 50-percent weight for 24 repetitions, 25-percent weight for 50 repetitions, maximum weight for an additional 12 repetitions.
   - 5 upper body exercises with selection from the following: 1 biceps curl, 1 upper shoulder pull-up, 1 triceps, 1 forward with reverse biceps curl, 1 deltoid, 1 forearm/wrist curl/extension, 1 back pull-down exercise, 1 forearm reverse curl, 1 neck exercise.
   - 4 lower body exercises with selection from the following: squats, lunges, quadriceps press, hamstring press.

- 2 to 3 core exercises with selection from the following: sit-ups, crunches, leg lifts, leg scissor crosses, leg extensions to chest.
2. Aerobic exercise for 20 minutes.

Chest Exercises

Push-up. Position yourself like a plank, on your hands and toes. The hands should be in alignment with the chest, fingers pointing straight forward, with the hands spaced a little wider than shoulder-width apart. The tummy is tucked in, and the butt muscle is down and straight, in alignment with the back. To work different areas of the muscles, the hands can be moved further apart (more chest) or closer together (more triceps)

Primary areas worked: chest
Secondary areas worked: triceps, shoulders

Bench Press. This uses the same movement as a push-up, except you lie on your back and push a weight up instead of raising the body. It can be done with dumbbells, a barbell, or on a machine. It can also be done in an incline or decline position.

Primary areas worked: chest
Secondary areas worked: triceps, shoulders

Dip. Best done on a "dipping bar", where the body is suspended and supported only by the arms. In the beginning position, the legs hang down, the body leans slightly forward, and the arms are fully extended. The elbows bend, and the arms lower the body down, then straighten to raise the body back to the starting position.

Primary areas worked: chest
Secondary areas worked: triceps, shoulders

Chest Fly. Can be performed using either dumbbells or a fly machine. Lie on your back on a bench with your arms out to the sides holding weights (or gripping the machine bars). Arc your arms up and in from the outstretched position until they are pressed together above the chest, with arms straight out and up. Keep a slight bend in the elbows the entire time. Get a good stretch at the bottom of the movement and a good squeeze at the top.

Primary areas worked: chest
Secondary areas worked: front deltoid

Cable Chest Fly. Done on a cable machine, standing, which each hand gripping a handle. Stand with one leg in front of the other, leaning slightly forward, the arms outstretched and back, with elbows slightly bent. The arms are then pulled forward until they are aligned directly in front of the chest.

Primary areas worked: chest
Secondary areas worked: front deltoid

Back Exercises

Back Row. Done with dumbbells (one or two), a barbell, or machines. Starting lying face down on a bench, with the arms hanging down straight and gripping the weights or the machine bars. The goal is to pull weight up toward the body. There are many variations, including close-grip rows performed on a pulley machine and the one-arm version done while leaning over a bench.

Primary areas worked: latissimi, rhomboids, trapezii
Secondary areas worked: biceps, rear deltoid

Pullover. Begin lying face up on a bench, with the arms extended beyond the head and down, holding the weight. Keeping the arms close together and elbows slightly bent, bring them up over the head in an arcing motion, then lower them back to the starting position. This is usually done with one dumbbell, although there are variations using a barbell, two dumbbells, or a pullover machine.

Primary areas worked: latissimi
Secondary areas worked: triceps, chest

Lat Pull-Down. Done using a weight machine. While sitting on the bench, reach up to grasp the bar and pull it down the top of the upper chest or upper back.

Primary areas worked: latissimi
Secondary areas worked: biceps

Dickerson. Done using either a lateral pull-down machine or a pulley system. Begin with the arms straight out in front of the body, slightly elevated to forehead level. Grasp the vertical bar. Keeping the arms stiff, elbows slightly bent and shoulder width (or a little wider apart), pull the bar down to just above the hips.

Primary areas worked: latissimi
Secondary areas worked: triceps

Pull-Up. This exercise uses body weight only. It is done hanging from a bar and pulling yourself up, or using a machine with a counterbalanced weight bar you stand on for a little help. This is ideal for those not yet strong enough to lift their own body weight.

Primary areas worked: latissimi
Secondary areas worked: biceps

Shrug. Standing and holding on to barbells or dumbbells, "shrug" your shoulders up toward your ears to pull the weight up. Arms remain straight the whole time.

Primary areas worked: trapezii
Secondary areas worked: none, or very minor shoulder action

Shoulder Exercises

Shoulder Press. The goal of this exercise is to push a weight up over the head. It can be done with dumbbells or barbells, or using a weight machine. Hand position can be varied. If a barbell or machine is used, the hands can be placed closer or farther apart on the bar. If dumbbells are used, the palms can be facing out (forward) or each other, or changed from one position to the other as the arms are raised.

Primary areas worked: deltoids
Secondary areas worked: triceps

Lateral Raise. Done with dumbbells, one held in each hand, arms hanging down to the sides, with the weights slightly in front of the body. The arms are raised to the side until the elbows reach just above the shoulder level, then are lowered back down. This exercise can be varied by keeping the elbows slightly bent, which makes it a bit harder, or completely bent, which is easier and safer for the shoulder joint.

Primary areas worked: side deltoids
Secondary areas worked: trapezii

Front Raise. Similar to the lateral raise, performed with dumbbells or a barbell. Begin with the weight(s) held in the hands, arms straight down in front of the body. The arms are lifted straight out and up in front of the body in an arcing movement, stopping just above shoulder level.

Primary areas worked: front deltoids
Secondary areas worked: trapezii, chest

Rear Fly. Performed with dumbbells, with the body leaning over the legs. It can be done either seated or standing with knees bent and leaning over. Start with weights hanging straight down from the body, level with the abdominal muscles. The weight is then lifted out to the side of the body, keeping a slight bend in the elbows.

Primary areas worked: rear deltoids
Secondary areas worked: rhomboids

Upright Row. Done with a barbell or two dumbbells. Begin standing, with the arms hanging down and weights in front of the body, palms facing in toward the body. The weight is lifted up to just below chin level, with the elbows kept high through the motion.

Primary areas worked: front deltoids
Secondary areas worked: trapezii

Arm Exercises

Biceps Curl. This exercise involves lifting a weight held in the hand by bending the elbow to bring the hands up toward the shoulders. There are many variations, including using dumbbells, a barbell, or a machine; you can stand or be seated; and the palms may be facing out, in, or rotating through the movement.

Primary areas worked: biceps
Secondary areas worked: front deltoid

Triceps Extension. The "reverse" of the biceps curl, with the goal being to straighten a bent arm while holding a weight, then releasing it back into the starting position. If you're using a dumbbell, begin leaning over a bench, supporting yourself with one hand, holding a dumbbell in the other, with the dumbbell arm pulled back and its elbow bent at a 90-degree angle. Keeping the upper arm stationary, extend the arm straight out so that the dumbbell moves backward and up. This can also be done using a pulley.

Primary areas worked: triceps
Secondary areas worked: shoulder, latissimi

Bench Dip. Similar to the dip but performed using a weight bench to support the upper body and with the feet on the ground. Begin with your hands on the edge of the bench, palms down and supporting your weight, with your rear end hanging just off the bench and your legs straight out in front, angling down to the floor so that your heels are resting on the floor. The fingers should be facing the body and the elbows close together. The arms are then bent, lowering the body toward the floor, then straightened so the body is raised.

Primary areas worked: triceps
Secondary areas worked: shoulders, chest

Leg Exercises

Squat. The goal of this exercise is to "sit" and stand back up while holding a weight. Starting in a standing position, push the gluteus muscles backwards and lower yourself as if you are going to sit in a chair until you are in a squatting position. The upper body leans slightly forward, toes are pointed straight out in front, feet are slightly wider than shoulder-width apart. This exercise can be done with a barbell held across the back, or using a squat machine, a Smith machine, a hack squat machine, or a Smith ball on the wall.

Primary areas worked: quadriceps, gluteals, hamstrings
Secondary areas worked: back

Leg Press. Similar to a bench press, but it works the leg rather than the chest muscles. Sit or lie down in a leg press machine, with knees bent toward the chest and feet against the weight platform. Push hard with the legs against the platform until the legs straighten but not entirely; the knees should be slightly bent.

Primary areas worked: quadriceps, gluteals, hamstrings
Secondary areas worked: back

Leg Extension. Done on a machine while sitting up, with the legs down and the ankles pressing up against a padded bar. The legs are lifted with the feet rising up in an arc, pushing the bar up, until the legs are straight out in front of the body.

Primary areas worked: quadriceps
Secondary areas worked: none, or minor hip flexor action

Leg Curl. The "reverse" of the leg extension, performed lying facedown on a bench with the Achilles tendons pressed up into a padded bar or sitting up, legs straight out in front, with the Achilles tendons resting on a padded bar. The legs are flexed, pushing against the bar until the knees are bent, with the heel toward the gluteals.

Primary areas worked: hamstrings
Secondary areas worked: none, or slight lower back action

Calf Raise. The goal is to stand up on tiptoes against resistance. This can be done various ways. The simplest is to stand straight up, feet flat on the ground, holding dumbbells by the sides. Stand up on tiptoes, then descend back to feet flat on the ground.

Primary areas worked: calves
Other benefit: full body stabilization

Lunge. The idea is to "dip" one leg, knee bent, similar to the way a fencer does when lunging with the foil. Begin in a standing position. Keeping the upper body erect, step forward so that one leg is in front of the body and one is behind. Lower the back knee toward the ground, bending the front leg as well, until the front leg is bent in a perfect 90-degree angle at the knee. The upper body is kept erect. There are many variations of this exercise, including holding the lunge while moving up and down, stepping out and pushing back, and walking. It can be done with dumbbells, barbells, and using a machine, such as the Smith machine.

Primary areas worked: gluteals, quadriceps, hamstrings
Secondary areas worked: low back, abductors, adductors

Step-up. Holding dumbbells at the sides or a barbell across the back, step up onto a step or low bench (as long as it is very secure and safe) and then back down. It can be done working one leg at a time and then the other, or with alternating legs.

Primary areas worked: gluteals, hamstrings, quadriceps
Secondary areas worked: low back, abductors, and adductors

Abductor and Adductor Toners. Performed with machines, these exercises work the "outside" and "inside" of the thighs. To work the muscles on the outsides of your thighs (abductors), you sit in the machine, legs straight forward and resting on pads attached to weights, then spread them out to the sides. To work the muscles on the insides of your thighs (adductors), you do the reverse, beginning in a seated position, legs straight out in front but spread, then squeeze them together.

Primary areas worked: abductors, adductors
Secondary areas worked: none

Dead Lift. Performed with dumbbells or barbells. Begin in a kneeling position, with the butt pushed back, upper body leaning forward, and the toes pointed straight out. The back is kept in alignment, without rounding. Hands grip the weights, which rest on the floor. Stand up, using only the legs, with the arms and hands acting only as hooking and carrying mechanisms. After reaching the standing position, lower the weight in the same fashion.

Primary areas worked: trapezius, latissimi, erector spinae, gluteals, hamstrings, quadriceps, and psoas

ABCT SUMMARY

- Exercise in the morning on an empty stomach after an eight- to 12-hour fast.
- Drink 6 ounces of water mixed with 10 grams of whey protein before starting warm-up or doing any exercise.
- Warm up for five minutes with stretching, flexing, and extension exercises of the upper and lower body.
- Start the resistance portion of ABCT based on one's present exercise level, time commitment, and the desired intensity of the workout. Pick ABCT 1, 2, 3, 4, or 5, with varied mixing and matching to accomplish the goals of bulk, contour, and tone. Do the intense exercise to the muscle burn. Alternate muscle groups each day for both upper and lower body, and do core exercises.
- Exercise two or three upper body and two or three lower body muscle groups per session, and increase the number of muscle groups exercised with your desired intensity and training time. Do two or three core exercises with some flexibility and balance work as well.
- Ten minutes into the resistance workout, start drinking 4 ounces of the ABCT Energy Shake at intervals after each exercise set. Rest sixty seconds between each repetition unless doing supersets or taking minimal rest periods.
- Take three-second rests as needed to combat muscle fatigue.

- After completing the resistance exercises, start the aerobic program, utilizing cross-training.
- Keep the ratio of resistance training to aerobic exercise at 2:1.
- Exercise at least 6 times per week, but daily is best with one day off every week if necessary.
- Compress the exercise time and increase training intensity by doing core exercises or other lower intensity exercises during the sixty-second rest periods. Alternatively, you can do the core exercises as part of the resistance exercise session.
- Eat the recommended post-exercise breakfast.

ABCT CONCLUSION

Optimal exercise with the ABCT combined AE and RT, with proper nutrition and the energy drink, results in numerous health benefits, including lowered blood pressure, reduced cardiovascular disease, and slowed aging. Skeletal muscle is an endocrine organ that secretes over 400 myokines, hormones, and mediators that result in gene expression patterns and intracellular and extracellular signaling to reduce inflammation, oxidative stress, and modulate immune function during chronic adaptive training. An understanding of the nutrient-gene-muscle interconnections, as it relates to proteomics and metabolomics will allow for a more scientific recommendation for the types and duration of exercise coupled with nutritional support to enhance good health outcomes, reduce morbidity, and mortality.

Here are the main elements of ABCT:

- Resistance Training. Weight-lifting is modified to encourage the muscles to "talk" to the body in such a way as to encourage the better heart and full-body health. ABCT uses graduated weights and variable repetitions. In brief, lift the heaviest weight you can twelve times to get the burn, then decrease the weight with each subsequent set—but keep increasing the number of times you lift that weight. This is done using five variable weights and repetitions as follows:
- If you do a bench press at 100 pounds for 12 repetitions then do the following sequence with a 60-second rest between each set.
- 75% of first weight or 75 lbs. for 18 reps.
- 50% of first weight or 50 lbs for 24 reps.
- 25% of first weight or 25 lbs. for 50 reps.
- 100% of first weight or 100 lbs. for 12 reps.

This maximizes post-exercise oxygen consumption, depletes glycogen, and increases the production of lactic acid to achieve all the muscle-, hormone-, cytokine-, and interleukin-stimulating effects that lead to the health benefits of exercise.

- Aerobic Training in Intervals. Jogging, swimming, biking, and other forms of continual movement that set the heart beating at an elevated rate and keep it there for a predetermined amount of time. However, the standard

TABLE 21.2
ABCT Exercise Program Health Benefits
Summary of the Aerobics, Build, Contour, and Tone Exercise Program

The ABCT Exercise Program has numerous positive effects on body and mind—much more than the typical aerobic-based programs. Among other things, it:
- Reduces risk of heart disease and heart attack, and lowers risk of recurrent heart attack
- Improves heart function
- Lowers blood pressure and reduces risk of developing hypertension
- Reduces total cholesterol, triglycerides, and LDL
- Increases HDL
- Reduces body weight and body fat
- Reduces clotting tendencies
- Lowers blood sugar and decreases risk of diabetes
- Improves insulin sensitivity
- Improves all abnormalities of metabolic syndrome
- Improves immune function
- Reduces risk of stroke
- Reduces risk of certain cancers, such as colon, breast, and prostate
- Improves memory and focus and reduces risk of Alzheimer's disease and dementia
- Improves skin tone and, elasticity and decreases wrinkles
- Improves depression, stress, anxiety, and overall psychological well-being
- Improves sleep

approach—keeping the heart beating at a certain elevated rate for 20, 30, or even 60 minutes—is faulty. The best technique is aerobic interval training, which consists of short periods ranging from 20 to 60 seconds of "burst" aerobic training at 80–90% of your maximal aerobic capacity (MAC) or heart rate for your age, then dropping to 50% of the MAC or heart rate for three times the length of the initial burst of activity. You then repeat this six times. The length of your burst of activity and resting period will depend on your present level of exercise conditioning. For example, if you run for 20 seconds at 80% MAC then you would do 60 seconds at 50% MAC. This more closely mimics the natural activities we evolved to perform and benefit from and strings together several periods of intense and semi-intense activity into a single, longer exercise period that still burns calories and builds endurance.
- Proper Ratio of Aerobic Training to Resistance Training. The optimal ratio of resistance to interval aerobic training should be 2:1. For example, during a 60-minute workout, you would perform 40 minutes of resistance training and 20 minutes of interval aerobics, with the aerobics coming after the resistance training. This should be done at least four days per week but daily is the best to achieve cardiovascular conditioning and optimal body composition with increases in lean muscle mass and reduction in body fat (Table 21.3).

TABLE 21.3
Recommended Exercises for ABCT Fitness with Aerobic Intervals and Resistance Training Based on Time Schedule

15 MINUTES: Aerobic intervals for 5 minutes
    Resistance training for 10 minutes
        Two upper body exercises with ABCT 1
        Two lower body exercises
        One core exercise

30 MINUTES: Aerobic intervals for 10 minutes
    Resistance training for 20 minutes
        Three upper body exercises with ABCT 1 AND 2
        Two lower body exercises
        One core exercise
        One flexibility exercise

45 MINUTES: Aerobic intervals for 15 minutes
    Resistance training for 30 minutes
        Three upper body exercises with ABCT 1, 2, and 3
        Three lower body exercises
        Two core exercises
        One flexibility exercise

60 MINUTES: Aerobic intervals for 20 minutes
    Resistance training for 40 minutes
        Four upper body exercises with ABCT 1, 2, 3, and 4
        Three lower body exercises
        Two core exercises
        One flexibility exercise

90 MINUTES: Aerobic intervals for 30 minutes
    Resistance training for 60 minutes
        Five upper body exercises with ABCT 1, 2 3, 4, and 5
        Three lower body exercises
        Two core exercises
        One flexibility exercise

120 MINUTES: Aerobic intervals for 40 minutes
    Resistance training for 80 minutes
        Six upper body exercises with ABCT 1, 2, 3, 4 and 5
        Four lower body exercises
        Three core exercises
        Two flexibility exercises

WEIGHT MANAGEMENT, BODY FAT, AND VISCERAL FAT

Obesity by itself is a well-known risk factor for CHD and MI, and losing weight and body fat will help to lower your blood pressure, serum lipids, blood sugar, and other cardiovascular risks. Total body fat, especially visceral or belly fat produces over 45 chemicals, called "adipokines" that can elevate blood pressure, cause inflammation, oxidative stress, cardiovascular disease, and diabetes mellitus. About 60% of hypertensive patients are at least 20% over ideal body weight. A weight loss of about 10–12 pounds will result in a reduction in blood pressure of 7/5 mm Hg in obese and nonobese patients. Weight loss also potentiates the effects of other lifestyle modifications and drug therapy. Weight reduction should decrease adipose tissue, not lean muscle mass. Reduction in visceral obesity, which is measured by the waist circumference and by a machine called body impedance analysis (BIA), is particularly important in reducing blood pressure and cardiovascular risk (Figure 21.1). Weight loss with fat loss improves the output of the heart, decreases blood volume, sodium and water retention, and swelling in the legs (edema). In addition, weight and fat loss reduce insulin levels, improves insulin sensitivity, lowers adrenalin levels, dilates the arteries, reduces the sympathetic nervous system activity, plasma renin activity (PRA), and serum aldosterone levels. Weight and fat loss reduce inflammation and oxidative stress. Here are some important and practical points about weight and body fat reduction.

1. It is not safe to lose over 3.3 pounds per week, and the preferred weight loss is 1–2 pounds per week.
2. Body fat is more important than body weight.
   Males should be < 15% body fat.
   Females should be < 22% body fat.
3. The number of calories needed per day to maintain the same weight is your weight in pounds times 10.
   That is, 160 lbs. = 1600 calories.
   1600 calories is your basal metabolic rate (BMR).
4. It takes a 3500-calorie deficit to lose one pound.
2. If the body mass index (BMI) is over 27, there is a danger of developing significant health problems.
6. Four major factors contribute to obesity:
   - Genetics.
   - Metabolic factors.
   - Diet.
   - Physical inactivity.
7. Waist circumference, taken around the natural waistline, over the value indicated below is associated with a high risk of disease and may be the single best predictor of obesity-related cardiovascular disease and overall morbidity and mortality:
   men over 40 inches
   and
   women over 35 inches.
8. Neck circumference also correlates with high disease risk.
   Men over 15.6 inches
   Women over 14.4 inches
9. Obesity will increase the risk of morbidity and mortality from the following diseases:
   - Hypertension.

- Dyslipidemia.
- Type II diabetes and insulin resistance.
- Coronary heart disease and heart attack.
- Stroke.
- Gallbladder disease.
- Osteoarthritis.
- Sleep apnea.
- Respiratory problems.
- Cancer of breast, prostate, colon, and endometrium.
- Chronic kidney disease.
- Microalbuminuria and proteinuria.

10. The ideal body weight calculation: (Depends on the body frame size)
    - Women: 100 lbs. first 5 feet, then 5 lbs. for each additional inch of height.
    - Men: 106 lbs. first 5 feet, then 6 lbs. for each additional inch of height.
    - 10% is added for large frame and you delete 10% for a small frame.

SUMMARY OF THE CHD NUTRITION PROGRAM

1. **Vegetables**: eat 8–12 servings per day of non-starchy vegetables such as broccoli, cauliflower, leafy greens, asparagus.
2. **Fruit**: limit to four servings. ½ cup of pieces = one serving. 1 small whole fruit = one serving.
3. **Fats**: eat good fats, avoid some saturated fats, and avoid all trans fats. Omega-3 fatty acids (1–2 grams per day), extra virgin olive oil (four tablespoons per day), and olive products.
4. **Protein**: 25–30% clean-sourced fish, turkey, chicken, and proteins. Organic meat and grass-fed beef. Avoid any animal product with pesticides, toxins, or hormones. Avoid farm-raised fish. Wild coldwater fish, like cod, mackerel, salmon, halibut, and tuna, are best.
5. **Water**: 50% of body weight (pounds) in ounces of filtered or distilled water per day. If you weigh 150 pounds, that is 75 ounces of water, herbal teas, broths daily. (Do not consume from plastic.)
6. **Tea**: 16 ounces of decaffeinated green tea per day.
7. **Pomegranate**: ¼ cup of pomegranate seeds per day.
8. **Detox + elimination diet**: followed by an anti-inflammatory, blood sugar regulating protocol. Emphasize high-fiber foods and add servings of cruciferous vegetables, such as broccoli, cauliflower, Brussels sprouts, and cabbage, etc.
9. **Caloric restriction (CR)**: 12–14 hours overnight fast with 12.5% caloric restriction and 12.5% increase in energy expenditure with exercise. PREDIMED (Mediterranean) type diet.
10. **Fasting Mimicking Diet (PROLON)**: three months on, then three months off and repeat.

The Integrative CHD Prevention Program

Chapters 12 and 13 on Nutrition and CHD discusses all the effective nutrition recommendations to safely reduce body weight, body fat, visceral (belly) fat; preserve lean muscle mass; decrease CHD risk; lower blood pressure; reduce blood glucose and cholesterol and other cardiovascular and general health risks.

TOBACCO PRODUCTS, E-CIGARETTES, AND VAPING

Smoking is a major risk factor for heart attack, coronary heart disease, congestive heart failure, stroke, chronic obstructive lung disease, and lung cancer. Smoking accounts for over 500,000 deaths per year in the US. It is very addictive, so you must seek your doctor's help in stopping this very bad habit. Second-hand smoke, e-cigarettes, vaping, and other forms of tobacco should also be stopped. It is never too late to quit! Discontinuation of smoking will reduce vasoconstriction, improve endothelial function, and lower blood pressure. It also decreases sympathetic nervous system activity, norepinephrine levels, RAAS activity, carbon monoxide levels, platelet stickiness, clotting risk, oxidative stress, and inflammation. All tobacco products must be stopped, such as chewing tobacco, snuff, e-cigarettes, vaping, and others. Here are some encouraging facts to help you quit smoking.

- Quitting smoking decreases CHD and cardiovascular risk. At about one year after quitting smoking, your risk for a heart attack drops dramatically.
- Within two to five years after quitting smoking, your risk for stroke may fall to that of a nonsmoker's.
- If you stop smoking, your risks for cancers of the mouth, throat, esophagus, and bladder drop by 50% within 5 years.
- Ten years after you discontinue smoking, your risk for dying from lung cancer drops by 50%.

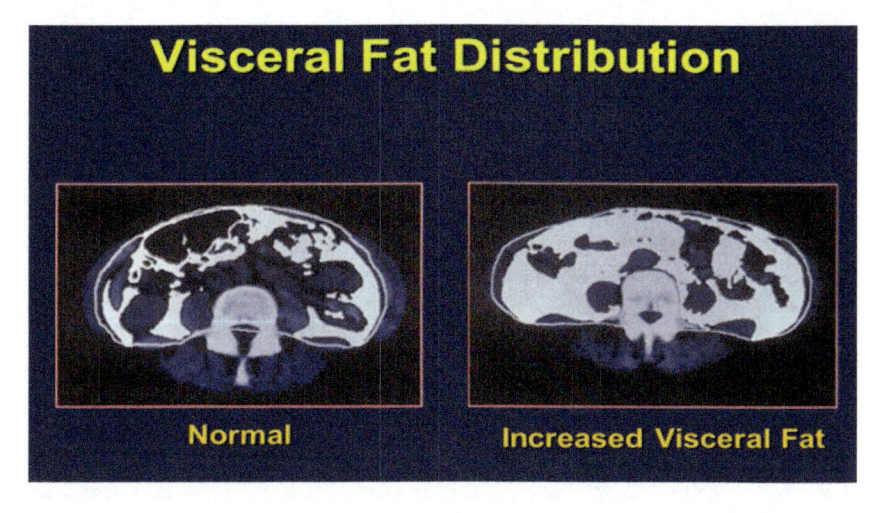

FIGURE 21.6 Visceral or belly fat is shown in the normal and the obese patient. The white represents the visceral fat.

CONTROL ALL RISK FACTORS FOR CORONARY HEART DISEASE

All CHD risk factors should be measured, identified early, and treated aggressively to the optimal levels. The top 5 and top 25 are the best starting risk factors to evaluate. See Chapter 18 for a detailed discussion.

These are the Top 5:

1. **Hypertension**: in addition to office and home BP monitory, the new gold standard to evaluate BP and CHD risk is the 24-hour ambulatory blood pressure monitoring (ABM) discussed in the Hypertension section. Goal BP is 120/80 mm Hg.
2. **Dyslipidemia**: an advanced blood lipid analysis that measures all the different lipid particle numbers and sizes as well a test to determine the function of HDL is much more accurate than the old standard lipid profiles. Goals are defined for each lipid value including LDL P to below 1000, LDL to below 100 mg/dL, TG to 75 mg/dL, HDL P to over 5000, HDL above 50 mg/dL and good HDL function.
3. **Hyperglycemia, metabolic syndrome, insulin resistance and diabetes mellitus**: evaluate all blood tests to determine the fasting glucose, two-hour GTT, insulin levels and HbA1C. Fasting blood sugar should be about 75 mg /dL with hemoglobin A1C below 5.5 units.

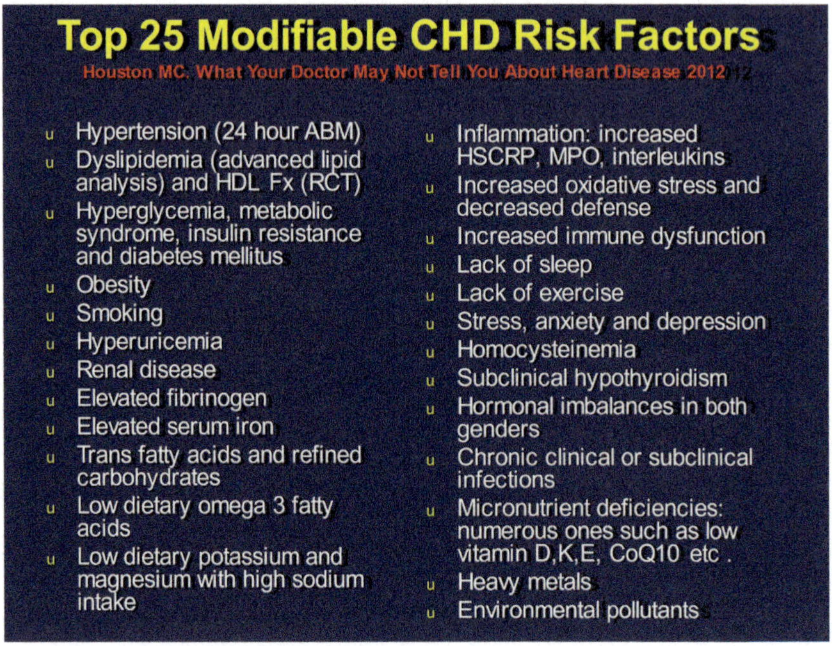

Top 25 Modifiable CHD Risk Factors
Houston MC. What Your Doctor May Not Tell You About Heart Disease 2012

- Hypertension (24 hour ABM)
- Dyslipidemia (advanced lipid analysis) and HDL Fx (RCT)
- Hyperglycemia, metabolic syndrome, insulin resistance and diabetes mellitus
- Obesity
- Smoking
- Hyperuricemia
- Renal disease
- Elevated fibrinogen
- Elevated serum iron
- Trans fatty acids and refined carbohydrates
- Low dietary omega 3 fatty acids
- Low dietary potassium and magnesium with high sodium intake
- Inflammation: increased HSCRP, MPO, interleukins
- Increased oxidative stress and decreased defense
- Increased immune dysfunction
- Lack of sleep
- Lack of exercise
- Stress, anxiety and depression
- Homocysteinemia
- Subclinical hypothyroidism
- Hormonal imbalances in both genders
- Chronic clinical or subclinical infections
- Micronutrient deficiencies: numerous ones such as low vitamin D,K,E, CoQ10 etc.
- Heavy metals
- Environmental pollutants

FIGURE 21.7 Top 25 CHD risk factors.

4. **Obesity**: measure the total body fat, regional fat, visceral fat, and lean muscle with the BIA and get to 16% in men and 22% in women.
5. **Smoking and all tobacco products, e-cigarettes, and vaping**: must be stopped immediately.

SUMMARY AND KEY TAKE AWAY POINTS

1. Nutrition is the foundation for the prevention and treatment of CHD and should follow the general guidelines of the Mediterranean diet. Also consider caloric restriction and fasting programs.
2. Numerous nutritional supplements have proven scientific validity to improve numerous CHD risk factors, noninvasive cardiovascular tests, coronary calcium, and plaque stabilization and reduction.
3. A combined aerobic and resistance exercise program is recommended for the prevention and treatment of CHD. This is the ABCT exercise program.
4. Achieving the ideal body weight, BMI, and body fat will improve CHD risk.
5. Stop all tobacco products, e-cigarettes, and vaping.
6. Control all risk factors for CHD to the new recommended goals.

BIBLIOGRAPHY

1. Sieve I, et al. Regulation and function of endothelial glycocalyx layer in vascular diseases. Vascul Pharmacol. 2018;100:26–33.11.
2. Dam H. Biochem. J. 1935;29(6);1273–85.
3. Vermeer C Food Nutr. Res. 2012;56. 5329. doi: 10.3402/fnr.v56i0.5329.
4. Hernlund E, et al. Arch. Osteoporos. 2013;8:136: doi: 10.1007/s11657-013-0136-1.
5. van Leeuwen JP, et al. Crit. Rev. Eukaryot. Gene Expr. 2001;11(1–3):199–226.
6. Booth SL. Annu. Rev. Nutr. 2009;29:89–110.
7. Crockett JC, et al. J. Cell Sci. 2011;124(7):991–8.
8. European Union. Official Journal of the European Union, Commission Directive (EU) 432/2012, L 285/9-12; 2008.
9. Personalized and Precision Integrative Cardiovascular Medicine. Houston, M, Editor and Contributor. Philadelphia and Chicago: Wolters Kluwer Publishers; November 2019.
10. Nicolaidou P, et al. Eur. J. Pediatr. 2006;165(8):540–5.
11. Suttie JS, Editor. Vitamin K in Health and Disease. Boca Raton: CRC Press; 2009.
12. Garber AK, et al. J. Nutr. 1999;129(6):1201–3.
13. Booth SL, et al. J. Clin. Endocrinol. Metab. 2008;93(4):1217–23.
14. Theuwissen E, et al. Br. J. Nutr. 2012;108(9):1652–7.
15. Sokoll LJ, et al. Am. J. Clin. Nutr. 1997;65(3):779–84.
16. Shea MK, et al. J. Nutr. 2011;141(8):1529–34.
17. Westenfeld R, et al. Am. J. Kidney Dis. 2012;59(2):186–95.
18. Theuwissen E, et al. Food Funct. 2014;5(2):229–34.
19. Knapen MH, et al. Osteoporos. Int. 2013;24(9):2499–507.
20. Yaegashi Y, et al. Eur. J. Epidemiol. 2008;23(3):219–25.

21. Iwamoto J, et al. Keio. J. Med. 2003;52(3):147–50.
22. Ushiroyama T, et al. Maturitas. 2002;41(3):211–21.
23. Ikeda Y, et al. J. Nutr. 2006;136(5):1323–8.
24. Kaneki M, et al. Nutrition. 2001;17(4):315–21.
25. Fujita Y, et al. Osteoporos. Int. 2012;23(2):705–14.
26. Schurgers LJ, et al. Blood. 2007;109(8):3279–83.
27. Iwamoto J, et al. Nutr. Res. 2009;29(4):221–8.
28. Cockayne S, et al. Arch. Intern. Med. 2006;166(12):1256–61.
29. www.eisai.jp/medical/products/di/EPI/GLA_SC_EPI.pdf; 2014.
30. von der Recke P, Hansen MA, Hassager C. Am. J. Med. 1999;106(3):273–8.
31. Demer LL. Int. J. Epidemiol. 2002;31(4):737–41.
32. Beulens JW, et al. Atherosclerosis. 2009;203(2):489–93.
33. Schurgers LJ, et al. Arterioscler. Thromb. Vasc. Biol. 2005;25(8):1629–33.
34. Knapen MH, et al. Thromb. Haemost. 2015;113(5):1135–44.
35. Kurnatowska I, et al. Pol. Arch. Med. Wewn. 2015;125(9):631–40.
36. Ganesan K, et al. J. Natl Med. Assoc. 2005;97(3):329–33.
37. Ginaldi L, et al. Immun. Ageing. 2005;2:14.
38. Abdel-Rahman MS, Alkady EA, Ahmed S. Eur. J. Pharmacol. 2015;761:273–8.
39. Houston M, Bell L. Controlling High Blood Pressure Though Nutrition, Nutrtional Supplement, Lifestyle and drugs. Boca Raton and Oxford: CRC Press; 2021.
40. Ross R Glomset J Harker L. Response to injury and atherogenesis. Am J Pathol. 1977;86(3):675–84.
41. Alphonsus CS, Rodseth RN. The endothelial glycocalyx: A review of the vascular barrier. Anaesthesia. 2014;69(7):777–84.
42. Fu BM, Tarbell JM. Mechano-sensing and transduction by endothelial surface glycocalyx: Composition, structure, and function. Wiley Interdiscip Rev Syst Biol Med. 2013;5(3):381–90.
43. Kolálová H Ambruzová B Šindlerová LŠ Klinke A, Kubala L. Modulation of endothelial glycocalyx structure under inflammatory conditions. Mediators of Inflammation. 2014; 62:Article ID 694312.
44. Reitsma S, Slaaf DW Vink H van Zandvoort MAMJ, oude Egbrink MGA. The endothelial glycocalyx: Composition, functions, and visualization. Eur J Physiol. 2007;454:345–59.
45. Becker BF, Chappell D, Jacob M. Endothelial glycocalyx and coronary vascular permeability: The Fringe Bene t. Basic Research in Cardiology. 2010;105:687–701.
46. Curry FE, Adamson RH. Endothelial glycocalyx: Permeability barrier and mechanosensor. Annals of Biomedical Engineering. 2012;40:828–39.
47. Mulivor AW, Lipowsky HH. Role of glycocalyx in leukocyte-endothelial cell adhesion. American Journal of Physiology: Heart and Circulatory Physiology. 2002;283(4):H1282–91.
48. Ebong EE, Lopez-Quintero SV, Rizzo V, Spray DC, Tarbell JM. Shearinduced endothelial NOS activation and remodeling via heparan sulfate, glypican-1, and syndecan1. Integr Biol. 2014;6(3):338–47. 9 10. Wanyi Yen, Bin Cai, Jinlin Yang, Lin Zhang, Min Zeng, John M. Tarbell & Bingmei M. Fu. Endothelial Surface Glycocalyx Can Regulate Flow-Induced Nitric Oxide Production in Microvessels In Vivo. PLoS One (2015) 10(1):e0117133.
49. Nieuwdorp M, van Haeften TW, Gouverneur MCLG, Mooij HL, van Lieshout MHP, Levi M, Meijers JCM, Holleman F, Hoekstra JBL, Vink H, Kastelein JJP, Stroes ESG. Loss of endothelial glycocalyx during acute hyperglycemia coincides with endothelial dysfunction and coagulation activation in vivo. Diabetes. 2006;55(2):480–6.

50. Vink H Constantinescu AA, Spaan JAE. Oxidized lipoproteins degrade the endothelial surface layer: implications for platelet-endothelial cell adhesion. Circulation. 2000;101:1500–2
51. Eskens BJM Leurgans TM, Vink H, VanTeeffelen JWGE. Early impairment of skeletal muscle endothelial glycocalyx barrier properties in diet induced obesity in mice. Physiol Rep. 2014;2(1):e00194.
52. Nussbaum C Heringa ACF, Mormanova Z, Puchwein-Schwepcke AF Pozza SBD, Genzel-Boroviczény O. Early microvascular changes with loss of the glycocalyx in children with type 1 diabetes. J Pediatr. 2014;164(3):584589.
53. Lopez-Quintero SV Cancel LM Pierides A Antonetti D Spray DC, Tarbell JM. High glucose attenuates shear-induced changes in endothelial hydraulic conductivity by degrading the glycocalyx. PLoS One. 2013;8(11):e78954.
54. Salmon AHJ, Satchell SC. Endothelial glycocalyx dysfunction in disease: Albuminuria and increased microvascular permeability. J Pathol. 2012;226:562–74.
55. Mulders TA Nieuwdorp M Stroes ES Vink H, Pinto-Sietsma S-J. Noninvasive assessment of microvascular dysfunction in families with premature coronary artery disease. Int J Cardiol. 2013;168(5):5026–8.
56. Vlahu CA Lemkes BA, Struijk DG, Koopman MG, Krediet RT, Vink H. Damage of the endothelial glycocalyx in dialysis patients. J Am Soc Nephro. 2012;123:1900–8.
57. Martens RJH, Vink H, van Oostenbrugge RJ, Staals J. Sublingual microvascular glycocalyx dimensions in lacunar stroke patients. Cerebrovasc Dis (2013) 35:451–4.
58. Rahbar E, Cardenas JC, Baimukanova G, Usadi B, Bruhn R Pati S, Ostrowski SR, Johansson PI, Holcomb JB, Wade CE. Endothelial glycocalyx shedding and vascular per-meability in severely injured trauma patients. J Transl Med. (2015) 13:117.
59. Drake-Holland AJ, Noble MI. The important new drug target in cardiovascular medicine: The vascular glycocalyx. Cardiovascular & Haematological Disorders-Drug Targets (2009) 9:118–23.
60. Drake-Holland AJ, Noble MI. Update on the important new drug target in cardiovascular medicine the vascular glycocalyx. Cardiovascular & Haematological Disorders-Drug Targets (2012) 12:76–81.
61. Tarbell JM, Cancel LM. The glycocalyx and its significance in human medicine. J Intern Med. 2016;280(1):97–113. doi: 10.1111/joim.12465. Epub 2016 Jan 8.
62. Cecelja M, Chowienczyk P. Role of arterial stiffness in cardiovascular disease. JRSM Cardiovascular Disease. 2012;1(4):4–11.
63. Grey E, Bratteli C, Glasser SP, Alinder C, Finkelstein SM, Lindgren BR, Cohn JN. Reduced small artery but not large artery elasticity is an independent risk marker for cardiovascular events. American Journal of Hypertension. 2003;16(4):265–9.
64. Tao J, Jin Y-F, Yang Z, Wang L-C, Gao X-R, Lei L, Ma H. Reduced arterial elasticity is associated with endothelial dysfunction in persons of advancing age. American Journal of Hypertension. 2004;17(8): 654–9.
65. Houston M and Hays L. Acute effects of an oral nitric oxide supplement on blood pressure, endothelial function, and vascular compliance in hypertensive patients. Journal of Clinical Hypertension. 2014;21:1–6.
66. Ren RMD, Liu JMD, Cheng, GMD, Tan, JMD, PhD. Editor(s). Găman., Mihnea-Alexandru vitamin K2 (Menaquinone-7) supplementation does not affect vitamin K-dependent coagulation factors activity in healthy individuals. Medicine. 2021June 11;100(23):e26221.
67. Derosa G, Pasqualotto S, Catena G, D'Angelo A, Maggi A, Maffioli P. A randomized, double-blind, placebo-controlled study to evaluate the effectiveness of a food supplement containing creatine and D-ribose combined with a physical exercise program in increasing stress tolerance in patients with ischemic heart disease. Nutrients. 2019 Dec 17;11(12):3075.

68. Herrick J, St Cyr JJ. Ribose in the heart. Diet Suppl. 2008;5(2):213–7.
69. Shecterle LM, Terry KR, St Cyr JA. The patented uses of D-ribose in cardiovascular diseases. Recent Pat Cardiovasc Drug Discov. 2010 Jun;5(2):138–42.

1. Priyadarsini KI. The chemistry of curcumin: From extraction to therapeutic agent. Molecules. 2014;19:20091–112. doi: 10.3390/molecules191220091. [PMC free article] [PubMed] [CrossRef] [Google Scholar].
2. Gupta SC, Patchva S, Aggarwal BB. Therapeutic roles of curcumin: Lessons learned from clinical trials. AAPS J. 2013;15:195–218. doi: 10.1208/s12248-012-9432-8. [PMC free article] [PubMed] [CrossRef] [Google Scholar].
3. Aggarwal BB, Kumar A, Bharti AC. Anticancer potential of curcumin: Preclinical and clinical studies. Anticancer Res. 2003;23:363–98. [PubMed] [Google Scholar].
4. Lestari ML, Indrayanto G. Curcumin. Profiles Drug Subst. Excip. Relat. Methodol. 2014;39:113–204. [PubMed] [Google Scholar].
5. Mahady GB, Pendland SL, Yun G, Lu ZZ. Turmeric (Curcuma longa) and curcumin inhibit the growth of Helicobacter pylori, a group 1 carcinogen. Anticancer Res. 2002;22:4179–81. [PubMed] [Google Scholar].
6. Reddy RC, Vatsala PG, Keshamouni VG, Padmanaban G, Rangarajan PN. Curcumin for malaria therapy. Biochem. Biophys. Res. Commun. 2005;326:472–4. doi: 10.1016/j.bbrc.2004.11.051. [PubMed] [CrossRef] [Google Scholar].
7. Vera-Ramirez L, Perez-Lopez P, Varela-Lopez A, Ramirez-Tortosa M, Battino M, Quiles JL. Curcumin and liver disease. Biofactors. 2013;39:88–100. doi: 10.1002/biof.1057. [PubMed] [CrossRef] [Google Scholar].
8. Wright LE, Frye JB, Gorti B, Timmermann BN, Funk JL. Bioactivity of turmeric-derived curcuminoids and related metabolites in breast cancer. Curr. Pharm. Des. 2013;19:6218–25. doi: 10.2174/1381612811319340013. [PMC free article] [PubMed] [CrossRef] [Google Scholar].
9. Aggarwal BB, Harikumar KB. Potential therapeutic effects of curcumin, the anti-inflammatory agent, against neurodegenerative, cardiovascular, pulmonary, metabolic, autoimmune and neoplastic diseases. Int. J. Biochem. Cell Biol. 2009;41:40–59. doi: 10.1016/j.biocel.2008.06.010. [PMC free article] [PubMed] [CrossRef] [Google Scholar].
10. Panahi Y, Hosseini MS, Khalili N, Naimi E, Simental-Mendia LE, Majeed M, Sahebkar A. Effects of curcumin on serum cytokine concentrations in subjects with metabolic syndrome: A post-hoc analysis of a randomized controlled trial. Biomed. Pharmacother. 2016;82:578–82. doi: 10.1016/j.biopha.2016.05.037. [PubMed] [CrossRef] [Google Scholar].
11. Kuptniratsaikul V, Dajpratham P, Taechaarpornkul W, Buntragulpoontawee M, Lukkanapichonchut P, Chootip C, Saengsuwan J, Tantayakom K, Laongpech S. Efficacy and safety of Curcuma domestica extracts compared with ibuprofen in patients with knee osteoarthritis: A multicenter study. Clin. Interv. Aging. 2014;9:451–8. doi: 10.2147/CIA.S58535. [PMC free article] [PubMed] [CrossRef] [Google Scholar].
12. Mazzolani F, Togni S. Oral administration of a curcumin-phospholipid delivery system for the treatment of central serous chorioretinopathy: A 12-month follow-up study. Clin. Ophthalmol. 2013;7:939–45. doi: 10.2147/OPTH.S45820. [PMC free article] [PubMed] [CrossRef] [Google Scholar].
13. Allegri P, Mastromarino A, Neri P. Management of chronic anterior uveitis relapses: Efficacy of oral phospholipidic curcumin treatment. Long-term follow-up. Clin. Ophthalmol. 2010;4:1201–6. doi: 10.2147/OPTH.S13271. [PMC free article] [PubMed] [CrossRef] [Google Scholar].

14. Trujillo J, Chirino YI, Molina-Jijón E, Andérica-Romero AC, Tapia E, Pedraza-Chaverrí J. Renoprotective effect of the antioxidant curcumin: Recent findings. Redox Biol. 2013;1:448–56. doi: 10.1016/j.redox.2013.09.003. [PMC free article] [PubMed] [CrossRef] [Google Scholar].
15. Anand P, Kunnumakkara AB, Newman RA, Aggarwal BB. Bioavailability of curcumin: Problems and promises. Mol. Pharm. 2007;4:807–18. doi: 10.1021/mp700113r. [PubMed] [CrossRef] [Google Scholar].
16. Han HK. The effects of black pepper on the intestinal absorption and hepatic metabolism of drugs. Expert Opin. Drug Metab. Toxicol. 2011;7:721–9. doi: 10.1517/17425255.2011.570332. [PubMed] [CrossRef] [Google Scholar].
17. Shoba G, Joy D, Joseph T, Majeed M, Rajendran R, Srinivas PS. Influence of piperine on the pharmacokinetics of curcumin in animals and human volunteers. Planta Med. 1998;64:353–6. doi: 10.1055/s-2006-957450. [PubMed] [CrossRef] [Google Scholar].
18. Basnet P, Skalko-Basnet N. Curcumin: An anti-inflammatory molecule from a curry spice on the path to cancer treatment. Molecules. 2011;16:4567–98. doi: 10.3390/molecules16064567. [PMC free article] [PubMed] [CrossRef] [Google Scholar].
19. Lao CD, Ruffin MT, Normolle D, Heath DD, Murray SI, Bailey JM, Boggs ME, Crowell J, Rock CL, Brenner DE. Dose escalation of a curcuminoid formulation. BMC Complement. Altern. Med. 2006;6:10. doi: 10.1186/1472-6882-6-10. [PMC free article] [PubMed] [CrossRef] [Google Scholar].
20. Kunnumakkara AB, Bordoloi D, Harsha C, Banik K, Gupta SC, Aggarwal BB. Curcumin mediates anticancer effects by modulating multiple cell signaling pathways. Clin. Sci. 2017;131:1781–99. doi: 10.1042/CS20160935. [PubMed] [CrossRef] [Google Scholar].
21. Lin YG, Kunnumakkara AB, Nair A, Merritt WM, Han LY, Armaiz-Pena GN, Kamat AA, Spannuth WA, Gershenson DM, Lutgendorf SK, et al. Curcumin inhibits tumor growth and angiogenesis in ovarian carcinoma by targeting the nuclear factor-κB pathway. Clin. Cancer Res. 2007;13:3423–30. doi: 10.1158/1078-0432.CCR-06-3072. [PubMed] [CrossRef] [Google Scholar].
22. Marchiani A, Rozzo C, Fadda A, Delogu G, Ruzza P. Curcumin and curcumin-like molecules: From spice to drugs. Curr. Med. Chem. 2014;21:204–22. doi: 10.2174/092986732102131206115810. [PubMed] [CrossRef] [Google Scholar].
23. Sahebkar A, Serbanc MC, Ursoniuc S, Banach M. Effect of curcuminoids on oxidative stress: A systematic review and meta-analysis of randomized controlled trials. J. Funct. Foods. 2015;18:898–909. doi: 10.1016/j.jff.2015.01.005. [CrossRef] [Google Scholar].
24. Banach M, Serban C, Aronow WS, Rysz J, Dragan S, Lerma EV, Apetrii M, Covic A. Lipid, blood pressure and kidney update 2013. Int. Urol. Nephrol. 2014;46:947–61. doi: 10.1007/s11255-014-0657-6. [PMC free article] [PubMed] [CrossRef] [Google Scholar].
25. Menon VP, Sudheer AR. Antioxidant and anti-inflammatory properties of curcumin. Adv. Exp. Med. Biol. 2007;595:105–25. [PubMed] [Google Scholar].
26. Panahi Y, Alishiri GH, Parvin S, Sahebkar A. Mitigation of systemic oxidative stress by curcuminoids in osteoarthritis: Results of a randomized controlled trial. J. Diet. Suppl. 2016;13:209–20. doi: 10.3109/19390211.2015.1008611. [PubMed] [CrossRef] [Google Scholar].
27. Priyadarsini KI, Maity DK, Naik GH, Kumar MS, Unnikrishnan MK, Satav JG, Mohan H. Role of phenolic O-H and methylene hydrogen on the free radical reactions and antioxidant activity of curcumin. Free Radic. Biol. Med. 2003;35:475–84. doi: 10.1016/S0891-5849(03)00325-3. [PubMed] [CrossRef] [Google Scholar].

28. Biswas SK. Does the Interdependence between Oxidative Stress and Inflammation Explain the Antioxidant Paradox? Oxid. Med. Cell. Longev. 2016;2016:5698931. doi: 10.1155/2016/5698931. [PMC free article] [PubMed] [CrossRef] [Google Scholar].
29. Jurenka JS. Anti-inflammatory properties of curcumin, a major constituent of Curcuma longa: A review of preclinical and clinical research. Altern. Med. Rev. J. Clin. Ther. 2009;14:141–53. [PubMed] [Google Scholar].
30. Recio MC, Andujar I, Rios JL. Anti-inflammatory agents from plants: Progress and potential. Curr. Med. Chem. 2012;19:2088–103. doi: 10.2174/092986712800229069. [PubMed] [CrossRef] [Google Scholar].
31. Hunter DJ, Schofield D, Callander E. The individual and socioeconomic impact of osteoarthritis. Lancet Nat. Rev. Rheumatol. 2014;10:437–41. doi: 10.1038/nrrheum.2014.44. [PubMed] [CrossRef] [Google Scholar].
32. Vos T, Barber RM, Bell B, Bertozzi-Villa A, Biryukov S, Bolliger I, Charlson F, Davis A, Degenhardt L, Dicker D, et al. Global, regional, and national incidence, prevalence, and years lived with disability for 301 acute and chronic diseases and injuries in 188 countries, 1990–2013: A systematic analysis for the Global Burden of Disease Study. Lancet. 2013;386:743–800. doi: 10.1016/S0140-6736(15)60692-4. [PMC free article] [PubMed] [CrossRef] [Google Scholar].
33. Goldring MB. Osteoarthritis and cartilage: The role of cytokines. Curr. Rheumatol. Rep. 2000;2:459–65. doi: 10.1007/s11926-000-0021-y. [PubMed] [CrossRef] [Google Scholar].
34. Rahimnia AR, Panahi Y, Alishiri A, Sharafi M, Sahebkar A. Impact of supplementation with curcuminoids on systemic inflammation in patients with knee osteoarthritis: Findings from a randomized double-blind placebo-controlled trial. Drug Res. 2015;65:521–5. doi: 10.1055/s-0034-1384536. [PubMed] [CrossRef] [Google Scholar].
35. Sahebkar A. Molecular mechanisms for curcumin benefits against ischemic injury. Fertil. Steril. 2010;94:e75–e76. doi: 10.1016/j.fertnstert.2010.07.1071. [PubMed] [CrossRef] [Google Scholar].
36. Henrotin Y, Priem F, Mobasheri A. Curcumin: A new paradigm and therapeutic opportunity for the treatment of osteoarthritis: Curcumin for osteoarthritis management. SpringerPlus. 2013;2:56. doi: 10.1186/2193-1801-2-56. [PMC free article] [PubMed] [CrossRef] [Google Scholar].
37. Belcaro G, Cesarone MR, Dugall M, Pellegrini L, Ledda A, Grossi MG, Togni S, Appendino G. Product-evaluation registry of Meriva®, a curcumin-phosphatidylcholine complex, for the complementary management of osteoarthritis. Panminerva Med. 2010;52(Supplement 1):55–62. [PubMed] [Google Scholar].
38. Belcaro G, Hosoi M, Pellegrini L, Appendino G, Ippolito E, Ricci A, Ledda A, Dugall M, Cesarone MR, Maione C, et al. A controlled study of a lecithinized delivery system of curcumin (meriva®) to alleviate the adverse effects of cancer treatment. Phytother. Res. 2014;28:444–50. doi: 10.1002/ptr.5014. [PubMed] [CrossRef] [Google Scholar].
39. Chandran B, Goel A. A randomized, pilot study to assess the efficacy and safety of curcumin in patients with active rheumatoid arthritis. Phytother. Res. 2012;26:1719–25. doi: 10.1002/ptr.4639. [PubMed] [CrossRef] [Google Scholar].
40. Panahi Y, Rahimnia AR, Sharafi M, Alishiri G, Saburi A, Sahebkar A. Curcuminoid treatment for knee osteoarthritis: A randomized double-blind placebo-controlled trial. Phytother. Res. 2014;28:1625–31. doi: 10.1002/ptr.5174. [PubMed] [CrossRef] [Google Scholar].
41. Francesco DP, Giuliana R, Eleonora ADM, Giovanni A, Federico F, Stefano T. Comparative evaluation of the pain-relieving properties of a lecithinized formulation of curcumin (Meriva®), nimesulide, and acetaminophen. J. Pain Res. 2013;6:201–5. doi: 10.2147/JPR.S42184. [PMC free article] [PubMed] [CrossRef] [Google Scholar].

42. Daily JW, Yang M, Park S. Efficacy of turmeric extracts and curcumin for alleviating the symptoms of ioint arthritis: A Systematic review and meta-snalysis of randomized clinical trials. J. Med. Food. 2016;19:717–29. doi: 10.1089/jmf.2016.3705. [PMC free article] [PubMed] [CrossRef] [Google Scholar].
43. Na LX, Li Y, Pan HZ, Zhou XL, Sun DJ, Meng M, Li XX, Sun CH. Curcuminoids exert glucose-lowering effect in type 2 diabetes by decreasing serum free fatty acids: A double-blind, placebo-controlled trial. Mol. Nutr. Food Res. 2013;57:1569–77. doi: 10.1002/mnfr.201200131. [PubMed] [CrossRef] [Google Scholar].
44. Chuengsamarn S, Rattanamongkolgul S, Luechapudiporn R, Phisalaphong C, Jirawatnotai S. Curcumin extract for prevention of type 2 diabetes. Diabetes Care. 2012;35:2121–7. doi: 10.2337/dc12-0116. [PMC free article] [PubMed] [CrossRef] [Google Scholar].
45. Bradford PG. Curcumin and obesity. Biofactors. 2013;39:78–87. doi: 10.1002/biof.1074. [PubMed] [CrossRef] [Google Scholar].
46. Hlavackova L, Janegova A, Ulicna O, Janega P, Cerna A, Babal P. Spice up the hypertension diet: Curcumin and piperine prevent remodeling of aorta in experimental L-NAME induced hypertension. Nutr. Metab. 2011;8:72. doi: 10.1186/1743-7075-8-72. [PMC free article] [PubMed] [CrossRef] [Google Scholar].
47. Sahebkar A. Are curcuminoids effective C-reactive protein-lowering agents in clinical practice? Evidence from a meta-analysis. Phytother. Res. 2013;28:633–42. doi: 10.1002/ptr.5045. [PubMed] [CrossRef] [Google Scholar].
48. Ak T, Gulcin I. Antioxidant and radical scavenging properties of curcumin. Chem. Biol. Interact. 2008;174:27–37. doi: 10.1016/j.cbi.2008.05.003. [PubMed] [CrossRef] [Google Scholar].
49. Sahebkar A, Mohammadi A, Atabati A, Rahiman S, Tavallaie S, Iranshahi M, Akhlaghi S, Ferns GA, Ghayour-Mobarhan M. Curcuminoids modulate pro-oxidant-antioxidant balance but not the immune response to heat shock protein 27 and oxidized LDL in obese individuals. Phytother. Res. 2013;27:1883–8. doi: 10.1002/ptr.4952. [PubMed] [CrossRef] [Google Scholar].
50. Mohammadi A, Sahebkar A, Iranshahi M, Amini M, Khojasteh R, Ghayour-Mobarhan M, Ferns GA. Effects of supplementation with curcuminoids on dyslipidemia in obese patients: A randomized crossover trial. Phytother. Res. 2013;27:374–9. doi: 10.1002/ptr.4715. [PubMed] [CrossRef] [Google Scholar].
51. DiSilvestro RA, Joseph E, Zhao S, Bomser J. Diverse effects of a low dose supplement of lipidated curcumin in healthy middle-aged people. Nutr. J. 2012;11:79. doi: 10.1186/1475-2891-11-79. [PMC free article] [PubMed] [CrossRef] [Google Scholar].
52. Sahebkar A. Curcuminoids for the management of hypertriglyceridaemia. Nat. Rev. Cardiol. 2014;11:123. doi: 10.1038/nrcardio.2013.140-c1. [PubMed] [CrossRef] [Google Scholar].
53. Soni KB, Kuttan R. Effect of oral curcumin administration on serum peroxides and cholesterol levels in human volunteers. Indian J. Physiol. Pharmacol. 1992;36:273–5. [PubMed] [Google Scholar].
54. Panahi Y, Khalili N, Hosseini MS, Abbasinazari M, Sahebkar A. Lipid-modifying effects of adjunctive therapy with curcuminoids-piperine combination in patients with metabolic syndrome: Results of a randomized controlled trial. Complement. Ther. Med. 2014;22:851–7. doi: 10.1016/j.ctim.2014.07.006. [PubMed] [CrossRef] [Google Scholar].
55. Panahi Y, Hosseini MS, Khalili N, Naimi E, Simental-Mendia LE, Majeed M, Sahebkar A. Antioxidant and anti-inflammatory effects of curcuminoid-piperine combination in subjects with metabolic syndrome: A randomized controlled trial and an updated meta-analysis. Clin. Nutr. 2015;34:1101–8. doi: 10.1016/j.clnu.2014.12.019. [PubMed] [CrossRef] [Google Scholar].

56. Ganjali S, Sahebkar A, Mahdipour E, Jamialahmadi K, Torabi S, Akhlaghi S, Ferns G, Parizadeh SMR, Ghayour-Mobarhan M. Investigation of the effects of curcumin on serum cytokines in obese individuals: A randomized controlled trial. Sci. World J. 2014;2014:898361. doi: 10.1155/2014/898361. [PMC free article] [PubMed] [CrossRef] [Google Scholar].
57. Strimpakos A, Sharm R. Curcumin: Preventive and therapeutic properties in laboratory studies and clinical trials. Food Chem. Toxicol. 2008;10:511–45. doi: 10.1089/ars.2007.1769. [PubMed] [CrossRef] [Google Scholar].
58. Epstein J, Sanderson I, Macdonald T. Curcumin as a therapeutic agent: The evidence from in vitro, animal and human studies. Br. J. Nutr. 2010;103:1545–57. doi: 10.1017/S0007114509993667. [PubMed] [CrossRef] [Google Scholar].
59. Cox KH, Pipingas A, Scholey AB. Investigation of the effects of solid lipid curcumin on cognition and mood in a healthy older population. J. Psychopharmacol. 2015;29:642–51. doi: 10.1177/0269881114552744. [PubMed] [CrossRef] [Google Scholar].
60. Chilelli NC, Ragazzi E, Valentini R, Cosma C, Ferraresso S, Lapolla A, Sartore G. Curcumin and Boswellia serrata modulate the glyco-oxidative status and lipo-oxidation in master sthletes. Nutrients. 2016;8:745. doi: 10.3390/nu8110745. [PMC free article] [PubMed] [CrossRef] [Google Scholar].
61. McFarlin BK, Venable AS, Henning AL, Sampson JN, Pennel K, Vingren JL, Hill DW. Reduced inflammatory and muscle damage biomarkers following oral supplementation with bioavailable curcumin. BBA Clin. 2016;5:72–8. doi: 10.1016/j.bbacli.2016.02.003. [PMC free article] [PubMed] [CrossRef] [Google Scholar].
62. Drobnic F, Riera J, Appendino G, Togni S, Franceschi F, Valle X, Pons A, Tur J. Reduction of delayed onset muscle soreness by a novel curcumin delivery system (Meriva®): A randomised, placebo-controlled trial. J. ISSN. 2014;11:31. doi: 10.1186/1550-2783-11-31. [PMC free article] [PubMed] [CrossRef] [Google Scholar].
63. Delecroix B, Abaïdia AE, Leduc C, Dawson B, Dupont G. Curcumin and piperine supplementation and recovery following exercise induced muscle damage: A randomized controlled trial. J. Sports Sci. Med. 2017;16:147–53. [PMC free article] [PubMed] [Google Scholar].
64. Esmaily H, Sahebkar A, Iranshahi M, Ganjali S, Mohammadi A, Ferns G, Ghayour-Mobarhan M. An investigation of the effects of curcumin on anxiety and depression in obese individuals: A randomized controlled trial. Chin. J. Integr. Med. 2015;21:332–8. doi: 10.1007/s11655-015-2160-z. [PubMed] [CrossRef] [Google Scholar].
65. Kocaadam B, Şanlier N. Curcumin, an active component of turmeric (Curcuma longa), and its effects on health. Crit. Rev. Food Sci. Nutr. 2017;57:2889–95. doi: 10.1080/10408398.2015.1077195. [PubMed] [CrossRef] [Google Scholar].
66. Sharma RA, Euden SA, Platton SL, Cooke DN, Shafayat A, Hewitt HR, Marczylo TH, Morgan B, Hemingway D, Plummer SM. Phase I clinical trial of oral curcumin: Biomarkers of systemic activity and compliance. Clin. Cancer Res. 2004;10:6847–54. doi: 10.1158/1078-0432.CCR-04-0744. [PubMed] [CrossRef] [Google Scholar].

1. Houston, MC. What Your Doctor May Not Tell You About Heart Disease. The Revolutionary Book that Reveals the Truth Behind Coronary Illness: And How You Can Fight Them. New York: Grand Central Life and Style, Hachette Book Group; 2012.
2. Vina J, Sanchis-Gomar F, Martinez-Bello V, Gomez-Cabrera MC. Exercise acts a drug; the pharmacological benefits of exercise Br J Pharmacol. 2012;167(1):1–12.
3. Meka N, Katragadda S, Cherian B, Arora RR. Endurance exercise and resistance training in cardiovascular disease. Ther Adv Cardiovasc Dis. 2008;2(2):115–21.
4. Coffey VG, Hawley JA. The molecular bases of training adaptation. Sports Med. 2007;37(9):737–63.

5. Radom-Aizik S, Hayek S, Shahar I, Rechavi G, Kaminski N, Ben-Dov I. Effects of aerobic training on gene expression in skeletal muscle of elderly men Med Sci Sports Exerc. 2005;37(10):1680–96.
6. Ostrowski K, Schjerling P, Pedersen BK. Physical activity and plasma interleukin-6 in humans: Effect of intensity of exercise Eur J Appl Physiol. 2000;83(6):512–5.
7. Pedersen BK, Ostrowski K, Rohde T, Bruunsgaard HCJ. The cytokine response to strenuous exercise. Physiol Pharmacol. 1998;76(5):505–11.
8. Steensberg A. The role of IL-6 in exercise-induced immune changes and metabolism. Exerc Immunol Rev. 2003;9:40–7.
9. Petersen AM, Pedersen BK. The anti-inflammatory effect of exercise J Appl Physiol. 2005;98(4):1154–62.
10. Bruunsgaard H. Physical activity and modulation of systemic low-level inflammation. J Leukoc Biol. 2005;78(4):819–35.
11. Scott JP, Sale C, Greeves JP, Casey A, Dutton J, Fraser WD. Cytokine response to acute running in recreationally-active and endurance-trained men. Eur J Appl Physiol. 2013;22:34–46. Mar 6. [Epub ahead of print].
12. Welc SS, Clanton TL. The regulation of interleukin-6 implicates skeletal muscle as an integrative stress sensor and endocrine organ. Exp Physiol. 2013;98(2):359–71.
13. Pratley R, Nicklas B, Rubin M, Miller J, Smith A, Smith M, Hurley B, Goldberg A. Strength training increases resting metabolic rate and norepinephrine levels in healthy 50- to 65-yr-old men. J Appl Physiol. 1994;76(1):133–7.
14. Anton MM, Cortez-Cooper MY, DeVan AE, Neidre DB, Cook JN, Tanaka H. Resistance training increases basal limb blood flow and vascular conductance in aging humans. J Appl Physiol. 2006;101(5):1351–5.
15. Okamoto T. Combined aerobic and resistance training and vascular function: effect of aerobic exercise before and after resistance training J Appl Physiol. 2007;103(5):1655–61.
16. McCartney N, McKelvie RS, Haslam DR, Jones NL. Usefulness of Weight lifting training in improving strength and maximal power output in coronary artery disease Am J Cardiol. 1991;67(11):939–45.
17. Morra EA, Zaniqueli D, Rodrigues SL, El-Aourar LM, Lunz W, Mill JG, Carletti L. Long- term intense resistance training in men is associated with preserved cardiac structure /function, decreased aortic stiffness, and lower central augmentation pressure. J of Hypertension. 2014;32:286–293.
18. Pu CT, Johnson MT, Forman DE, Hausdorff JM, Roubenoff R, Foldvari M, Fielding RA, Singh MA. Randomized trial of progressive resistance training to counteract the myopathy of chronic heart failure. J Appl Physiol. 2001;90(6):2341–50.
19. Church TS, Blair SN, Cocreham S, Johannsen N, Johnson W, Kramer K, Mikus CR, Myers V, Nauta M, Rodarte RQ, Sparks L, Thompson A, Earnest CP. Effects of aerobic and resistance training on hemoglobin A1c levels in patients with type 2 diabetes: a randomized controlled trial. JAMA. 2010;304(20):2253–62.
20. Mujica V, Urzúa A, Leiva E, Díaz N, Moore-Carrasco R, Vásquez M, Rojas E, Icaza G, Toro C, Orrego R,Palomo IJ. Intervention with education and exercise reverses the metabolic syndrome in adults. Am Soc Hypertens. 2010;4(3):148–53.
21. Warner SO, Linden MA, Liu Y, Harvey BR, Thyfault JP, Whaley-Connell AT, Chockalingam A, Hinton PS,Dellsperger KC, Thomas TR. The effects of resistance training on metabolic health with weight regain J Clin Hypertens. 2010;12(1):64–72.
22. Lamina S. Effects of continuous and interval trainingprograms in the management of hypertension: A randomized controlled trial. J Clin Hypertens. 2010;12(11):841–9.
23. Marzolini S, Oh PI, Brooks D. Effect of combined aerobic and resistance training versus aerobic training alone in individuals with coronary artery disease: a meta-analysis. Eur J Prev Cardiol. 2012;19(1):81–94.

24. Rossi A, Dikareva A, Bacon SL, Daskalopoulou SS. The impact of physical activity on mortality in patients with high blood pressure: a systematic review. J Hypertens. 2012;30(7):1277–88.
25. Sesso HD, Paffenbarger RS Jr, Lee IM. Physical activity and coronary heart disease in men: The Harvard Alumni Health Study. Circulation. 2000;102(9):975–80.
26. Hawley JA, Burke LM, Phillips SM, Spriet LL. Nutritional modulation of training-induced skeletal muscle adaptations. J Appl Physiol. 2011;110(3):834–45.
27. Pennings B, Koopman R, Beelen M, Senden JM, Saris WH, van Loon LJ. Exercising before protein intake allows for greater use of dietary protein-derived amino acids for de novo muscle protein synthesis in both young and elderly men. Am J Clin Nutr. 2011;93(2):322–31.
28. Lecoultre V, Benoit R, Carrel G, Schutz Y, Millet GP, Tappy L, Schneiter P. Fructose and glucose co-ingestion during prolonged exercise increases lactate and glucose fluxes and oxidation compared with an equimolar intake of glucose. Am J Clin Nutr. 2010;92(5):1071–9.
29. Stephens BR, Braun B. Impact of nutrient intake timing on the metabolic response to exercise. Nutr Rev. 2008;66(8):473–6.
30. Peake JM, Suzuki K, Coombes JS. The influence of antioxidant supplementation on markers of inflammation and the relationship to oxidative stress after exercise. J Nutr Biochem. 2007;18(6):357–71.
31. Fukuda DH, Smith AE, Kendall KL,Stout JR. The possible combinatory effects of acute consumption of caffeine, creatine, and amino acids on the improvement of anaerobic running performance in humans. Nutr Res. 2010;30(9):607–14.
32. Wray DW, Uberoi A, Lawrenson L, Bailey DM, Richardson RS. Oral antioxidants and cardiovascular health in the exercise-trained and untrained elderly: A radically different outcome. Clin Sci. 2009;116(5):433–41.
33. Panza VS, Wazlawik E, Ricardo Schütz G, Comin L, Hecht KC, da Silva EL. Consumption of green tea favorably affects oxidative stress markers in weight-trained men. Nutrition. 2008;24(5):433–42
34. Karanth J, Jeevaratnam K. Effect of carnitine supplementation on mitochondrial enzymes in liver and skeletal muscle of rat after dietary lipid manipulation and physical activity. Indian J Exp Biol. 2010;48(5):503–10.
35. Huang A, Owen K. Role of supplementary L carnitine in exercise and exercise recovery.Med Sport Sci. 2012;59:135–42.
36. Broad EM, Maughan RJ, Galloway SDR. Effects of exercise intensity and altered substrate availability on cardiovascular and metabolic responses to exercise after oral carnitine supplementation in athletes Int J Sport Nutr Exerc Metab. 2011;21(5):385–97.
37. Addis P, Shecterle LM, St Cyr JA. Cellular protection during oxidative stress: A potential role for D-ribose and antioxidants. J Diet Suppl. 2012;9(3):178–82.
38. Cramer JT, Housh TJ, Johnson GO, Coburn JW. Stout JREffects of a carbohydrate-, protein-, and ribose-containing repletion drink during 8 weeks of endurance training on aerobic capacity, endurance performance, and body composition. J Strength Cond Res. 2012;26(8):2234–42.
39. Seifert JG, Subudhi AW, Fu MX, Riska KL, John JC, Shecterle LM, St Cyr JA. The role of ribose on oxidative stress during hypoxic exercise: a pilot study. J Med Food. 2009;12(3):690–3.
40. Hoffman JR, Williams DR, Emerson NS, Hoffman MW, Wells AJ, McVeigh DM, McCormack WP, Mangine GT, Gonzalez AM, Fragala MS. L-alanyl-L-glutamine ingestion maintains performance during a competitive basketball game. J Int Soc Sports Nutr. 2012;9(1):4.

41. Rowlands DS, Clarke J, Green JG, Shi X. L-Arginine but not L-glutamine likely increases exogenous carbohydrate oxidation during endurance exercise . Eur J Appl Physiol. 2012 Jul;112(7):2443–53.
42. Ra SG, Miyazaki T, Ishikura K, Nagayama H, Suzuki T, Maeda S, Ito M, Matsuzaki Y, Ohmori H. Additional effects of taurine on the benefits of BCAA intake for the delayed-onset muscle soreness and muscle damage induced by high-intensity eccentric exercise. Adv Exp Med Biol. 2013;776:179–87.
43. Tang FC, Chan CC, Kuo PL. Contribution of creatine to protein homeostasis in athletes after endurance and sprint running. Eur J Nutr. 2013;19:66–73. Feb 8. [Epub].
44. Howatson G, Hoad M, Goodall S, Tallent J, Bell PG, French DN. Exercise-induced muscle damage is reduced in resistance-trained males bybranched chain amino acids: a randomized, double-blind, placebo controlled study. J Int Soc Sports Nutr. 2012 May 8;9(1):20. [Epub ahead of print].
45. Breen L, Phillips SM. Nutrient interaction for optimal protein anabolism in resistance exercise. Curr Opin Clin Nutr Metab Care. 2012 May;15(3):226–32.
46. Pasiakos SM, McClung HL, McClung JP, Margolis LM, Andersen NE, Cloutier GJ, Pikosky MA, Rood JC,Fielding RA, Young AJ. Leucine-enriched essential amino acid supplementation during moderate steady state exercise enhances postexercise muscle protein synthesis. Am J Clin Nutr. 2011;94(3):809–18.
47. Chen S, Li Z, Krochmal R, Abrazado M, Kim W, Cooper CBJ. Effect of Cs-4 (Cordyceps sinensis) on exercise performance in healthy older subjects: A double-blind, placebo-controlled trial. Altern Complement Med. 2010;16(5):585–90.
48. Kumar R, Negi PS, Singh B, Ilavazhagan G, Bhargava K, Sethy NK. Cordyceps sinensis promotes exercise endurance capacity of rats by activating skeletal muscle metabolic regulators. J Ethnopharmacol. 2011;136(1):260–6.
49. Noreen EE, Buckley JG, Lewis SL, Brandauer J, Stuempfle KJ. The Effects of an Acute Dose of Rhodiola rosea on Endurance ExercisePerformance. J Strength Cond Res. 2013;27(3):839–47.
50. Xu J, Li Y. Effects of salidroside on exhaustive exerciseinduced oxidative stress in rats Mol Med Rep. 2012;6(5):1195–8.
51. Noreen EE, Buckley JG, Lewis SL, Brandauer J, Stuempfle KJ. The Effects of an Acute Dose of Rhodiola Rosea on Endurance ExercisePerformance. J Strength Cond Res. 2012;44:16–24. May 24. [Epub ahead of print].
52. Parisi A, Tranchita E, Duranti G, Ciminelli E, Quaranta F, Ceci R, Cerulli C, Borrione P, Sabatini S. Effects of chronic Rhodiola Rosea supplementation on sport performance and antioxidant capacity in trained male: preliminary results. J Sports Med Phys Fitness. 2010;50(1):57–63.
53. Evdokimov VG. Effect of cryopowder Rhodiola rosae L. on cardiorespiratory parameters and physical performance of humans]. Aviakosm Ekolog Med. 2009;43(6):52–6.
54. Churchward-Venne TA, Breen L, Di Donato DM, Hector AJ, Mitchell CJ, Moore DR, Stellingwerff T, Breuille D, Offord EA, Baker SK, Phillips SM. Leucine supplementation of a low-protein mixed macronutrient beverage enhances myofibrillar protein synthesis in young men: a double-blind, randomized trial. Am J Clin Nutr. 2014;99(2):276–86.
55. Osterberg KL, Melby CL. Effect of acute resistance exercise on postexercise oxygen consumption and resting metabolic rate in young women. Int J Sport Nutr Exerc Metab. 2000;10(1):71–81.

1. Whelton PK, et al. ACC/AHA/AAPA/ABC/ACPM/AGS/APhA/ASH/ASPC/NMA/PCNA guideline for the prevention, detection, evaluation, and management of high blood pressure in adults: A report of the American College of Cardiology/American Heart Association Task Force on clinical practice guidelines. *Hypertension*. 2018 Jun;71(6):e13–e115.

2. Houston M. The role of nutrition and nutraceutical supplements in the treatment of hypertension. *World Journal of Cardiology.* 2014;6(2):38–66.
3. Houston M. Nutrition and nutraceutical supplements for the treatment of hypertension: Part 1. *Journal of Clinical Hypertension.* 2013;15:752–7.
4. Houston M. Nutrition and nutraceutical supplements for the treatment of hypertension: Part II. *Journal of Clinical Hypertension.* 2013;15:845–51.
5. Houston M. Nutrition and nutraceutical supplements for the treatment of hypertension: Part III. *Journal of Clinical Hypertension.* 2013;15:931–7.
6. Borghi C, Cicero AF Nutraceuticals with a clinically detectable blood pressure-lowering effect: a review of available randomized clinical trials and their meta-analyses. *Br J Clin Pharmacol.* 2017;83(1):163–71.
7. Sirtori CR, Arnoldi A, Cicero AF. Nutraceuticals for blood pressure control. *Review. Ann Med.* 2015;47(6):447–56.
8. Cicero AF, Colletti A. Nutraceuticals and blood pressure control: Results from clinical trials and meta-analyses. *High Blood Press Cardiovasc Prev.* 2015;22(3):203–13.
9. Turner JM, Spatz ES. Nutritional supplements for the treatment of hypertension: A practical guide for clinicians. *Curr Cardiol Rep.* 2016;18(12):126. Review.
10. Caligiuri SP, Pierce GN. A review of the relative efficacy of dietary, nutritional supplements, lifestyle and drug therapies in the management of hypertension. *Crit Rev Food Sci Nutr.* 2016 Aug 5;89:120–137. [Epub ahead of print].
11. Houston MC, Fox B, Taylor N. *What Your Doctor May Not Tell You About Hypertension. The Revolutionary Nutrition and Lifestyle Program to Help Fight High Blood Pressure.* New York: AOL Time Warner, Warner Books; September 2003.
12. Houston M. Treatment of hypertension with nutrition and nutraceutical supplement: Part 1. *Alternative and Complimentary Medicine.* 2019;24:260–75.
13. Houston M. Treatment of hypertension with nutrition and nutraceutical supplement: Part 2. *Alternative and Complimentary Medicine.* 2019;25:23–36.
14. Sinatra S and Houston M, Editors. *Nutrition and Integrative Strategies in Cardiovascular Medicine.* Boca Raton: CRC Press; 2015.

1. U.S. Department of Health and Human Services. *The Health Consequences of Smoking—50 Years of Progress: A Report of the Surgeon General.* Atlanta: U.S. Department of Health and Human Services, Centers for Disease Control and Prevention, National Center for Chronic Disease Prevention and Health Promotion, Office on Smoking and Health; 2014.
2. U.S. Department of Health and Human Services. *How Tobacco Smoke Causes Disease: What It Means to You.* Atlanta: U.S. Department of Health and Human Services, Centers for Disease Control and Prevention, National Center for Chronic Disease Prevention and Health Promotion, Office on Smoking and Health; 2010.
3. Centers for Disease Control and Prevention. Quick stats: Number of deaths from 10 leading causes—national vital statistics system, United States, 2010. *Morbidity and Mortality Weekly Report.* 2013:62(8);155.
4. Mokdad AH, Marks JS, Stroup DF, Gerberding JL. Actual causes of death in the United States. *JAMA: Journal of the American Medical Association.* 2004;291(10):1238–45.
5. U.S. Department of Health and Human Services. *Women and Smoking: A Report of the Surgeon General.* Rockville: U.S. Department of Health and Human Services, Public Health Service, Office of the Surgeon General; 2001 [accessed 2017 Apr 20].
6. U.S. Department of Health and Human Services. *Reducing the Health Consequences of Smoking: 25 Years of Progress. A Report of the Surgeon General External.* Rockville: U.S. Department of Health and Human Services, Public Health Service, Centers for Disease Control, National Center for Chronic Disease Prevention and Health Promotion, Office on Smoking and Health; 1989.

22 Medicines for CHD

Many different medicines are used to treat coronary heart disease (CHD) by reducing the CHD risk factors. You should consult with your personal physician regarding all drug treatments for CHD.

BLOOD-THINNING MEDICINES

Blood thinners are a type of medicine that can help reduce the risk of a heart attack by thinning your blood and preventing it from clotting.

Antithrombotic drugs in routine use include

1. Antiplatelet drugs (aspirin, clopidogrel, and glycoprotein IIb/IIIa receptor antagonists).
2. Anticoagulants (unfractionated and low molecular weight heparin, warfarin, and direct thrombin inhibitors).
3. Factor X inhibitors.

Factor X Inhibitors

Xarelto
Generic name: rivaroxaban

Eliquis
Generic name: apixaban

Arixtra
Generic name: fondaparinux

Savaysa
Generic name: edoxaban

Bevyxxa
Generic name: betrixaban

STATINS

Statins (also called HMG-CoA reductase inhibitors) block an enzyme called HMG-CoA reductase (3-hydroxy-3-methylglutaryl coenzyme A reductase) that is involved in the synthesis of mevalonate, a naturally occurring substance that is then used by the body to make sterols, including cholesterol.

By inhibiting this enzyme, cholesterol and LDL-cholesterol production are decreased. Statins also increase the number of LDL receptors on liver cells, which enhances the uptake and breakdown of LDL cholesterol. Most of the effects of statins (including the blocking of the HMG-CoA reductase enzyme) occur in the liver.

Lowering cholesterol and other types of fats is important because research has shown that elevated levels of total cholesterol, LDL cholesterol, triglycerides, and apolipoprotein B increase a person's risk of developing heart disease or having a stroke.

Crestor
Generic name: rosuvastatin
Lipitor
Generic name: atorvastatin
Livalo
Generic name: pitavastatin
Zocor
Generic name: simvastatin
Pravachol
Generic name: pravastatin
Mevacor
Generic name: lovastatin

BETA BLOCKERS

Beta blockers, including atenolol, bisoprolol, metoprolol, and nebivolol, are often used to prevent angina and after an MI.

NITRATES

Nitrates are used to dilate your blood vessels and treat angina. They are available in a variety of forms, including tablets, sprays, and skin patches, such as glyceryl trinitrate and isosorbide mononitrate.

RANOLAZINE

This medication may help people with chest pain (angina). It may be prescribed with a beta blocker or instead of a beta blocker if you can't take it.

ACE inhibitors are commonly used to treat high blood pressure but also may improve CHD.

Generic Name of Drug	Brand Name of Drug
Benazepril hydrochloride	Lotensin
Captopril	Capoten
Enalapril maleate	Vasotec
Fosinopril sodium	Monopril
Lisinopril	Prinivil, Zestril
Moexipril	Univasc
Perindopril	Aceon
Quinapril hydrochloride	Accupril
Ramipril	Altace
Trandolapril	Mavik

Angiotensin II receptor blockers are used to treat hypertension and CHD.

Generic name	Common brand names
Candesartan	Atacand
Eprosartan mesylate	Teveten
Irbesartan	Avapro
Losartan potassium	Cozaar
Telmisartan	Micardis
Valsartan	Diovan
Olmesartan	Benicar
Azilsartan medoxomil	Edarbi

Calcium channel blockers are used to treat hypertension, CHD, and angina.

Generic name	Common brand names
Amlodipine besylate	Norvasc, Lotrel
Bepridil	Vasocor
Diltiazem hydrochloride	Cardizem CD, Cardizem SR, Dilacor XR, Tiazac
Felodipine	Plendil
Isradipine	DynaCirc, DynaCirc CR
Nicardipine	Cardene SR
Nifedipine	Adalat CC, Procardia XL
Nisoldipine	Sular
Verapamil hydrochloride	

23 Future Perspectives

CHD GENETICS AND INFORMATICS

The next big phase of coronary heart disease (CHD) genetics will be informatics, bringing together masses of data and trying to understand how hundreds of small genetic differences combine to put people at risk for CHD. As part of these genetics, we will provide whole DNA sequences or sequences of particular genes.

STEM CELLS

Another fascinating area is inducible pluripotent stem cells (iPSCs). We can study patient heart cells, understand the problem, and correct it. In addition, we can remove stem cells from adipose tissue or bone marrow (mesenchymal stem cells), grow them in culture, and inject them into the patient to grow new coronary arteries and heart muscle. Your stem cells contain your entire DNA. What we can do is basically change those cells in the laboratory into stem cells, and then change those stem cells into heart cells so in effect what you end up with is the patient's own heart cells in a dish. That means we can test heart cells to find out what the fundamental problems are, then test different therapies and strategies to find something that modifies the way the heart cells work in a culture dish.

NANOTECHNOLOGY

Nanotechnology for the delivery of drugs and supplements. Nanotechnology is the manipulation of matter on a near-atomic scale to produce new structures, materials, and devices. It is generally defined as engineered structures, devices, and systems. Nanomaterials are defined as those things that have a length scale between 1 and 100 nanometers.

INFLAMMATION TREATMENTS AND IMMUNOTHERAPY

The identification of novel drug targets and the development of novel therapeutics that block atherosclerosis-specific inflammatory pathways and exhibit safe and limited immune-suppressive side effects.

VACCINES

There is active research to develop effective vaccines that will reduce the formation of coronary artery plaque through a variety of mechanisms. However, it may be decades before these are available clinically.

BIBLIOGRAPHY

1. Lutgens E, Atzler D, Döring Y, Duchene J, Steffens S, Weber C. Immunotherapy for cardiovascular disease. *Eur Heart J.* 2019 Dec 21;40(48):3937–46. doi: 10.1093/eurheartj/ehz283.
2. Kandaswamy E, Zuo L. Recent Advances in Treatment of Coronary Artery Disease: Role of Science and Technology. *Int J Mol Sci.* 2018 Jan 31;19(2):424. doi: 10.3390/ijms19020424.
3. Katsi V, Antoniou CK, Manolakou P, Toutouzas K, Tousoulis D.What's in a prick? Vaccines and the cardiovascular system. *Hellenic J Cardiol.* Jul–Aug 2020;61(4):233–40. doi: 10.1016/j.hjc.2019.09.002. Epub 2019 Nov 15.

24 Grand Summary and Conclusions

Coronary heart disease (CHD) is the number-one cause of death in the US and worldwide. We have reached a limit in our ability to prevent and treat CHD with current concepts, prevention, and treatment strategies. Coronary heart disease is preventable and treatable. It starts in utero or at a very early age. Endothelial dysfunction and glycocalyx dysfunction are the earliest vascular findings in CHD and precede the clinical symptoms and events, such as MI by decades. Coronary heart disease is the result of atheromatous changes and plaque formation in the coronary arteries supplying the heart. Approximately 80% of the cases of coronary heart disease can be prevented by achieving optimal nutrition, optimal exercise, optimal weight, and body fat; stress and anxiety reduction; and avoiding all tobacco products and smokeless tobacco. The three finite responses of the coronary arteries and cardiovascular system to the infinite insults are vascular inflammation, vascular oxidative stress, and vascular immune dysfunction. The top five CHD treatable risk factors are hypertension, diabetes mellitus, dyslipidemia (abnormal cholesterol), obesity, and smoking. These must be redefined, measured accurately, and treated early and aggressively. It is also important to measure and treat the top 25 CHD risk factors. However, there are over 400 known risk factors. Genetics and family history are most important as discussed in Chapters 9 and 10.

Coronary heart disease is caused by a reduction in the blood supply due to obstruction by a plaque in one or more of the coronary arteries to the heart muscle (myocardium), which results in decreased delivery of fresh blood, oxygen, and nutrients. Less commonly, vasospasm may cause CHD. The plaque is made up of fatty material, oxidized cholesterol and fats, inflammatory cells, white blood cells, immune cells, smooth muscle cells, and other substances. The top five CHD risk factors (hypertension, diabetes mellitus, dyslipidemia (abnormal cholesterol), obesity, and smoking) must be redefined based on new information, and the top 25 CHD risk factors need to be assessed as well. There are actually over 400 risk factors for CHD. There are an infinite number of insults to the coronary artery to cause CHD but only three finite responses: inflammation, oxidative stress, and vascular immune dysfunction. Seventy-five percent of myocardial infarctions are caused by unstable plaque rupture and a blood clot (thrombus) in a coronary artery without previous angina symptoms.

The symptoms and signs of CHD and MI can be mild or severe, typical or atypical, and will vary depending on your gender. If you develop any of these then go to the nearest emergency room or hospital. Time is critical in dealing with an MI and may truly save your heart and your life.

DOI: 10.1201/b22808-25

The arteries have five parts: lumen, glycocalyx, endothelium, muscle (media), and adventitia. The endothelium and the glycocalyx are like the air traffic control systems of the coronary arteries (as well as other arteries) and determine what happens in the blood and the subendothelial lining and the artery muscle, such as clotting, leakage or contraction of the muscle, hypertension, vasodilation, growth, inflammation, oxidative stress or immune dysfunction of the artery, atherosclerosis, arteriosclerosis, and CHD. The endothelium makes nitric oxide and many other substances which helps to reduce blood pressure, atherosclerosis, arteriosclerosis, CHD, and MI. Glycocalyx dysfunction (GD) and endothelial dysfunction (ED) of the coronary arteries may precede the development of CHD by decades. There are many treatments for endothelial dysfunction and glycocalyx dysfunction that are discussed.

The coronary arteries have only three finite or final responses that are generated from a large number of environmental and internal insults, coupled with our genetics and our nutrition. These are vascular inflammation, oxidative stress, and immune dysfunction. Inflammation in the coronary arteries is part of the injury process that repairs and heals the acute damage to the glycocalyx and endothelium. But if it is chronic, then the arteries become innocent bystanders to more damage leading to CHD. Inflammation can be measured in the blood using HS-CRP, interleukins, TNF alpha, and other inflammatory markers. There are many foods and supplements that reduce inflammation. Oxidation is a normal process in the body that uses oxygen to make energy that is needed for life. But if there is an imbalance with more oxidative stress than oxidative defense, then CHD is more common. Oxidative stress can be measured in the blood with many tests. There are many effective foods and supplements to provide an oxidative defense. Vascular immune dysfunction involves types of WBCs such as leukocytes, lymphocytes, monocytes, T cells, and macrophages that attempt to control the invasion of the glycocalyx and endothelium of the artery by LDL cholesterol.

All three finite vascular responses occur at the same time and lead to foam cells, fatty streaks, and atherosclerotic plaque that may rupture and form a clot (thrombus) in the artery leading to MI.

CHD starts early in life, but it can be identified with new cardiovascular testing which will allow early and aggressive prevention and treatment. Assessment of the three finite vascular responses, removing the vascular insult(s), and measuring correctly all the CHD risk factors coupled with personalized and precision cardiovascular program and genetics will help to reduce CHD and MI.

There are many causes of chest pain. Some of these are serious and life threatening. Angina chest pain or MI can have variable types of chest pain depending on age, gender, and other issues. If you have chest pain, with or without other symptoms, you need to get an immediate medical evaluation by your doctor, the emergency department, or hospital. Do not wait and assume that the chest pain is not serious. Your life can depend on rapid action and evaluation.

Our modern nutrition is not healthy. We consume too much sodium, refined sugars, bad fats (trans fats and some saturated fats), omega-6 fatty acids, but not enough potassium, magnesium, fruit, fiber, vegetables, high-quality protein, omega-3 fatty acids, and monounsaturated fats. These nutritional habits contribute to CHD and

MI. Your genes interact with your nutrition and your environment (gene expression) to determine your risk for CHD and MI. You can test many CHD genes with CARDIA-X from Vibrant America Labs (see the Sources section).

If one of your parents had a MI at an early age, you have a higher chance of having an MI. Numerous genes that cause CHD and hypertension can have their genetic expression altered with nutrition, supplements, exercise, drugs, and lifestyle. Evaluate specific genetic SNPs for CHD and CHD risk factors using the CARDIA-X genetic profile. The traditional Mediterranean diet (TMD) with five tablespoons EVOO/day (50 grams), nuts, and CoQ10 reduces the risk for CHD. Omega-3 fatty acids should be given to all patients, dose dependent (1–5 grams per day) to lower blood pressure and reduce CHD and MI risk. Avoid caffeine in CYP 1A2 SNP, if a slow metabolizer of caffeine, to reduce the risk of CHD and MI. Selective use of ASA and vitamin E depending on COMT phenotype. Methyl folate and B vitamins depend on the MTHFR genotype for methylation. Selenium should be given with the GSH-Px gene if defective.

All of the genes listed in the Vibrant CARDIA-X testing have specific treatments that will modify the gene expression to reduce CHD risk and CHD risk factors, such as hypertension, dyslipidemia, blood glucose, and the three finite responses. These include treatments with nutrition, nutritional supplements, lifestyle, and medications.

The gut microbiome has more genes than are in the human genome. The gut microbes help with the digestion of nutrients, prevent significant colonization of pathogens, and promote gut immunity, while the host provides a favorable environment for microbial survival. The microbiome has direct effects on the risk for CHD, MI, hypertension, high cholesterol, diabetes mellitus, and obesity. A damaged or leaky gut allows bacteria and other products into the blood and causes endotoxemia, inflammation, and arterial damage. CHD is a post-meal disease. Eating healthy foods and taking specific supplements before eating will reduce arterial damage, higher blood sugar, and triglycerides. The treatments to repair the gut include proper nutrition with a low intake of long-chain SFA, TFA, and sugars, such as glucose and fructose. You should consume more fruits, vegetables, fiber, prebiotics, probiotics, omega-3 fatty acids, olive products, specific vitamins, minerals, and glutamine. Avoid long-term use of medications that block acid production in your stomach, such as proton pump inhibitors (PPIs) and H2 blockers, as they will increase the risk for endothelial dysfunction, lower nitric oxide, and increase the risk of CHD and MI.

The role of nutrition in the prevention and treatment of CHD has been clearly demonstrated in published clinical trials. The top five cardiovascular risk factors, as presently defined, are not an adequate explanation for the current limitations to prevent and reduce CHD. The three finite responses and sound nutritional advice and evaluation based on scientific studies will be required to effect an improvement in risk for CHD.

Early detection of CHD coupled with aggressive prevention and treatment of all cardiovascular risk factors will diminish the progression of functional and structural cardiovascular abnormalities and clinical CHD. Utilization of targeted personalized and precision treatments that apply genetics with optimal nutrition coupled with exercise, ideal weight, and body composition, and discontinuation of all tobacco use can prevent approximately 80% of CHD. Nutritional studies provide evidence that CHD can be reduced with a weighted plant-based diet with ten servings of fruits and vegetables per day, MUFA, PUFA, nuts, whole grains, cold water fish, the DASH diets, PREDIMED-TMD diet; reduction of refined carbohydrates and sugars, sucrose, sugar substitutes, high fructose corn syrup, long-chain SFAs, processed foods, and elimination of all TFA.

Eggs and dairy products are not associated with CHD.

Coconut oil is neutral for CHD risk.

Organic grass-fed beef and wild game may reduce CHD.

High intakes of potassium and magnesium are recommended in conjunction with sodium restriction.

Caffeine intake should be adjusted, depending on the genetic ability to metabolize it via the CYP 1A2 system.

Alcohol is associated with a U-shaped curve and CHD.

The role of gluten, soy, and caloric restriction and CHD in humans will require more studies.

EATING A HEALTHY HEART DIET IS EASY, TASTES GOOD, AND CAN PREVENT CHD

1. **Vegetables**. Eat 8–12 servings per day of non-starchy vegetables, such as broccoli, cauliflower, leafy greens, asparagus.
2. **Fruit**. Limit to four servings. ½ cup of pieces = one serving. One small whole fruit = one serving.
3. **Fats**. Eat good fats and avoid some saturated fats and all trans fats. Omega-3 fatty acids (1–2 grams per day), extra virgin olive oil (four tablespoons per day), and olive products.
4. **Protein**: 25–30% clean-sourced fish, turkey, chicken, and proteins. Organic meat, grass-fed beef. Avoid any animal product with pesticides, toxins, or hormones. Avoid farm-raised fish. Wild cold water fish, like cod, mackerel, salmon, halibut, and tuna, are best.
5. **Water**. Fifty percent of body weight (pounds) in ounces of filtered or distilled water per day. If you weigh 150 pounds that is 75 ounces of water, herbal teas, broths daily. Do not drink from plastic containers or bottles.
6. **Tea.** Sixteen ounces of decaffeinated green tea per day.
7. **Pomegranate.** One-quarter cup of pomegranate seeds per day.
8. **Detox + elimination diet**, followed by an anti-inflammatory, blood sugar-regulating protocol. Emphasize high-fiber foods and add servings of cruciferous vegetables, such as broccoli, cauliflower, Brussels sprouts, and cabbage, etc.
9. **Caloric restriction (CR)**. Fasting 12–14 hours overnight fast with 12.5% caloric restriction and 12.5% increase in energy expenditure with exercise. PREDIMED (Mediterranean)-type diet.
10. **FMD (Fasting Mimicking Diet)** (PROLON). Three months on, then three months off, and repeat.

Grand Summary and Conclusions

Chronic stress, anxiety, and depression increase the risk of CHD and MI through an imbalance of the ANS with increased SNS activity. The autonomic nervous system is a balance of the SNS, PNS, and ENS. Increased and chronic activation of the SNS is a CHD risk factor and needs to be reduced and controlled. Increased SNS activity causes heart and arterial inflammation, oxidative stress, and vascular immune dysfunction. Increased SNS via NE activity elevates blood pressure, heart rate, other cardiovascular problems, and many endocrine problems, such as elevated blood sugar, adrenaline, and cortisol. The PNS via Ach reduces inflammation, oxidative stress, and vascular immune dysfunction. The SNS and PNS must be balanced to reduce the risk of CHD. There are many treatments to balance the autonomic nervous system including lifestyle, nutrition, exercise, herbs, and supplements.

Myocardial infarction is due to plaque in the coronary arteries. The final stage of MI is a thrombosis in the coronary artery. The primary types of plaque are hard plaque (or stable plaque), which is more obstructive, and soft rupture-prone plaque that is not as obstructive. They have different compositions and clinical presentations. About 75% of MIs are caused by soft rupture-prone plaque. The formation of plaque starts as glycocalyx and endothelial dysfunction and progresses over decades through many stages to become complicated plaque. Most MIs occur without any previous symptoms of chest pain, shortness of breath, or other cardiovascular symptoms. Early subendothelial plaque is extraluminal and may not be detectable with a routine coronary arteriogram but can be seen with CAC, MRA of the heart, IVUS, and CTA of the heart. There are many supplements and medications that will reduce plaque size and stabilize the plaque by stabilizing and reducing the lipid core.

In addition to severe obstructive CHD, patients may have NO-CHD or CA-VS that cause anginal chest pain, tightness, pressure, shortness of breath, and MI. The underlying cause for both NO-CHD and CA-VS is a reduction in arterial nitric oxide and possible vascular inflammation, oxidative stress, and immune dysfunction. The diagnosis is made with cardiac monitors and by injection of certain drugs into the coronary artery during a coronary angiogram. The treatment is with nitrates, calcium channel blockers, statins, alpha blockers, and some natural compounds that increase arterial nitric oxide.

CHD and MI are the number-one causes of death in women in the US. The symptoms of CHD and MI in women are often different than they are in men. The CHD risk factors may also be different in women than in men. Women are often diagnosed later than men and may have a poorer prognosis. Women tend to have worse results from PTCA, stents, and CABG compared to men. After menopause the risk for CHD and MI increase due to a reduction in estrogen, increased inflammation, changes in blood lipids, and other factors.

The top five CHD risk factors must be redefined and evaluated accurately and treated in all patients. The top modifiable 25 CHD risk factors should now be included in the evaluation of patients with any risk of CHD or known CHD. Other CHD risk factors, metabolomics, and proteomics in blood and urine are now emerging and will be important in the future evaluation of CHD risk. Early measurement of the top 25 modifiable CHD risk factors and early, aggressive, and appropriate prevention and treatment will help prevent and reduce CHD and MI.

Several noninvasive vascular tests will predict the risk for CHD or MI decades before an event occurs. These noninvasive vascular tests will predict the presence and severity of CHD and the risk for MI. Numerous supporting blood tests also predict CHD and MI. All of these tests are available clinically. Immediate prevention and treatment programs can be started to correct these abnormal tests and prevent the progression of CHD or an MI.

Nutrition is the foundation for the prevention and treatment of CHD and should follow the general guidelines of the Mediterranean diet. Also consider caloric restriction and fasting programs. Numerous nutritional supplements have proven scientific validity to improve numerous CHD risk factors, noninvasive cardiovascular tests, coronary calcium, and plaque stabilization and reduction. A combined aerobic and resistance exercise program is recommended for the prevention and treatment of CHD. This is the ABCT exercise program. The ideal body weight, BMI, and body fat will improve CHD risk. Stop all tobacco products, e-cigarettes, and vaping. Control all risk factors for CHD and adhere to the new recommended goals.

Sources

GENERAL HYPERTENSION EDUCATION AND INFORMATION

Hypertension Institute: www.hypertensioninstitute.com

RECOMMENDED NUTRITIONAL SUPPLEMENT COMPANIES

Company name: Biotics Research Corporation
Phone number: 800-231-5777
Email: biotics@bioticsresearch.com
Website: www.bioticsresearch.com

Company name: Designs for Health
Phone number: 860-623-6314
Email: info@designsforhealth.com
Website: www.designsforheath.com

Company name: Metagenics and Metagenics Institute
Phone number: 800-692-9400
Email: info@metagenicsinstitute.com
Website: Metagenics.com and MetagenicsInstitute.com (nonbranded educational content)

Company name: Human N
Phone number: 512-488-4477
Email: Professional@humann.com
Website: https://humannpro.com/drhouston

Company name: AC Grace Company
Phone number: 800-833-4368 or 903-636-4368
Email: Info@acgrace.com
Website: www.acgrace.com

Company name: Ortho Molecular Products
Phone number: 800-332-2351
Email: contactus@ompimail.com
Website: www.orthomolecularproducts.com

Company: Calroy Health Sciences, LLC
Product: Arterosil HP
Phone number: 800-609-6409

289

Email: support@arterosil.com
Website: arterosil.com

Company name: MitoQ Ltd.
Phone number: +649-379-8222
Email: customerservice@mitoq.com
Website: www.mitoq.com

ADDITIONAL NUTRITIONAL SUPPLEMENT COMPANIES

Xymogen
Carlson Labs
Douglas Labs
Dr Sinatra
Pure Encapsulations
Standard Process
Swanson
Klaire Labs
Juice Plus (NSA)

LAB TESTING COMPANIES, CARIDOVASCULAR TESTING, AND MEDICAL EQUIPMENT COMPANIES

Company name: Vibrant America Clinical Lab
Phone number: 866-364-0963
Email: support@vibrant-america.com
Website: www.vibrant-america.com

Company name: SpectraCell Laboratories, Inc.
Phone number: 800-227-5227
Email: support@spectracell.com
Website: www.spectracell.com

Company name: AtCor Medical (cardiovascular and blood pressure testing company)
Phone number: 630-228-8871
Email: info@atcormedical.com
Website: www.atcormedical.com

Company name: Predictive Health Diagnostics, Inc: Puls Test
Phone number: 866-299-8998
Email: Info@pulstest.com
Website: www.pulstest.com
Fax: 888-424-7505

Celine Peters, RN, MN
Vice President, Clinical Development
11335 NE 122nd Way Ste 105
Kirkland, WA 98034
Cell: 619.889.8539

Prevencio utilizes Machine Learning (Artificial Intelligence) + Multi-Proteomic Biomarkers + Proprietary Algorithms to deliver cardiovascular diagnostic & prognostic tests that are significantly more accurate than standard-of-care stress tests, individual biomarkers, genetic markers, and clinical risk scores. The six blood tests in our product pipeline are:

1. HART CVE™ – blood test for one-year risk of heart attack, stroke, or cardiac death
2. HART CAD™ / HART CADhs™ – blood tests for obstructive coronary artery disease diagnosis
3. HART PAD™ – blood test for peripheral artery disease diagnosis
4. HART AS™ – blood test for aortic valve stenosis
5. HART AMP™ – blood test for risk of amputation
6. HART AKI™ – blood test for risk of acute kidney injury in cardiac catheterization

To view our video, log onto www.prevenciomed.com and scroll down to click on the video.

CORUS

Smart Heart Labs
1301 Finks Hideaway Rd
Monroe, LA 71203
318-605-3060
service@smartheartlabs.com

Index

A

ACE I/D (DD allele), 71, 73, 80
Acetylcholine (Ach), 151, 168
Acute myocardial infarction, 11, 186
Adenosine triphosphate (ATP), 42, 70, 221, 246
Adipokines, 40, 178, 260
ADR B2 gene, 73
Advanced glycation end products (AGE), 44, 105–106
Advanced oxidation protein products (AOPP), 44
Adventitia, 4, 25, 26, 35, 284
Aerobics, Build, Contour, and Tone (ABCT) exercise program, 237–239
    aerobic training in intervals, 244
    arm exercises, 255
    back exercises, 253–254
    chest exercises, 252–253
    core exercises, 245
    elements, 240–242
    energy drink, 246
    exercise on an empty stomach, 246–247
    health benefits, 259
    hybrids/supersets/rapid sets, 243
    leg exercises, 255–257
    nutrient availability, 247–248
    nutrition/water, 246
    resistance training, 242–243
    resistance training hints, 243–244
    shoulder exercises, 254–255
    time-intensive exercise, 245
    time schedule, 260
    training schedules, 248–252
Aged garlic (Kyolic garlic), 224–225
Albumin creatinine ratio (ACR), 181
Alpha-blockers, 169
Alpha-linolenic acid (ALA), 96, 225
Alternate-day (ADF) fasting, 109
Amlodipine, 20, 169
Aneurysm, 70, 200–201
Angiotensin-converting enzyme inhibitors (ACEI), 73, 80, 164, 181
Angiotensin receptor blockers (ARB), 73, 164, 181
Antigens, 51
Anti-thrombin III (AT-III), 31
Aortic aneurysm, 60, 70, 195
Aortic ultrasound, 200–201
Arm exercises, 255
Arterial inflammation, 39, 152, 287
Arterosil, 80, 164, 186, 220–221
Asymptomatic atherosclerosis, 2
Atheroma, 158, 283
Atherosclerosis, 89, 155–159
    fatty streak, 158
    foam cells, 158
    initial lesion, 158
Augmentation index, 201, 220, 221
Autonomic nervous system (ANS), 149–152, 183, 287

B

Back exercises, 253–254
Basophils, 27, 30
Beta blockers, 180, 231, 278
Bioidentical hormone replacement (BIHRT), 185
Blood pressure (BP), 20, 34–35, 56, 73, 81, 98, 151, 201
Blood-thinning medicines, 277
Body impedance analysis (BIA), 177, 261
Body mass index (BMI), 5, 40, 106, 109, 177
Brain inflammation, 151
Brain natriuretic peptide (BNP), 210

C

C-1 compliance, 195
C-2 compliance, 195
Calcium channel blockers (CCBs), 169, 279, 287
Carboxylation, 221
Cardiac stress test, 20, 21
Cardiopulmonary Exercise Testing (CPET), 201, 238
Cardiovascular genetics
    CHD and CHD risk factor genes, 70–71
    COMT polymorphisms, 71
    glutathione-related SNPs, 71
Cardiovascular laboratory testing, 201
Cardiovascular SNPs, 73–74
Carotid artery ultrasound, 196–198
Catalase (CAT), 44, 234
Catechol-O-methyltransferase (COMT), 71
Cell adhesion molecules (CAMs), 45, 192, 234
Cerebrovascular accidents (CVA), 96, 100
CHAN2T 3 CHD scoring system, 211
Chest exercises, 252–253
Chronic kidney disease (CKD), 180, 201

Chronic obstructive pulmonary disease (COPD), 59
Clean-sourced proteins, 126–127
Coenyzme Q10 (CoQ10), 69
    clinical physiology, 232–236
    endothelial function, 236
    functions, 232
    hypertension, 236
    pharmacokinetics, 231–232
Colony-forming units (CFU), 93
Computerized Arterial Pulse Wave Analysis (CAPWA), 195, 210
Computerized tomographic angiogram (CTA), 155–156, 159
Conduit arteries, 26
Congestive heart failure (CHF), 69, 96, 107, 196, 231, 263
Consortium of Southeastern Hypertension Centers (COSEHC), 209–210
Contrast dye, 159, 162
Coronary arteries, 164, 167, 171–173, 186, 198, 284
Coronary artery bypass graft (CABG), 21, 171, 196
Coronary artery bypass graft occlusion (CABG occlusion), 100
Coronary artery calcification (CAC), 106, 108, 159, 185, 198–199
Coronary artery disease (CAD), 2
Coronary artery dissection, 60
Coronary artery plaque, 10
    anatomy, 160–161
    characteristics, 162–164
    prevention and treatment, 164
    types, 159
Coronary artery spasm, 9, 11, 20, 167, 168, 185, 220
Coronary artery vasospasm (CAVS), 167
    causes, 168
    diagnosis, 168–169
    partial obstruction, 168
    treatment, 169
Coronary heart disease (CHD), 1
    avoid/eat in moderation
        legumes, 131
        oils, 131
        sodium, 132–133
        sweeteners, 131
    cardiovascular and genetics program, 57
    case studies, 19–23
    causes, 12
    chest pain, 59
    chronic clinical infections, 185
    chronic clinical/subclinical infections, 185–186
    control risk factors, 264–265

diabetes mellitus, 177
dietary fats
    MUFA, 100
    omega-3 fatty acids, 99–100
    SFA, 100–101
    TFA, 102–103
dysglycemia, 177
dyslipidemia, 176–177
early age, 3–4
environmental pollutant, 186
exercise, 237–239
exercise, twist, 239–240
family history, 67
heavy metals, 186
hormonal imbalances in both genders, 185
hyperglycemia, 177
hypertension, 175–176
insulin resistance, 177
Korean War, 3
lack of exercise, 183
lack of sleep, 183
metabolic syndrome, 177
micronutrient deficiencies, 186
modifiable risk factors, 179–180
nutrition
    alcohol consumption, 109–110
    animal protein diets, 106–107
    caffeine, 108–109
    caloric restriction, 109
    DASH diet, 98–99
    dietary magnesium, 111
    dietary sodium, 111
    fermented foods, 130
    fish, 107
    flours, 129
    fruits, 128
    gluten, 110
    herbs, 129
    Mediterranean diet, 96, 99
    nuts, 110–111
    nuts/seeds, 130
    pastured poultry, 129
    protein, 106–107
    refined carbohydrates, 105
    sodium–potassium ratio, 111
    soy protein, 107
    starches, 130
    sweet, 128
    sweeteners, 129
obesity and body fat, 177–178
PDAY study, 3
prevention, 4–5
recipes, 133–146
removing the vascular insult(s), 55
risk factors, 9, 56, 95, 99, 189–193
SFA

Index   295

coconut oil, 103
milk products, 103
peptides, 103
smoking, 179
stress/anxiety/depression, 183
subclinical hypothyroidism, 184–185
subclinical infections, 185
symptoms, 59
symptoms and signs, 15–17
three finite vascular responses, 55
United States, 2–3
whey protein, 104–105
women
   menopause, 172–173
   risk factors, 172
   symptoms, 171–172
Coronary heart disease prevention program (CHDPP), 5
Costochondritis, 59
Curcumin, 108, 229–230
CYP 11 B2, 73
CYP4A11, 73
CYP-450-1A2, 72
Cytomegalovirus (CMV), 186

D

Deoxyribonucleic acid (DNA), 43, 68
Diabetes mellitus, 3, 22, 51, 67, 73–74, 177, 283, 285
Dietary acid load, 107–108
Dietary approaches to stop hypertension (DASH) diets, 69, 98–99
Dietary fats
   MUFA, 100
   omega-3 fatty acids, 99–100
   SFA, 100–101
   TFA, 102–103
Dietary magnesium, 111
Dietary sodium, 111
D-Ribose, 221, 241
Dysbiosis, 88
Dysglycemia, 177
Dyslipidemia, 102–103, 176–177, 264, 283

E

Echocardiography (ECHO), 199–200
E-cigarettes, 263
Ecosystem, 87
Ejection fraction (EF), 200, 233, 235
Electrocardiogram (EKG), 3, 22, 23, 169, 201, 210
Electrolytes, 69
Electron transport chain (ETC), 231–233
Elevated blood fibrinogen, 181
Elevated serum iron, 181–182

Endogenous antioxidants, 44–45
Endothelial dysfunction (ED), 33, 35, 42, 43, 56, 89, 111, 157, 181, 186, 196, 220, 234, 284
Endothelial glycocalyx (EGC), 30–33
Endotoxemia, 88, 89, 101, 285
Energy production, 233, 234
Enteric nervous system (ENS), 149
Enterocyte responses, 90
Enterocytes, 88, 93
Environmental pollutant, 186
Eosinophils, 27, 30
Epigenetics, 68–69
Epithelial sodium channel (ENaC), 73
Eplerenone, 73, 79
Epstein Barr virus (EBV), 186
Especially extra virgin olive oil (EVOO), 219
European Prospective Investigation into Cancer and Nutrition (EIPC), 106
Exogenous antioxidants, 44
Extra cellular matrix (ECM), 216
Extracellular superoxide dismutase (ec-SOD), 31
Extraluminal disease, 159
Extra virgin olive oil (EVOO), 40, 91, 96, 219

F

Fasting blood sugar (FBS), 177, 264
Fasting mimicking diet (FMD), 5, 40, 90
Fatty streak, 3, 50, 158, 284
Ferric reducing antioxidant power (FRAP), 45
Fibroatheroma, 158
Fibromyalgia, 60
Finite vascular responses, 12, 38, 49, 57, 70, 95, 284
Foam cells, 50, 158, 284
Folate metabolism, 184
Follicle-stimulating hormone (FSH), 185
Foreign invader, 29, 37, 51
Framingham Risk Score (FRS), 207–209
Free radicals, 42

G

Gamma-aminobutyric acid (GABA), 153
Gamma carboxyglutamate, 221
Gamma-linolenic acid (GLA), 100
Gastroesophageal reflux (GERD), 60
Gastrointestinal tract (GI), 28, 87, 110
Gene expression, 64, 67, 69
Gene expression score (GES), 214
Gene expression testing
   CORUS, 214
   GES, 214
   measurements, 214–216
Genes and gene expression
   1q25, 79

4q25, 80–81
6p 24.1, 80
9p21, 79
ACE I/D, 80
ADRB2, 81
AGTR1, 81
APO A 1, 80
APO A 2, 80
APO C3, 80
APO E, 81
CORIN, 81
CYP 11 B2, 79
CYP 1A2, 81
CYP 4 F2, 80
GSH PX, 80
MTHFR, 80
NOS 3, 80
SCARB1, 81
Genetic hypertension, 73
Genetics, 69, 107, 281, 283
Genetic testing, 63, 67, 72–73
Glucose tolerance test (GTT), 177
Glutathione peroxidase (GSH-Px), 44, 45, 71
Gluten, 110
Glycocalyx dysfunction (GD), 33, 57, 58, 219, 284
Gramnegative bacteria, 88
Green tea extract (EGCG), 164, 186, 237
Gut microbiome (GM), 87
　　effects on enterocytes and arteries, 93
　　treatment, 91–93

H

Hard plaque, 161, 287
Health Professionals Follow-Up Study (HPFS), 101
Health-promoting fats, 127–128
Healthy heart diet, 286–288
Heart-healthy diet, 147
Heat shock proteins (HSPs), 50
Heavy metals, 39, 43, 186
Hemoglobin A1C (HbA1C), 22, 23, 177, 264
Herpes simplex virus (HSV), 186
Hiatal hernia, 60
High-density cholesterol (HDL-C), 104
High-density lipoprotein (HDL), 96, 164, 172
High fructose corn syrup (HFCS), 105, 219
High-sensitivity C reactive protein (HS-CRP), 38, 39, 96, 179
HMG-CoA reductase inhibitors, 277–278
Homocysteinemia, 183–184
Hormonal Imbalances, 185
Human circulatory system
　　adventitia, 35
　　arterial muscle/media, 35
　　EGC, 30–33
　　endothelium, 33–35
　　lumen, 26–30
Human genome, 63, 285
Human intestine, 88
8-Hydroxydeoxyguanosine (8-OHdG), 43
Hyperglycemia, 89, 177, 181, 264
Hypertension, 72–73, 175–176
Hypertriglyceridemia, 89
Hypertrophic cardiomyopathy, 60
Hypertrophy, 73, 200
Hyperuricemia, 180
Hypothyroidism, 184–185

I

Immune dysfunction, 28, 49, 50, 151, 284
Immunotherapy, 281
Inducible pluripotent stem cells (iPSCs), 281
Inflammation, 28, 37
　　artery, 39
　　measure, 38–40
　　treatments, 281
　　vascular, 40
Informatics, 281
Innocent bystander, 38, 50, 284
Insulin resistance, 88, 100, 177, 180, 264
Interleukin 6 (IL-6), 38, 39, 96
Interleukin I B (IL-1B), 38
Intermittent fasting, 40, 90, 109
Intimal medial thickness (IMT), 157
Intravascular ultrasound (IVUS), 156, 157, 163
Isocaloric (ISC) replacement, 102

K

Kaplan–Meier Curve, 67, 68
Krebs cycle (TCA), 233

L

LDL cholesterol, 43, 49, 51, 101, 111, 178, 231
LDL lipoprotein, 49, 51
LDL particle number (LDL P), 96
Left anterior descending artery (LAD), 9, 10, 20
Left circumflex (LCX) artery, 9, 10
Left main (LM) artery, 9, 10
Leg exercises, 255–257
Lipid peroxidation, 43–44, 231
Lipoic acid, 89
Lipoprotein saccharide (LPS), 91
Long-chain fatty acids (LCFA), 101
Low-density cholesterol (LDL-C), 104
Low-density lipoprotein (LDL), 96, 100, 172, 177
Lung disease, 59
Luteinizing hormone (LH), 185
Lymphocytes, 29, 284

Index

M

Macrophages, 29, 38, 50, 151, 158
Macrovasculature, 26
Magnetic resonance angiogram (MRA), 156, 159
Magnetic resonance imaging (MRI), 156
Major adverse CV event (MACE), 214
Malondialdehyde (MDA), 43–44
Matrix metalloproteinases (MMPs), 216
Maximum aerobic capacity (MAC), 244
Maximum heart rate (MHR), 244
Mediastinum, 60
Mediterranean diet (MedDiet), 46, 69, 96–98
Medium chain fatty acids (MCFA), 101
Menopause, 172–173
Metabolic memory, 89
Metabolic syndrome, 104, 177, 180
Methylmercury, 107
Micronutrient deficiencies, 186
Micronutrients, 63, 187–188, 219, 237
Micronutrient testing (MNT), 3, 219, 232
Microvascular angina, 22
Microvasculature, 26
Mineral absorption, 91
Mitochondria, 42, 70, 222, 231–237
Mitoquinone (Mito Q), 237
Modern diet, 65
Modified LDL, 49, 50, 176
Monocyte chemotactic protein (MCPs), 50, 96
Monocytes, 29, 51, 58, 160
Monounsaturated fatty acids (MUFA), 56, 65, 100, 219, 247
MTHFR gene, 73
Multi-nutrient cardiovascular protection supplement, 237
Myeloperoxidase (MPO), 43, 231
Myocardial infarction (MI), 1
    atherosclerosis, 155–159
    blockage, 11
    case studies, 1–2
    causes, 12
    symptoms, 155
    thrombosis, 287
    unstable plaque rupture, 12, 283
Myocarditis, 60

N

N acetyl cysteine (NAC), 164, 184, 186
Nanotechnology, 281
Natural compounds, 169, 287
NEO 40, 73, 80, 164, 186, 220
Neoantigens, 51
Neutrophils, 29–30
Nitrates, 169, 278
Nitration, 44

Nonalcoholic fatty liver disease (NAFLD), 105
Noninvasive vascular tests, 195
    retina/fundus examination, 198
    risk of CHD, 288
Nonobstructive coronary heart disease (NO-CHD), 167, 168
Non-starchy vegetables, 126
Norepinephrine (NE), 71, 151, 263
NOS 3, 71, 73, 80
Nuclear power plants, 70
Nurses' Health Study (NHS), 99, 100
Nutrient gene interactions, 64
Nutrients, 68, 97–98
    electrolytes, 69
    monounsaturated fat, 70
    Omega-3 fatty acids, 70
Nutrition
    alcohol consumption, 109–110
    animal protein diets, 106–107
    caffeine, 108–109
    caloric restriction, 109
    DASH diet, 98–99
    dietary magnesium, 111
    dietary sodium, 111
    fermented foods, 130
    fish, 107
    flours, 129
    fruits, 128
    gluten, 110
    herbs, 129
    Mediterranean diet, 96, 99
    nuts, 110–111
    nuts/seeds, 130
    pastured poultry, 129
    protein, 106–107
    refined carbohydrates, 105
    sodium–potassium ratio, 111
    soy protein, 107
    starches, 130
    sweet, 128
    sweeteners, 129
Nutritional supplements, 219–220, 288
Nuts, 47, 70, 96, 110–111

O

Obesity, 39, 177–178, 260–262
Omega-3 fatty acids, 70, 99–100
    cardiovascular benefits, 226–228
    ingredients, 229
    pathway, 226
    RCT, 228–229
    safety, 229
    types, 225
Oxidative stress, 49, 50
    antioxidants

endogenous, 44–45
exogenous, 44
foods and supplements, 46–47
reduce CHD/MI, 45
sources of food, 47
supplements, 48
DNA/RNA damage, 43
lipid peroxidation, 43–44
measure, 43
nitration, 44
oxidation, 42–43
protein oxidation, 44
tests to measure defense, 46
treatment, 46–47
Oxidized LDL (oxLDL), 38, 49, 96
Oxygen radical antioxidant capacity (ORAC), 45

P

Parasympathetic nervous system (PNS), 149
Pattern recognition receptors (PRR), 41, 52
Pericarditis, 60
Peripheral arterial disease (PAD), 196
Plaque, 2, 37
anatomy, 160–161
arterial inflammation, 39
atherosclerotic, 50, 155
causing MI, 2
coronary artery, 10, 159
LAD, 11
MI, 287
obstruction, 9
prevention and treatment, 164
rupture-prone, 163
soft, 162
stabilizing effect, 229
types, 161
Plasma albumin, 28
Plasma renin activity (PRA), 261
Postprandial blood glucose, 90
Prebiotics, 93
Premature ventricular beats (PVCs), 201
Pritikin diet, 69
Probiotics, 89, 93
Protein oxidation, 44
Proteoglycans, 30
Proton pump inhibitors (PPIs), 91, 285
Pulmonary embolus (PE), 59
Pulmonary hypertension, 60
PULS cardiac test (CHL), 211–214

Q

Quantitative coronary artery angiogram (QCA), 214
Quercetin, 231

R

Radical nitrogen species (RNS), 48, 63
Radical oxygen species (ROS), 48, 63
Randomized clinical trials (RCTs), 228–229
Ranolazine, 278–279
Rasmussen center CHD scoring, 210–211
Reactive nitrogen species (RNS), 43
Reactive oxygen species (ROS), 43
Recipes, 133–146
Red blood cell (RBC), 26
Refined carbohydrates, 92, 105
Remnant lipoproteins (RLP), 49
Renal disease, 180–181
Reverse cholesterol transport (RCT), 231
Ribonucleic acids (RNA), 43, 68
Right coronary arteries (RCA), 9, 10

S

Saturated fatty acids (SFA), 65, 88, 91–92, 99–101
coconut oil, 103
milk products, 103
peptides, 103
Security system, 49
Sex hormone-binding globulin (SHBG), 185
Short chain fatty acids (SCFA), 88, 101
Shoulder exercises, 254–255
Single nucleotide polymorphisms (SNPs), 67
Slugging, 26
Smoking, 179, 263
Soft plaque, 161
Soluble proteoglycans (SP), 31
Spironolactone, 79
Starchy and root vegetables, 128
Statins, 169, 277–278
Stem cells, 281
Stroke, 26
Sudden cardiac death (SCD), 100
Sulfated polysaccharide (SSP), 220
Superoxide dismutase (SOD), 44
Supplements, 40–42
Sympathetic nervous system (SNS), 149, 183
treatment, 152–153

T

T cells, 38, 49
T-helper cells, 49
Thrombomodulin (TM), 31
Thrombosis, 287
Thyroid binding globulin (TBG), 184
Thyroid stimulating hormone (TSH), 184
Tietze's syndrome, 59
Time-intensive exercise, 245

Index

Tissue inhibitor of matrix metalloproteinase (TIMP), 216
Toll-like receptors (TLR), 41, 52
Total cholesterol (TC), 96
Total iron-binding capacity (TIBC), 181
Traditional Mediterranean diet (TMD), 96, 285
Trans fatty acid (TFA), 65, 89, 92, 101–103, 219
Treadmill test, 20
Triglycerides (TG), 89, 96
Trimethylamine N-oxide (TMAO), 88
Trolox equivalent antioxidant capacity assay (TEAC), 45
Tumor necrosis factor alpha (TNF alpha), 38
Turmeric, 229–230
2D echocardiography, 199–200
Type 2 diabetes mellitus (T2DM), 96, 104

U

Ubiquinone (CoQ10), 231

V

Vaccines, 281
Vaping, 263
Vascular immune dysfunction, 49–51
Vascular inflammation, 40, 49–51
VasculoSirt, 237
Vegetarian diets, 106
Very low-density lipoprotein (VLDL), 177
Visceral fat, 89
Vitamin absorption, 91
Vitamin K2 MK 7, 221–224

W

White blood cells (WBCs), 26–28
    functions, 29–30
    types, 29–30

Taylor & Francis eBooks

www.taylorfrancis.com

A single destination for eBooks from Taylor & Francis with increased functionality and an improved user experience to meet the needs of our customers.

90,000+ eBooks of award-winning academic content in Humanities, Social Science, Science, Technology, Engineering, and Medical written by a global network of editors and authors.

TAYLOR & FRANCIS EBOOKS OFFERS:

- A streamlined experience for our library customers
- A single point of discovery for all of our eBook content
- Improved search and discovery of content at both book and chapter level

REQUEST A FREE TRIAL
support@taylorfrancis.com

Routledge — Taylor & Francis Group
CRC Press — Taylor & Francis Group

Printed in Great Britain
by Amazon